Sexual Segregation in Vertebrates
Ecology of the Two Sexes

Males and females often differ in developmental patterns, adult morphology, ecology and behaviour, and in many mammals males are often larger. Size dimorphism results in divergent nutritional and energetic requirements or reproductive strategies by the sexes, which in turn sometimes causes them to select different forage, use different habitats and express differing social affinities. Such divergent lifestyles often lead males and females to live large parts of their lives separately. Sexual segregation is widespread in animals. Males and females may share the same habitat but at different times, for example, or they might use different habitats entirely. Why did sexual segregation evolve and what factors contribute to it? *Sexual Segregation in Vertebrates* explores these questions by looking at a wide range of vertebrates, and is aimed as a synthesis of our current understanding and a guide for future research.

KATHREEN RUCKSTUHL is currently Assistant Professor at the University of Calgary, Ecology Division, in the Department of Biological Sciences.

PETER NEUHAUS is part-time Research Assistant at the University of Neuchâtel, and Adjunct Professor at the University of Calgary.

Sexual Segregation in Vertebrates

Ecology of the Two Sexes

Edited by

K. E. RUCKSTUHL
*University of Calgary, Canada
and University of Cambridge,
UK*

P. NEUHAUS
*Université de Neuchâtel,
Switzerland, University of
Calgary, Canada and University
of Cambridge, UK*

CAMBRIDGE UNIVERSITY PRESS
Cambridge, New York, Melbourne, Madrid, Cape Town, Singapore,
São Paulo, Delhi, Dubai, Tokyo, Mexico City

Cambridge University Press
The Edinburgh Building, Cambridge CB2 8RU, UK

Published in the United States of America by Cambridge University Press, New York

www.cambridge.org
Information on this title: www.cambridge.org/9780521184212

© Cambridge University Press 2005

First published 2005
First paperback edition 2010

A catalogue record for this publication is available from the British Library

ISBN 978-0-521-83522-0 Hardback
ISBN 978-0-521-18421-2 Paperback

Contents

Contributors

John D. Altringham
School of Biology, University of Leeds, Leeds, LS2 9JT, UK

Vernon C. Bleich
California Department of Fish and Game, 407 W. Line St. Bishop, CA 93514, USA

Richard Bon
Centre de Recherches sur la Cognition Animale, CNRS – Université Paul Sabatier – Toulouse III, 118 Route de Narbonne, 31062 Toulouse cedex 4, France

Paulo Catry
Unidade de Investigação em Eco-Etologia, Instituto Superior de Psicologia Aplicada, Rua Jardim do Tabaco 44, 1149-041 Lisboa, Portugal

Tim H. Clutton-Brock
Department of Zoology, University of Cambridge, Downing Street, Cambridge, CB2 3EJ, UK

Larissa Conradt
School of Biological Sciences, University of Sussex, Falmer, Brighton BN1 9QG, UK

Graeme Coulson
Department of Zoology, University of Melbourne, Victoria 3010, Australia

Darren P. Croft
School of Biology, University of Leeds, Leeds, LS2 9JT, UK

John P. Croxall
British Antarctic Survey, High Cross, Madingley Road, Cambridge CB3 0ET, UK

Jean-Louis Deneubourg
Service d'Ecologie Sociale, CP 231, Université Libre de Bruxelles, Boulevard du Triomphe, 1050 Bruxelles, Belgium

Johan T. du Toit
Department of Forest, Range, and Wildlife Sciences, Utah State University, 5230 Old Main Hill, Logan UT 84322-5236, USA

Jean-François Gerard
Institut de Recherche sur les Grands Mammifères, INRA, BP 27, 31326 Castanet-Tolosan, France

Jacob Gonzáles-Solís
Departamento Biologia Animal (Vertebrats), Av Diagonal 645, Barcelona 08028, Spain

Richard James
Department of Physics, University of Bath, Bath, BA2 7AY, UK

Jens Krause
School of Biology, University of Leeds, Leeds, LS2 9JT, UK

Jeffrey D. Long
Educational Psychology, University of Minnesota, 204 Burton Hall, 178 Pillsbury Drive SE, Minneapolis, MN 55455-0208, USA

Abigail M. MacFarlane
Department of Zoology, University of Melbourne, Victoria 3010, Australia

Martin B. Main
University of Florida, Southwest Florida Research & Education Center, 2686 State Road 29 North, Immokalee, FL 34142, USA

Robert Michaud
Group de Recherche et d'éducation sur les mammifères marin, 295 Chemin Sainte-Foy, Quebec G1R 1T5, Canada

Pablo Michelena
Centre de Recherches sur la Cognition Animale, CNRS – Université Paul Sabatier – Toulouse III, 118 Route de Narbonne, 31062 Toulouse cedex 4, France

Elizabeth A. Mizerek
Center for Applied Research & Educational Improvement, University of Minnesota, 204 Burton Hall, 178 Pillsbury Drive SE, Minneapolis, MN 55455-0208, USA

Peter Neuhaus
Department of Zoology, University of Cambridge, Downing Street, Cambridge CB2 3EL, UK

and
Institute de Zoology, Université de Neuchâtel, Emile-Argand 11, Case
postale 2, 2007 Neuchâtel, Switzerland

Anthony D. Pellegrini
Educational Psychology, University of Minnesota, 204 Burton Hall, 178
Pillsbury Drive SE, Minneapolis, MN 55455-0208, USA

Richard A. Phillips
British Antarctic Survey, High Cross, Madingley Road, Cambridge CB3 0ET,
UK

Esther S. Rubin
Zoological Society of San Diego, Center for Reproduction of Endangered
Species, P.O. Box 120551, San Diego, CA 92112-051, USA

Kathreen E. Ruckstuhl
Department of Zoology, University of Cambridge, Downing Street,
Cambridge CB2 3EL, UK
and
Department of Biological Sciences, University of Calgary, 2500 University
Drive NW, Calgary AB, T2N 1N4, Canada

Paula Senior
School of Biology, University of Leeds, Leeds, LS2 9JT, UK

Richard Shine
School of Biological Sciences A08, University of Sydney, NSW 2006,
Australia

David W. Sims
Marine Biological Association of the United Kingdom, The Laboratory,
Citadel Hill, Plymouth PL1 2PB, UK

Iain J. Staniland
British Antarctic Survey, High Cross, Madingley Road, Cambridge CB3 0ET,
UK

Mike Wall
School of Biological Sciences A08, University of Sydney, NSW 2006,
Australia

David P. Watts
Department of Anthropology, Yale University, P.O. Box 208277, New Haven,
CT 06520-8277, USA

José C. Xavier
British Antarctic Survey, High Cross, Madingley Road, Cambridge CB3 0ET,
UK
and
Centre of Marine Sciences (CCMAR), University of Algarve, Campus de
Gambelas, 8000–117 Faro, Portugal

Preface

In September 2002, we invited researchers to discuss their work, but also to consolidate our knowledge of sexual segregation in vertebrates, during a three-day workshop in the Zoology Department of the University of Cambridge. The book stems from this workshop, but is much more than a compilation of different research chapters. At that workshop we gave each author the task to integrate their knowledge of a particular system into the framework of sexual segregation. Many of the authors had not even previously worked on sexual segregation but, as you will see, they all did an excellent job in synthesizing information on sexual segregation and the ecology of the two sexes in different taxa. We're very fortunate to have been able to attract so many outstanding scientists who either contributed to the book itself or commented on the chapters. Our special thanks also go to Lotti and Hans Neuhaus who flew over from Switzerland to look after Anna May, so that we did not have to segregate for parental duties. We are also very grateful to the staff at Gonville and Caius College, Cambridge, who provided accommodation for the workshop; to the University of Cambridge, who provided us with tea and biscuits during tea breaks; and, in particular, we would like to thank the staff in the Zoology Department of Cambridge and all our colleagues who supported us in various stages during the writing and editing of this book. The whole process, from organizing the workshop to editing the book, has been achieved while Kathreen and I were postdoctoral fellows at the University of Cambridge. Anne Carlson deserves a special thank you for all her help, particularly with the book proposal. We are also greatly indebted to Nick Davis (Cambridge), Peter Jarman (Australia) and an anonymous referee who commented on the book proposal and improved it considerably. A special thank you also goes to Tracey Sanderson who initially was our book editor, Martin Griffiths, Carol Miller, Kath Pilgrem and Clare Georgy of Cambridge

University Press for comments, editing and getting the book published. We greatly enjoyed working with all of them and are very grateful for their help and encouragement. Before we acknowledge all the people who have reviewed chapters in this book we would like to mention that we dedicate this book to our daughter, Anna May, and to my late father, Frederic Ruckstuhl. To both of them the wonders of nature are forever puzzling and attractive, just as to us.

This book would never have been realized without the encouragement and support of so many people all of whose names we cannot list here – a big, big thank you! Special thanks go to the numerous reviewers of the chapters in this book. All chapters were peer-reviewed by at least two specialists. We highly appreciate the excellent and helpful comments with which the chapters in this book have been considerably improved. Finally, we would like to thank all the authors for their significant contributions. It was a pleasure to work with all of you!

Part I Overview

KATHREEN E. RUCKSTUHL AND TIM H. CLUTTON-BROCK

1

Sexual segregation and the ecology of the two sexes

According to Greek mythology, Amazons were female warriors who lived on an island. They occasionally met with men of another people to mate, keeping female offspring and sending male offspring back to their fathers. In many animals too, males and females live apart for most of the year, only gathering for mating. These include many fish species (Sims et al., 2001a), birds (Myers, 1981), lizards and snakes (Parmelee & Guyer, 1995; Shine et al., 2003a), and most mammals with a pronounced sexual dimorphism in body size (Ruckstuhl & Neuhaus, 2000). This book aims at synthesizing our current understanding of the evolution of sexual segregation in different vertebrates, focusing on taxa in which there is sufficient evidence to investigate causes of sexual segregation.

Sexual segregation has caused confusion in the literature, as it can occur at different levels (see Chapter 2 for details). Why, for example, do house mice, Mus musculus, embryos segregate by sex in their mother's womb (Terranova & Laviola, 1995)? Why do African ground squirrels, Xerus inauris, live in separate male and female groups within the same area (Waterman, 1997)? Why do female dogfish, Scyliorhinus canicula, use different feeding and resting areas than males (Chapter 8)? Or why do humans prefer to socialize with same-sex peers (Chapter 12; Maccoby, 1998)? Some species show social segregation of the sexes: males and females are found in different groups outside the mating season, but use the same areas and habitat types (see Chapters 9 and 10). In other cases, males and females may use different habitats, either within the same or in different areas. This type of segregation is referred to as habitat segregation.

What causes segregation of the sexes, and what are its adaptive advantages? Sexual segregation has been an important research focus in ungulate ecology, but until recently only a few studies had been done on other vertebrate species. There may be more than one reason why the sexes segregate, and the factors responsible for segregation may vary within and across the different vertebrate taxa. It has been suggested that social segregation is caused by habitat segregation but this need not necessarily be so (see Chapters 2, 8, 10 and 11). Let us start by looking at what could cause habitat segregation and social segregation separately, and then we will discuss how they could affect or be independent of each other.

The causes of habitat segregation may vary. In some cases, for example, sex differences in body size may be responsible for habitat segregation. Dimorphism in body size could result in divergent nutritional and energetic requirements, reproductive strategies, activity budgets and social affinities (Main & Coblentz, 1990; Bon, 1991; Ruckstuhl & Neuhaus, 2002). Comparative analyses on ungulates confirm that size-monomorphic social species generally live in mixed-sex groups or pairs displaying similar social and habitat preferences and behaviours, while sexually dimorphic species segregate outside the breeding season (Mysterud, 2000; Ruckstuhl & Neuhaus, 2002). Sexual dimorphism in body length or body size could also allow males and females to access different food items. Male giraffe, *Giraffa camelopardalis*, or elephants, *Loxodonta africana*, are able to feed higher up in the canopy than the smaller females (Ginnett & Demment, 1999; see also Chapter 2). Other examples including diving birds (i.e. some penguins, see Chapter 18), Antarctic fur seals, *Arctocephalus gazella* (Chapter 4), or larger whales (Chapter 16). Males in these species might be able to dive deeper and exploit other food sources than their smaller female counterparts, who are limited to foraging closer to the surface. In such cases sexual segregation in diet would not be adaptive, but a by-product of sexual differences in body size.

Body size dimorphism seems a likely candidate to explain sex differences in ecology. There are, however, notable exceptions, which include some species of bats (see Chapter 15), in which sexes segregate despite a lack of size dimorphism, chimpanzees, *Pan troglodytes* (Chapter 17), many social carnivores, and killer whales, *Orcinus orca* (Chapter 16), where size-dimorphic males and females are found in mixed-sex groups. Resource limitations (Oli, 1996), the distribution and receptivity of females (chimpanzees and whales), physiological limitations (bats), predator avoidance and other factors may well explain these departures from the rule.

What causes social segregation? Although habitat segregation will invariably lead to social segregation, social segregation can occur in the absence of habitat segregation (see Chapters 2, 10, 11). Differences in social affinity (Chapter 11), sexual avoidance (Geist & Bromley, 1978), sexual differences and asynchrony in activity budgets (Chapter 10) have all been suggested to lead to social segregation in the absence of habitat segregation. Ruckstuhl (1998) and Conradt (1998a) have independently suggested that sexual differences in activity budgets could lead to social segregation. If male and female activities differ considerably the cost of synchronizing activities could be too high, leading to the fission of mixed-sex into uni-sex groups. According to this hypothesis social segregation could even occur on different levels and lead to groups assorted by sex, age and reproductive status. Bon (1991), on the other hand, suggested that innate social preferences for the same sex lead to social segregation in adulthood. Geist and Bromley (1978) further suggested that male deer, for example, would segregate from females when they had antlers and only return into female groups once they had shed their antlers. They argued that males with antlers are conspicuous to predators, while males without antlers are using female mimicry to avoid predation. Lastly, males and females might avoid each other due to increased aggression in mixed-sex groups. Increased intra-sexual aggression was shown in the Roosevelt elk, *Cervus elaphus roosevelti*, where females were more aggressive towards each other in the presence than in the absence of males in their group (Weckerly *et al.*, 2001).

What then, if at all, is the relationship between social and habitat segregation? There are three equally likely scenarios one could propose. Firstly, habitat segregation leads to social segregation (Main *et al.*, 1996). Secondly, it is possible that habitat and social segregation are two independent phenomena (Conradt, 1999). Thirdly, social segregation leads to habitat segregation, as suggested in the model by Ruckstuhl and Kokko (2002). We will not investigate these different scenarios here, as they will be dealt with in more detail in Parts II to VI of the book.

The factors that cause or constrain sexual segregation appear to vary and are difficult to differentiate. Sexual segregation could be caused by a multitude of factors. Or is it not? Is there a single or a defined set of factors that could explain sexual segregation in all vertebrates? Although it may seem unlikely, there are at least some factors that can explain social and habitat segregation across taxa. These potential candidates are sexual differences in energy requirements and sensitivity to temperatures (Chapters 2, 3, 4, 15), sexual differences in reproductive strategies and predator vulnerability (Chapters 7, 8, 9, 13 and 18) and sexual differences in social affinities and activity budgets (Chapters 10 and 11), all of which are discussed in depth in this book.

What are the ecological consequences of sexual segregation? The study of sexual segregation has important implications for studies on population viability, gene flow, management of economically relevant species and the conservation of rare species (Clemmons & Buchholz, 1997; Komdeur & Deerenberg, 1997). The abundance and spatial distribution of food may affect the spacing or habitat use of males and females differently. Hence, the further male and female ranges are apart the less likely an encounter for breeding will occur. It is therefore conceivable that gene flow in small or reintroduced populations is limited if only subgroups of males and females of adjacent home ranges meet during breeding cycles (Komdeur & Deerenberg, 1997). While differences in body size may allow males and females to exploit resources differently (as mentioned earlier), this might also expose one sex to a higher predation risk than the other. Asian elephant bulls, *Elephas maximus*, for example, raid crops in India and are sometimes killed because of their transgressions, while females avoid human settlements (Sukumar & Gadgil, 1988). Female wandering albatross, *Diomedea exulans* (see Chapter 5), or females of some fish species (Wirtz & Morato, 2001) are at higher risk than males, because their foraging range overlaps with fishery activities, where they often get trapped and killed. Such cases of sex-biased mortality could lead to local extinctions of a species, particularly if female numbers are low or populations are small (see Chapters 5 or 18). It is therefore imperative to better understand the requirements of each sex of a species in order to protect the species effectively. Chapters 18 and 19 will focus on these conservation aspects. To summarize, it is possible that sexual segregation has many different ecological consequences that need to be taken into account by conservation groups and management, such as affecting sex specific carrying capacity, sex ratio in space and time, population distribution, variability in emigration and immigration and gene flow.

What does this book cover? This book reviews and discusses the different ecological factors affecting sexual segregation in fish (Chapters 7 and 8), reptiles (Chapter 13), birds (Chapters 5, 6 and 18), bats (Chapter 15), ungulates (Chapters 2, 3, 9, 10, 11 and 19), seals

(Chapter 4), whales (Chapter 16), marsupials (Chapter 14), monkeys, apes (Chapter 17) and humans (Chapter 12), and discusses important conservation aspects. The book is divided into eight parts. Part I is meant as an introduction to the topic, while Part II deals with definitions and measures of segregation. Part III describes in detail sexual differences in foraging ecology, while Part IV describes cases where predator avoidance and reproductive strategies may cause ecological segregation. Part V discusses sex related activities and social preferences and how these could cause social segregation. Part VI gives an overview of sexual differences in ecology within different taxa of vertebrates, while Part VII deals with implications for conservation and management. Part VIII includes a short conclusion and directions for future research. With this book, we portray and discuss sexual segregation in its many different forms and possible causes. By doing so, we will not only clarify ideas and concepts, but also initiate and focus on promising future research on this topic.

Part II Concepts and methodology

LARISSA CONRADT

2

Definitions, hypotheses, models and measures in the study of animal segregation

OUTLINE

In this chapter, the term 'segregation' is clarified, the different types of segregation are defined, and the importance of using terms accurately and consistently is illustrated. The rationale of the key hypotheses relating to causes of different types of segregation is briefly explained and a measure of the degree of segregation (the 'segregation coefficient') is presented that is suitable (i) to test hypotheses relating to segregation; and (ii) to make comparisons between populations, species and types of segregation. The chapter concludes with a model to predict the degree of social segregation in a population and ends with three brief empirical examples.

DEFINITIONS OF DIFFERENT TYPES OF SEGREGATION
AND DISTINCTION BETWEEN THEM

In many group-living animals, individuals of different classes such as, for example, males and females, subadults and adults, or large and small individuals tend to form separate social groups (e.g. Croft et al., 2003). This is termed 'social segregation' (Villaret & Bon, 1995, 1998; Bon & Campan, 1996). Further, classes of animals often differ in their habitat use, which is called 'habitat segregation', and/or they differ in their area use, which is termed 'spatial segregation' (e.g. Clutton-Brock et al., 1982), whereby 'spatial segregation' should be treated as an auxilliary concept (see later). Additionally, the terms 'diet segregation' (the classes differ in diet choice) and 'temporal segregation'

Sexual Segregation in Vertebrates: Ecology of the Two Sexes, eds. K. E. Ruckstuhl and P. Neuhaus.
Published by Cambridge University Press. © Cambridge University Press 2005.

(the classes use the same area but at different times of the year) are used by many authors (e.g. see Chapter 14). In the past, many authors have failed to distinguish social segregation clearly from habitat segregation (see Main et al., 1996). In particular, the wide and indiscriminate use of the unfortunate term 'sexual segregation' (referring to either social or habitat segregation between males and females) has significantly confounded theoretical arguments (Villaret & Bon, 1995; Bon & Campan, 1996; Conradt, 1997, 2000). The term 'ecological segregation' has also been used when referring to 'habitat segregation' (e.g. Bon & Campan, 1996). However, for reasons of consistent terminology (see Ruckstuhl & Neuhaus, 2000) the term should be avoided in future.

Habitat segregation describes differences between animal classes in their spatial distribution and, thus, in their use of the physical environment (e.g. Clutton-Brock et al., 1982). Its calculation is based on the categorization of habitat types, whereby habitat categorization carries the danger of not being adequate from the point of view of the animals. Therefore, habitat segregation should be studied complementarily with spatial segregation (Conradt et al., 2001), whereby spatial segregation is based on grid references (e.g. Clutton-Brock et al., 1982). However, it has to be remembered that the concept of 'spatial segregation' is an auxilliary concept, since, lastly, social segregation leads also to small-scale spatial segregation.

Social segregation needs to be clearly distinguished from habitat segregation (Villaret & Bon, 1995; Conradt, 1998a) for the following reasons. Firstly, social segregation could occur independently of habitat segregation. For example, animals of different classes could use the same habitats but at different times, and, thus, be segregated socially, but not with respect to habitat use (Fig. 2.1(c): Conradt, 1999). Alternatively, animals of different classes could prefer different habitats, but when found within the same area they might randomly socialize with individuals of other sex or age classes (e.g. bats: Russo, 2002). They would then display habitat segregation but no social segregation beyond the social segregation caused as a by-product of habitat segregation (Fig. 2.1(b)). Secondly, social and habitat segregation are likely to have different causes and consequences (e.g. Ruckstuhl & Neuhaus, 2000). For example, social avoidance of aggressive, large individuals by small individuals might explain social segregation between size classes but not necessarily habitat segregation, since animals can often exploit the same habitat types at different times or in different areas (Fig. 2.1(c)). Differences in activity rhythms between animal classes might also explain social segregation (Conradt, 1998b; Ruckstuhl, 1998,

(a)   Inter-sexual habitat and
        social segregation

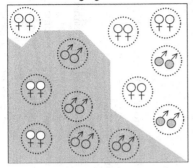

(c)   Inter-sexual social but no
        habitat segregation

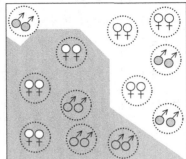

(b)   Inter-sexual habitat but no
        social segregation

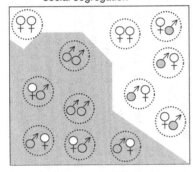

(d)   Neither inter-sexual social
        nor habitat segregation

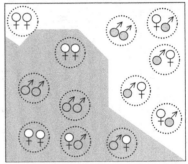

Figure 2.1 Relationships between social and habitat segregation. The diagrams depict the distribution of twelve social groups (circles) of two animals each between two habitats (grey and white). The degree of inter-sexual habitat segregation is the same in (a) and (b): females prefer the white and males the grey habitat (i.e. eight females and four males are in the white habitat and four females and eight males in the grey habitat). In (a) the sexes segregate into separate social groups *within* each habitat additionally to the segregation *between* habitats (all groups are uni-sex); while in (b) the sexes form groups at random *within* each habitat (two-thirds of groups are mixed-sex). In (c) and (d) the animals do not segregate between habitats (i.e. six males and six females in each habitat). However, in (c) the sexes segregate socially (all groups are uni-sex); while in (d) the sexes form groups at random within each habitat (half of the groups are mixed-sex).

Table 2.1 *Main predictions of the hypotheses that have been suggested to explain habitat segregation.*

Main environmental factor	Forage selection hypothesis (forage availability)	Scramble competition hypothesis (forage availability)	Predation risk hypothesis (predation risks)	Weather sensitivity hypothesis (weather conditions)
			Predictions	
Degree of segregation increases with:				
class difference in body size?	yes	yes	no prediction	yes
forage shortage?	yes	yes	no prediction	(yes)
predation risk and/or vulnerability?	no	no	yes	no
adverse weather conditions?	no	no	no	yes
population density?	no	yes, with density of smaller competitors	no	no obvious prediction
sex ratio?	no	yes	no	no
Differences between classes with respect to:				
use of high quality forage habitat	preferred by smaller sized class	preferred by smaller sized class	preferred by less predation sensitive class	preferred by less weather sensitive class
use of safe habitats	none	none	preferred by more predation sensitive class	none
use of sheltered habitats	none	none	none	preferred by more weather sensitive class
effect of weather conditions on habitat use	none	none	none	adverse weather conditions with stronger negative influence on high quality habitat use by more weather sensitive class

1999; Conradt & Roper, 2000) but not why animal classes should differ in habitat choice (see Cransac *et al.*, 1998; Ruckstuhl & Neuhaus, 2000; see also Chapter 10).

Diet segregation is often a consequence of habitat segregation, but can also occur when both classes use the same habitats, if they select different forage within a habitat (e.g. Chapter 3). Temporal segregation is also often a form of habitat segregation, whereby animals of different classes choose different habitats during only part of the year (e.g. wintering grounds of birds: Chapter 18). Therefore, their causal explanation is likely to be found in one or more of the hypotheses that aim to explain habitat segregation (see later).

POTENTIAL EXPLANATIONS OF SEGREGATION

Habitat segregation

Habitat segregation refers to differences between animal classes in their spatial distribution. Therefore, factors that affect the spatial distribution of animals could affect animal classes differently and cause habitat segregation. The major factors that affect the spatial distribution of most segregating animals are (i) forage and water (e.g. Clutton-Brock *et al.*, 1982); (ii) predation (e.g. Jakimchuk *et al.*, 1987; Lingle, 2002; Lingle & Pellis, 2002); and (iii) the physical environment, especially the availability of shelter against weather conditions (e.g. Jackes, 1973). The key hypotheses which have been suggested to explain habitat segregation are based on differences between animal classes in requirements with respect to one or more of these three factors. In the following, a brief summary of the rationale of the key hypotheses is given.

Forage selection hypothesis

Firstly, animal classes could differ in forage or water requirements. For example, lactating females, juveniles or small individuals often need access to higher quality forage or more water than do males or larger individuals, because of the allometric relationship between metabolic requirements and body size/gut capacity (Bell, 1971; Jarman, 1974; Van Soest, 1982; Cork, 1991; Oswald *et al.*, 1993; however, see also Conradt *et al.*, 1999a). Differences in forage or water requirements can lead to habitat segregation if different habitats offer dissimilar access to these resources (see Ruckstuhl & Neuhaus, 2002 for a review). Testable predictions for the hypothesis are summarized in Table 2.1. Forage-related hypotheses are treated in detail in the chapters of Part III.

Scramble competition hypothesis

Additionally to differences in forage requirements, asymmetry in forage competition could increase habitat segregation. For herbivores, Clutton-Brock et al. (1982, 1985, 1987a) suggested that larger individuals are less competitive feeders because of an allometric relationship between bite and body size. For example, large male ungulates have a lower ratio of incisor arcade width relative to absolute metabolic requirements than do the smaller females (Illius & Gordon, 1987, 1992). As a result, smaller herbivores tolerate lower plant biomass (Bell, 1971; Jarman, 1974). Therefore, at times of forage shortage, smaller herbivores can forage the plant biomass so low in high quality forage habitats that larger herbivores are no longer able to obtain sufficient forage intake rates and are, thus, forced by indirect competition into marginal habitats with lower forage quality but higher plant biomass (Clutton-Brock et al., 1982, 1987a). This hypothesis is termed the 'scramble competition hypothesis'. In the past it has also sometimes been referred to as 'body size hypothesis' (Main et al., 1996). However, for reasons of consistent terminology, the use of the second term should be avoided. Testable predictions for the hypothesis are summarized in Table 2.1.

Predation risk hypothesis

Secondly, animal classes could differ in predation sensitivity or risk. For example, females with young or juveniles are often more vulnerable to predation than are adult males (e.g. Bowyer, 1984; Jakimchuk et al., 1987; Lingle, 2000; Lingle & Wilson, 2001), and might therefore be more sensitive to predation risks. Such differences in predation sensitivity between animal classes could lead to habitat segregation if some habitats offer better protection from predators but are inferior with respect to other resources, such as forage. In such a case, one animal class might stay in a safe but nutritionally inferior habitat, while the other class followed a more risky strategy by using more dangerous habitats with better access to forage (e.g. Jakimchuk et al., 1987; du Toit, 1995; Bleich et al., 1997; Berger, 1991; Lingle, 2002). This hypothesis has been termed the 'reproductive strategy hypothesis' when it relates to intersexual habitat segregation (see Main et al., 1996) or, more generally, the 'predation risk hypothesis' (Ruckstuhl & Neuhaus, 2000, 2002). For reasons of consistent terminology, the use of the later, more general term is encouraged. Testable predictions of the hypothesis are summarized in Table 2.1. Predation-related hypotheses are treated in detail in the chapters of Part IV.

Physical conditions hypotheses

Thirdly, animal classes could differ in their choice of physical environments, namely temperature, humidity and weather conditions (e.g. see Chapters 8, 13, 15 and 18). Comparatively little empirical information exists on this topic.

Weather sensitivity hypothesis

The weather sensitivity hypothesis is a special case of the physical conditions hypothesis. It suggests that animal classes could differ in weather sensitivity (e.g. Jackes, 1973; Young & Isbell, 1991). Conradt et al. (2000) suggested that, for example in ungulates, larger individuals might be more sensitive to cold weather because their net energy gain decreases with absolute heat loss more steeply than that of smaller individuals. (Since this argument seems counter-intuitive, it is illustrated in more detail in a model later). Such differences in weather sensitivity between animal classes could lead to habitat segregation if most habitats offered either shelter or good foraging opportunities but rarely both, as is often the case (Ozoga, 1968; Staines, 1976, 1977). In such a situation, the more weather sensitive (e.g. larger) animals would prefer to stay in better sheltered but nutritionally inferior habitats, while the less weather sensitive (e.g. smaller) animals used more exposed habitats with better access to forage (Jackes, 1973; Conradt et al., 2000). This hypothesis has been termed the 'weather sensitivity hypothesis', and its testable predictions are summarized in Table 2.1. As no further chapters will address the weather sensitivity hypothesis, a brief empirical example is given in Box 2.1.

Social segregation

Four main hypotheses have been suggested to explain social segregation (see also chapters of Part IV). They are not mutually exclusive.

By-product hypothesis

The *first* hypothesis suggests that social segregation is a by-product of habitat segregation (e.g. Clutton-Brock et al., 1982; see also later): if animal classes use different habitats, they necessarily have to dwell to a certain extent in different social groups (Figure 2.1(b)).

Social avoidance hypothesis

The *second* hypothesis suggests that animals of different classes avoid each other socially, for example because of inter-class aggression (see Main *et al.*, 1996 for a review, but see also Weckerly *et al.*, 2001). Another reason for inter-class avoidance could be inter-class differences in activity budgets. Since this is such a potentially important mechanism, it is treated below as a separate hypothesis.

Social attraction hypothesis

The *third* hypothesis suggests that social segregation is due to an attraction between animals of the same class to facilitate social learning (Appleby, 1982, 1983; Bon & Campan, 1996; Villaret & Bon, 1995, 1998, see Chapter 11 for details) or lower conspicuousness to predators (Lingle, 2001; Croft *et al.*, 2003).

Activity budget hypothesis

The *fourth* hypothesis is the 'activity budget hypothesis'. Its rationale is as follows. A social group can only be spatially coherent if its members synchronize activities such as foraging and resting (Conradt, 1998b, Ruckstuhl, 1998). For example, if an animal wishes to remain with its group it cannot go off and forage while the rest of the group is sleeping, or stay behind and sleep while the rest of the group is foraging (Rook & Penning, 1991; Côté *et al.*, 1997). In order to synchronize their activities with those of other group members, an individual may have to compromise its own activity budget (Ruckstuhl, 1999). This will entail a cost (Conradt & Roper, 2003), which could influence the individual's decision to remain in the group and, as a consequence, group stability (Conradt & Roper, 2000). The costs of activity synchronization for individual group members are likely to be particularly high in groups that include members of different sex, size or age classes because the optimal allocation of time to various activities is likely to differ between such classes (e.g. Clutton-Brock *et al.*, 1982; Prins, 1987; Gompper, 1996; Ruckstuhl, 1999). These higher costs between group members of different classes could cause mixed-class groups to be less well synchronized in activity and, therefore, to be less stable, so that uni-class groups predominate in the population and the animal classes are socially segregated (Conradt, 1998b; Ruckstuhl & Neuhaus, 2000; see also later). The 'activity budget hypothesis' is treated in detail in Chapter 10.

MEASURES OF THE DEGREE OF SEGREGATION

Confounding factors when measuring segregation

The quantitative study of segregation and the testing of hypotheses require a suitable measure for the degree of segregation. In the past, authors have used a variety of ecological measures of overlap, e.g. the percentage of animals in uni-sex groups vs. mixed-sex groups (Nievergelt, 1981; Hillman, 1987; Owen-Smith, 1993; see Conradt, 1998a for a review). However, most of these measures of overlap have the problem that they are stochastically dependent on animal density, group sizes, class ratio (e.g. sex ratio) and similar variables. I illustrate the problem in the following, using hypothetical examples.

Confounding effects of sex ratio

Consider a population A with a sex ratio of $1♀:1♂$, in which the sexes do not segregate socially (i.e. degree of social segregation is zero) but associate randomly into groups of size two. In this population, the probability of observing a group with two females is $(0.5)^2 = 0.25$; with two males $(0.5)^2 = 0.25$; and with one female and one male $2 \cdot (0.5)^2 = 0.5$. Thus, we would measure a percentage of 50% of animals in uni-sex vs. 50% in mixed-sex groups. From these values, we could not easily infer that the actual tendency in the population for the sexes to segregate socially is zero. Moreover, let us compare the degree of social segregation in population A to that in a second population B with a sex ratio of $3♀:1♂$, in which the sexes also do not segregate socially and typical group size is two. In population B, the probability of observing a group with two females is $(0.75)^2 = 0.5625$; with two males $(0.25)^2 = 0.0625$; and with one female and one male $2 \cdot (0.75) \cdot (0.25) = 0.375$. Thus, we would measure a percentage of 62.5% of animals in uni-sex vs. 37.5% in mixed-sex groups. We would conclude that social segregation is larger in population B than in population A, although the actual degree of social segregation is the same in both populations (namely, zero). It is obvious that this measure would pose considerable problems if we were, for example, to investigate the influence of sex ratio on the degree of segregation (e.g. Clutton-Brock et al., 1987a) or to make comparisons between species with different degrees of sexual dimorphism, which also differ in sex ratios (e.g. Ruckstuhl & Neuhaus, 2002; Neuhaus & Ruckstuhl, 2002a).

Confounding effects of population density and group size

Variation in population density and group size can also lead to confounding results. To illustrate this, let us assume that in the next year the population density in population A rises and typical group size increases to four. Thus, the probability of observing a group with four females is $(0.5)^4 = 0.0625$; with four males $(0.5)^4 = 0.0625$; with three females and one male $4 \cdot (0.5)^3 \cdot (0.5) = 0.25$; with two females and two males $6 \cdot (0.5)^2 \cdot (0.5)^2 = 0.375$; and with one female and three males $4 \cdot (0.5) \cdot (0.5)^3 = 0.25$. The measured social segregation would be 12.5% of animals in uni-sex vs. 87.5% in mixed-sex groups. We would conclude that degree of social segregation has decreased with increasing population density in population A, while in reality it is still zero. Again, the problems of this measure are obvious for the study of the influence of population density on segregation (e.g. Clutton-Brock et al., 1987a) and comparative studies between populations or species (e.g. Ruckstuhl & Neuhaus, 2002). Other measures of overlap pose similar problems (see Conradt, 1998a for a review).

Conradt's segregation coefficient

Conradt (1998a) suggested, as a measure of the degree of segregation, the 'segregation coefficient' (SC), and showed that this measure is better at avoiding confounding effects of population density, group sizes or the relative number of animals in either class type than other ecological measures which have been used in the past. The segregation coefficient is therefore suitable for studying the influence of environmental factors and of population density, sex ratio, group sizes, etc. on the degree of segregation. It can further be used to test hypotheses and to make comparisons between populations, species and types of segregation. The segregation coefficient reflects the proportion of animals that decide to segregate (see Conradt, 1998a for mathematical details). If there is segregation, SC takes values between a maximum of one (complete segregation between animals of Class I and Class II) and zero (no segregation). However, note that if there is aggregation (i.e. the opposite of segregation) between animals of Classes I and II, the value in the square-root becomes negative and SC is no longer defined.

Thus, a negative value in the square-root indicates that there is no segregation but aggregation. The segregation coefficient cannot be used as a measure of aggregation because in the case of aggregation, it would be sensitive to confounding factors.

The suggested measure of the degree of segregation, SC, between two animal classes is (Conradt, 1998a):

$$SC = \sqrt{1 - \frac{X + Y - 1}{X \cdot Y} \cdot \sum_{i=1}^{k} \frac{x_i \cdot y_i}{x_i + y_i - 1}} \qquad (2.1)$$

where for

> **social segregation:** x_i is the number of animals of Class I in the ith group; y_i is the number of animals of Class II in the ith group; k is the number of groups with at least two animals (i.e. solitary animals are excluded); X is the total number of animals of Class I in all k groups; Y is the total number of animals of Class II in all k groups;
>
> **habitat segregation:** x_i is the number of animals of Class I in the ith habitat type; y_i is the number of animals of Class II in the ith habitat type; k is the number of habitat types which are used by at least two animals; X is the total number of animals of Class I in all k habitat types; Y is the total number of animals of Class II in all k habitat types (the measure is sensitive to the classification of habitat types);
>
> **spatial segregation:** x_i is the number of animals of Class I in the ith grid square; y_i is the number of animals of Class II in the ith grid square; k is the number of grid squares which are used by at least two animals; X is the total number of animals of Class I in all k grid squares; Y is the total number of animals of Class II in all k grid squares (the measure is sensitive to grid square size, i.e. spatial scale). It is important that sample sizes are large. Conradt (1998a) recommends that $X > 30$; $Y > 30$; and $k \geq 10$.

Quantitative relationships between different types of segregation

As well as being used to investigate the influence of environmental factors on segregation and to test hypotheses, Conradt's (1998a) social, habitat and spatial segregation coefficients can be used to examine the quantitative relationship between social, habitat and spatial

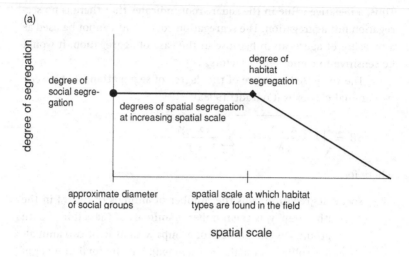

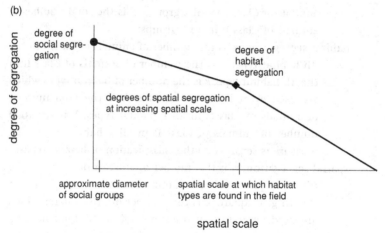

Figure 2.2 Theoretical relationships between degrees of social, spatial and habitat segregation. (a) If social segregation is only a by-product of habitat segregation: the degree of social segregation is not larger than the degree of habitat segregation. (b) If social segregation occurs additionally to habitat segregation: the degree of social segregation is larger than the degrees of habitat and spatial segregation at all spatial scales.

segregation. This is, for example, useful to test whether social segregation is just a by-product of habitat segregation. If animal classes use different habitats, they necessarily have to dwell to a certain extent in different social groups. However, there might, or might not, be social segregation additionally to the social segregation caused by habitat segregation (compare Figs. 2.1(a) and (b)). If animal classes form separate social groups to a larger extent than expected by differences in habitat use, social segregation can no longer be explained as a by-product of habitat segregation (Conradt, 1999; see Figs. 2.1(a) and (c)) but is likely to have different causes (Bon & Campan, 1996; Conradt, 1998b; Ruckstuhl, 1998; see also later). Conradt (1998a) has shown that if animal classes segregate with respect to habitat use but socialize randomly within habitats (i.e. there is no independent social segregation; Figs. 2.1(b) and 2.2(a)), then: $SC_{social} = SC_{habitat \, (for \, small \, spatial \, scales)}$. On the other hand, if animal classes segregate socially within habitats, additionally to segregation between habitats (Figs. 2.2(a) and 2.2(b)), then: $SC_{social} > SC_{habitat \, (for \, all \, spatial \, scales)}$ (Fig. 2.2.; see Conradt, 1998a for mathematical details). To illustrate this concept, a brief empirical example is given in Box 2.1.

Box 2.1 Empirical examples

A. Test of the weather sensitivity hypothesis

Male and female red deer (*Cervus elaphus* L.) on the Isle of Rum are segregated with respect to habitat use in winter (Clutton-Brock et al., 1982, 1987a). The weather sensitivity hypothesis predicts that (i) the larger males should forage in better-sheltered habitats than the smaller females; and (ii) males should be more strongly influenced by adverse weather conditions in their choice of good foraging habitats (Table 2.1). The most crucial weather factor for red deer on Rum is wind speed (Clutton-Brock et al., 1982, 1985). Conradt et al. (2000) found that (i) wind speed was significantly lower at male than at female foraging sites on windy days (Fig. 2.3); and (ii) the influence of wind, temperature and rain on the use of high quality foraging habitats was significantly stronger in males than in females (Table 2.2; Conradt et al., 2000). They concluded that the weather sensitivity hypothesis could explain inter-sexual habitat segregation in red deer on the Isle of Rum (Conradt et al., 2000).

B. Is social segregation a by-product of habitat segregation?

Conradt (1999) examined whether social segregation is a by-product of habitat segregation in red deer on Rum, and in a population of feral Soay sheep on the Isle of St Kilda. She

Table 2.2 *The influence of various weather factors on the use of high
quality forage habitats in winter by female and male red deer, and the
difference between the sexes.*

Weather factor	Males	Females	Difference (males–females)
Wind			
slope of regression line	−0.18	−0.11	−0.07*
t-value	−6.23	−4.61	2.00
p-value	<0.001	<0.001	<0.05
Temperature			
slope of regression line	0.08	0.24	0.16*
t-value	2.00	5.23	2.82
p-value	<0.05	<0.001	<0.01
Rainfall			
slope of regression line	−0.0006	−0.0017	−0.0011
t-value	−1.96	−3.27	−1.82
p-value	<0.05	<0.01	0.07

* Differences in regression slopes between the sexes, and, thus, differences
in the influence of the weather factor on habitat choice, were significant.
Shown are the results from a multiple logistic regression of weather factors
on high quality habitat use by female and male red deer in winter (sample
size: 546 census days (males) and 614 census days (females); census days
over a period of 20 years). Source: Conradt *et al.*, 2000.

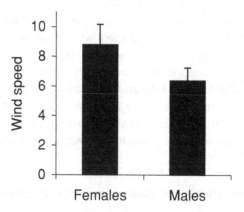

Figure 2.3   Average wind speed (+/−SE) at corresponding female and
male foraging sites. The differences between the sexes were significant
(Wilcoxon test: $T = 20$, $n = 14$, $p < 0.025$, one tailed).

(a)

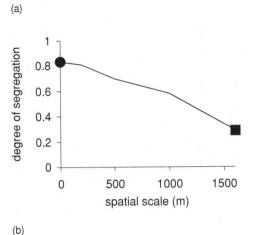

(b)

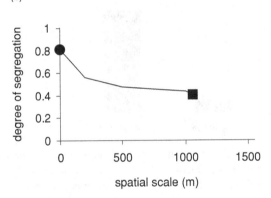

Figure 2.4 Average degrees of social (filled circle), habitat (filled square) and spatial (line) segregation at different spatial scales outside the rut in (a) red deer; and (b) feral Soay sheep.

predicted that (i) the degree of social segregation should not be larger than the degree of habitat segregation; and (ii) the degree of social segregation should be influenced by the environmental parameters that influence habitat segregation. However, Conradt (1999) found that the degree of social segregation outside the rutting season was significantly larger than the degree of habitat (and spatial) segregation for both species (results of t-tests for red deer: $t > 2.7$, $p < 0.01$ in all cases, $n = 20$ years of census data; for Soay sheep: $t > 12.0$, $p < 0.001$ in all cases, $n = 10$ years of census data; Fig. 2.4). The second prediction could only be tested in red deer. While the degree of habitat segregation depended significantly on weather conditions (wind speed: $F_{1,730} = 8.9$,

(a)

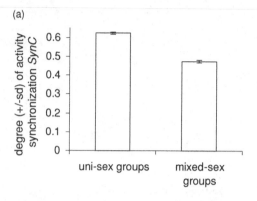

(b)

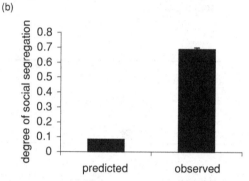

Figure 2.5   (a) Degrees (+/−SD) of activity synchronization in uni- and mixed-sex groups in a population of red deer; (b) predicted vs. observed degrees of social segregation in the same population.

$p < 0.01$, significant; temperature: $F_{1,729} = 18.2$, $p < 0.001$, significant; source: Conradt, 1997), degree of social segregation did not (wind speed: $F_{1,443} = 1.5$, $p = 0.21$, NS; temperature: $F_{1,443} = 1.1$, $p = 0.29$, NS; source: Conradt, 1999). Conradt (1999) concluded that social segregation was not a by-product of habitat segregation in either species.

C. Can differences in activity synchronization between uni-sex and mixed-sex classes explain inter-sexual social segregation?

Conradt and Roper (2000) used the activity synchronization model described earlier in this chapter, together with data on red deer from the Isle of Rum, to test whether lower activity synchronization in mixed-sex than in uni-sex groups could explain the inter-sexual social segregation observed in deer. Red deer were significantly better synchronized in uni-sex than in mixed-sex groups (t-test: $t = -8.5$, $p < 0.001$, $n = 153$ months in 20 years; Fig. 2.5(a); see also: Conradt, 1998b). Mean group size n was 2.8 adult deer and the proportion of male solitary animals was $\hat{p} = 0.31$ (Conradt and Roper, 2000). However, the predicted degree of social segregation, based on differences in activity synchronization between group types, was much lower than the observed degree of social segregation in the population (Fig. 2.5(b); see also: Conradt and Roper, 2000). Thus, although differences in activity synchronization existed between group types, they were not large enough to account fully for the observed degree of social segregation (see also: Ruckstuhl and Kokko, 2002).

MODELS RELEVANT TO THE STUDY OF SEGREGATION

Weather sensitivity model

At times of forage shortage and in cold weather conditions (e.g. in winter), energy intake can be very limited, heat loss considerable and energy gain a crucial factor for survival in ungulates (Clutton-Brock et al., 1982; Bubenik, 1984; Wagenknecht, 1986: Illius & Gordon, 1987, 1992). Ungulates should, thus, select a habitat in which their net energy gain is largest. The net energy gain in a habitat depends mainly on intake of digestible energy minus heat loss (Conradt, 1997). Energy intake and heat loss, in turn, are body size dependent (Clutton-Brock et al., 1987a; Illius & Gordon, 1987; Bubenik, 1984). The expected net energy gain per time (net_i) in a habitat i for an animal of body weight bw is (Conradt et al., 2000):

$$net_i = e_i \cdot bw^{0.48} - h_i \cdot bw^{0.63} - m \cdot bw^{0.66} \qquad (2.2)$$

where e_i is a habitat-specific rate of intake of digestible energy that depends on standing crop and quality of forage in habitat i; h_i is a habitat-specific rate of heat loss that depends on exposure in habitat i and on weather conditions; and m is the rate of movement expenditure (assuming that movement expenditure is similar in all habitats). The exponents are derived from published literature (for forage intake: Illius & Gordon, 1987; for heat loss and movement expenditure: Bubenik, 1984). Consider a habitat A, which offers better foraging opportunities than a habitat B (i.e. $e_A > e_B$), but is more exposed (i.e. $h_A > h_B$). An animal should switch from habitat A to habitat B, if $net_A < net_B$, and (using Eq. 2.2), thus, if:

$$(e_A - e_B) \cdot bw^{-0.15} < h_A - h_B \qquad (2.3)$$

Therefore, if the difference in site- and weather-specific rate of heat loss ($h_A - h_B$) exceeds a body weight dependent threshold ($(e_A - e_B) \cdot bw^{-0.15}$), animals are better off in a sheltered habitat with inferior foraging opportunities than in an exposed habitat with better foraging opportunities. For the larger animals this threshold is lower than for the smaller animals. This is because energy intake scales in a more strongly decelerating way with body weight than does heat loss. Furthermore, heat loss increases more steeply in exposed than in sheltered habitats with adverse weather conditions (Conradt et al., 2000). Therefore, the difference between habitats ($h_A - h_B$) increases with increasingly bad weather and reaches the threshold ($(e_A - e_B) \cdot bw^{-0.15}$) in milder weather conditions in large than in small animals (see Box 2.1).

Activity synchronization model

It has been suggested that differences between animal classes (i.e. the sexes) in activity rhythms/budgets result in higher asynchrony in activity in mixed-class compared to uni-class groups, and that this 'activity asynchrony' leads to relatively unstable mixed-class groups, to a predominance of uni-class groups in the population, and, thus, to social segregation ('activity budget hypothesis': Conradt, 1998b; Ruckstuhl, 1998, 1999; Conradt & Roper, 2000; Ruckstuhl & Neuhaus, 2001; see also Box 2.1). Conradt & Roper (2000) developed a model, which (i) predicts the degree of social segregation in a population based on observed differences in activity synchrony in mixed-class versus uni-class groups; and (ii) can be used to test the activity budget hypothesis (Conradt & Roper, 2000; Ruckstuhl & Neuhaus, 2000). The model is a system of differential equations that describe changes in the numbers of groups of different type (i.e. uni- and mixed-class groups) over time within a population. The system has a stable equilibrium (i.e. when the number of breaking-up groups equals the number of newly fusing groups of each type, giving no further net changes in group numbers). The number of groups of different types at equilibrium can, therefore, be used to calculate the expected degree of social segregation in the population, using Conradt's (1998a) social segregation coefficient SC_{social} (see above).

Model assumptions

The model assumes that (i) animals are free to leave or join groups ('fission-fusion groups': e.g. Clutton-Brock *et al.*, 1982; Prins, 1987; Albon *et al.*, 1992; Raman, 1997); (ii) group fission rates depend on the degree of activity synchronization, whereby mixed-class groups are less well synchronized than uni-class groups; (iii) fission rates are proportional to the proportion of group members that do not synchronize their activities (i.e. the less likely members are to synchronize activities, the more likely the group is to break apart); and (iv) individuals meet at random and fuse into a new group with a probability that is independent of animal class.

Differential equation system

Let $G_{i,n-i}(t)$ be the number of groups of size n that have i members of Class I and $(n - i)$ members of Class II (whereby $0 \leq i \leq n$) at time t. The change per time in number of groups $\left(\dfrac{dG_{i,n-i}(t)}{dt}\right)$ depends on the number of newly fusing groups minus the number of breaking-up groups per unit time. The number of newly fusing groups, in turn, depends on the probability that i animals of Class I and $(n - i)$ animals of Class II meet by chance (which is proportional to $\left(\dfrac{n!}{i!(n-i)!}\right) \cdot S_I(t)^i \cdot$ $S_{II}(t)^{n-i}$, whereby $S_I(t)$ is the number of solitary animals of Class I at time t, and $S_{II}(t)$ that of Class II) and fuse into a new group (with fusion rate: a). The number of breaking-up groups per time depends on the number of existing groups that could break up, and the rate at which they break up (i.e. fission rate: $b(i) = b$ for uni-class groups with $i = 0$ or $i = n$, and $b(i) = k \cdot b$ for mixed-class groups with $0 < i < n$ and $k > 1$). Then:

$$\frac{dG_{i,n-i}(t)}{dt} = a \cdot \left(\frac{n!}{i!(n-i)!}\right) \cdot S_I(t)^i \cdot S_{II}(t)^{n-i} - b(i) \cdot G_{i,n-i}(t)$$

for all $i : 0 \leq i \leq n$ \hfill (2.4)

whereby the total numbers of animals of Class I and II (T_I and T_{II}) are constant, and, therefore:

$$S_I(t) = T_I - \sum_{i=1}^{n} i \cdot G_{i,n-i}(t)$$ \hfill (2.5)

and:

$$S_{II}(t) = T_{II} - \sum_{i=0}^{n-1} (n - i) \cdot G_{i,n-i}(t)$$

Equilibrium

At equilibrium, the net changes per time in number of groups are zero (i.e. $\dfrac{dG_{i,n-i}(t)}{dt} = 0$ for all i), and the number of groups of each type at equilibrium ($\hat{G}_{i,n-i}$) can be calculated, using Eq. (2.4):

$$\hat{G}_{i,n-i} = \frac{a}{b(i)} \cdot \left(\frac{n!}{i!(n-i)!} \right) \cdot \hat{S}_I^i \cdot \hat{S}_{II}^{n-i} \tag{2.6}$$

where $\hat{S}_I$ is the number of solitary animals of Class I at equilibrium and $\hat{S}_{II}$ that of Class II. Using Eq. (2.6) in Eq. (2.5):

$$\hat{S}_I = T_I - \sum_{i=1}^{n-1} \frac{a}{b(i)} \cdot \left(\frac{n!}{i!(n-i)!} \right) \cdot \hat{S}_I^i \cdot \hat{S}_{II}^{n-i} \cdot i$$

$$= T_I - \frac{a}{b \cdot k} \cdot n \cdot \hat{S}_I \cdot (\hat{S}_I + \hat{S}_{II})^{n-1} - \frac{a}{b} \cdot n \cdot \hat{S}_I^n \cdot \left(1 - \frac{1}{k} \right) \tag{2.7}$$

and:

$$\hat{S}_{II} = T_{II} - \frac{a}{b \cdot k} \cdot n \cdot \hat{S}_{II} \cdot (\hat{S}_I + \hat{S}_{II})^{n-1} - \frac{a}{b} \cdot n \cdot \hat{S}_{II}^n \cdot \left(1 - \frac{1}{k} \right)$$

Using Eq. (2.1) and Eqs. (2.6) and (2.7), the degree of social segregation at equilibrium is:

$$\hat{SC}_{social} = \sqrt{1 - \frac{(T_I - S_I) + (T_{II} - S_{II})}{(T_I - S_I) \cdot (T_{II} - S_{II})} \cdot \sum_{i=1}^{n-1} \frac{i \cdot (n-i)}{n-1} \cdot \hat{G}_{i,n-i}}$$

$$= \sqrt{(k-1) \cdot \hat{p} \cdot (1 - \hat{p}) \cdot \frac{(k-1) \cdot [\hat{p}(1 - \hat{p})]^{n-2} + \hat{p}^{n-2} + (1 - \hat{p})^{n-2}}{[(k-1) \cdot \hat{p}^{n-1} + 1] \cdot [(k-1) \cdot (1 - \hat{p})^{n-1} + 1]}}$$

$$\tag{2.8}$$

where $\hat{p} = \dfrac{\hat{S}_I}{\hat{S}_I + \hat{S}_{II}}$ is the proportion of solitary animals that are of Class I, which can be easily obtained from observational data. k is the factor by which the fission rate of mixed-class groups is higher than that of uni-class groups (see earlier), and, thus, k is proportional to the proportion of group members which do not synchronize activities.

Degree of activity synchronization

Conradt (1998b) recommended as a measure of the proportion of group members, which do not synchronize activities, the synchronization coefficient $SynC$, which measures the degree of activity synchronization in groups:

$$SynC = \sqrt{1 - \sum_{h=\text{observation start}}^{\text{observation end}} \frac{N_h}{N} \cdot \frac{N_h - 1}{A_h \cdot R_h} \cdot \sum_{j=1}^{m_h} \frac{a_{h,j} \cdot r_{h,j}}{n_{h,j} - 1}} \qquad (2.9)$$

where N_h is the total number of non-solitary animals which have been observed in the h-hour since the observation started; N is the total number of non-solitary animals; A_h is the total number of non-solitary active animals that have been observed in the h-hour; R_h is the total number of non-solitary resting animals that have been observed in the h-hour; m_h is the number of groups that have been observed in the h-hour; $a_{h,j}$ is the number of active animals that have been observed in the h-hour in the jth group; $r_{h,j}$ is the number of resting animals that have been observed in the h-hour in the jth group; and $n_{h,j}$ is the size of the jth group in the h-hour. $SynC$ can be obtained from field data (e.g. Conradt, 1998b). It is (Conradt & Roper, 2000):

$$k = \frac{1 - SynC_{\text{mixed-class}}}{1 - SynC_{\text{uni-class}}} \qquad (2.10)$$

where $SynC_{\text{mixed-class}}$ and $SynC_{\text{uni-class}}$ are the observed degrees of activity synchronization in mixed-class and uni-class groups, respectively.

Predicted degree of social segregation at equilibrium

Using Eqs. (2.9) and (2.10) it follows:

$$\hat{S}\hat{C}_{social}$$

$$= \sqrt{\frac{\dfrac{(SynC_{\text{uni-class}} - SynC_{\text{mixed-class}}) \cdot \hat{p} \cdot (1 - \hat{p})}{(SynC_{\text{uni-class}} - SynC_{\text{mixed-class}}) \cdot \hat{p}^{n-1} + 1 - SynC_{\text{uni-class}}}}{\dfrac{(SynC_{\text{uni-class}} - SynC_{\text{mixed-class}}) \cdot [\hat{p}(1 - \hat{p})]^{n-2} + (1 - SynC_{\text{uni-class}}) \cdot (\hat{p}^{n-2} + (1 - \hat{p})^{n-2})}{(SynC_{\text{uni-class}} - SynC_{\text{mixed-class}}) \cdot (1 - \hat{p})^{n-1} + 1 - SynC_{\text{uni-class}}}}}$$

$$(2.11)$$

Therefore, the degree of social segregation at equilibrium depends on (i) the proportion $\hat{p}$ of solitary animals that is in Class I; (ii) the degree of activity synchronization $SynC_{\text{mixed-class}}$ in mixed-class groups in the population; (iii) the degree of activity synchronization $SynC_{\text{uni-class}}$ in uni-class groups; and (iv) the average group size n (including only animals of Class I or II and excluding solitary animals). In order to test the activity budget hypothesis, these four entities can be measured in the field, and the predicted equilibrium degree of social segregation can be calculated and compared to the actually observed degree of social segregation in the population (see Box 2.1 for an empirical example).

ACKNOWLEDGEMENTS

I thank Tim Roper and Tim Clutton-Brock for commenting on the manuscript, and Tim Clutton-Brock for allowing me to use his study site and for access to his long-term data set on red deer. I also thank the Royal Society of London, the Studienstiftung des Deutschen Volkes, the German Academic Exchange Service DAAD, the EU and NERC for financial support.

Part III Foraging ecology

3

Sex differences in the foraging ecology of large mammalian herbivores

INTRODUCTION

Adult males and females of many animal species differ in terms of the taxonomic range of food types they use, and/or the physical and chemical properties of the meals they ingest. Surprisingly, however, recognition and understanding of these differences has advanced slowly. For example, practitioners of wildlife production and conservation typically use total animal numbers for setting stocking rates, estimating area requirements, monitoring plant–animal interactions, etc., with no consideration of sex differences in feeding ecology. Yet the reason why textbooks on wildlife ecology and management (e.g. Caughley & Sinclair, 1994) seldom address sex differences is more the lack of conclusive published information than simple oversight. Hence, my purpose in writing this chapter will be served if it stimulates further research on this ecologically important topic.

My focus here is on sex differences in the foraging ecology of large mammalian herbivores (>5 kg), mostly because the principles underlying diet selection have been better studied in this group than in any other. It is obvious that substantial dietary differences will occur when male and female herbivores feed in separate plant communities, the possible reasons for which are discussed in Chapter 9. Here, however, I discuss sex differences in the consumption of plant material by large herbivores at the scale of the small patch, or feeding station (as defined by Senft et al., 1987). At this scale, if the diets of conspecifics of each sex differ consistently then we can assume it is due to sex differences in diet selection, which is the subject of this chapter.

Sexual Segregation in Vertebrates: Ecology of the Two Sexes, eds. K. E. Ruckstuhl and P. Neuhaus. Published by Cambridge University Press. © Cambridge University Press 2005.

In functional terms, large herbivores remove plant material from the available standing crop of vegetation in two ways: by grazing or browsing (reviewed by Hofmann, 1989). Grazers consume mainly grasses and sedges (monocots) from an essentially two-dimensional carpet close to the ground, while browsers consume forbs and foliage of shrubs and trees (dicots) from discrete three-dimensional structures supported by lignified stems and branches (McNaughton & Georgiadis, 1986). Various large herbivores are mixed feeders, in that they switch between grazing and browsing depending on the accessibility and quality of each food type, although in African savannas, for example, the proportion of mixed feeders in the large herbivore community is actually quite small. As described by Tieszen & Imbamba (1980) in Kenya, most trees and shrubs (dicots) employ the C_3 photosynthetic pathway while most grasses (monocots) employ the C_4 pathway. Because of this the relative proportions of grass and browse ingested by a herbivore can be estimated from the isotopic composition of its hard tissues, because the tissues of C_3 and C_4 plants differ with respect to the $^{13}C{:}^{12}C$ ratio. By determining this ratio in dentine and keratin, and analysing its distribution in the exceptionally diverse ungulate fauna of East Africa, Cerling et al. (2003) found that the 28 savanna bovid species separate out into two main clusters: 13 grazers and 12 browsers; only 3 species (10.7%) are mixed feeders. In this chapter I recognize the functional dichotomy between grazing and browsing guilds and consider how mechanisms underlying sex differences in feeding ecology may operate within each guild, thereby (hopefully) refining the links between observed patterns and suggested processes.

DIFFERENCES IN BODY SIZE

In large herbivorous mammals, increasing body size is associated with increasing absolute intake requirements per individual, which increasingly precludes dietary specialization on discretely distributed high quality foods (Jarman, 1974). Generalized feeding over large overlapping home ranges entails frequent interactions among conspecifics, which, together with common criteria for diet selection and predation avoidance, promotes gregariousness and polygyny (Jarman, 1974; Post et al., 1999; Badyaev, 2002; Pérez-Barbería et al., 2002). Among African savanna antelopes, for example, the transition between solitary and gregarious species occurs at ~30 kg, with the gazelles (Tribe Antilopini) being the smallest to occur in herds of >20 animals (Roberts et al., 2001). In polygynous ungulates it is to be expected that the larger males should (i) be more successful among male competitors for mates, and

(ii) be more attractive to females that exercise mate choice, so sexual selection should favour larger males (Loison *et al.*, 1999). Females, however, should begin reproducing as soon as they attain the body size and condition required for gestation and lactation rather than continuing to invest in their own growth (Post *et al.*, 1999).

So, for various reasons, adult males tend to be larger than adult females in most species of large herbivorous mammal (despite some notable exceptions, to be discussed later in this chapter). For various other reasons (discussed in Chapter 9 and elsewhere in this book), adult males often segregate away from female–young groups, so that the size frequency distribution of animals in a bachelor group of adult males can be quite distinct from that of a group comprising females and their offspring. As will be explained later, there are strong theoretical grounds for expecting this distinction to translate into sex differences in feeding ecology, even when both sexes share the same habitat.

The Jarman–Bell principle

In a landmark paper, Geist (1974) explained how two PhD studies concurrently crystallized an understanding of the ways in which body size affects the ecology and behaviour of ungulates. Jarman (1968) and Bell (1969), who both worked in African savannas, subsequently produced papers (Bell, 1971; Jarman, 1974) that have become the springboard for most analyses of ecological interactions in ungulate assemblages that include a range of body sizes. The Jarman–Bell principle applies to the ability of larger-bodied herbivores to ingest more ubiquitous and fibrous plant parts than the selectively feeding smaller members of grazing or browsing guilds. The basis for this is that gut capacity increases isometrically with body mass (i.e. as a constant proportion; $\alpha\ M^1$), while metabolic requirements increase allometrically with body mass (i.e. as a fractional power; $\alpha\ M^{0.75}$), so an animal's energetic demands from each unit of ingesta diminish across guild members of incremental size (Demment & van Soest, 1985). Hence, despite the absolutely greater intake requirements of larger herbivores, which prevent them from feeding as selectively as smaller ones, the range of food qualities they can tolerate increases with body size (Fig. 3.1).

Within a sexually size-dimorphic herbivore species, the Jarman–Bell principle predicts that adult males and females will differ in feeding ecology, with the males accepting diets higher in fibre. There is theoretical support for this prediction (Illius & Gordon, 1987; Pérez-Barbería & Gordon, 1998a; Barboza & Bowyer, 2000) and studies on Nubian ibex (*Capra ibex nubiana*), for example, provide some empirical

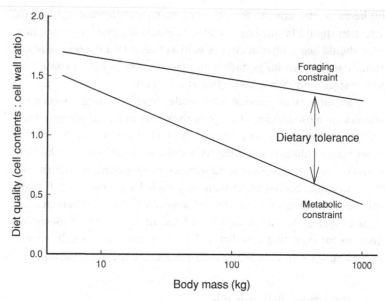

Figure 3.1   A schematic representation of the relationship between
dietary tolerance and body size in ungulates. Minimum diet quality
required for maintenance (bottom line) and maximum diet quality
that can be achieved (top line) both vary as functions of body mass.
All herbivores should eat the best food available, but the larger the
animal the more difficult it is to find, manipulate and ingest enough
of the best food items to meet intake requirements, so maximum diet
quality is limited by a foraging constraint. All herbivores should also
minimize searching time, but most plant tissue is too fibrous and/or
woody to meet energetic requirements, especially for the smaller
ungulates, so the use of abundant food is limited by a metabolic
constraint. As body size increases, minimum diet quality decreases
more steeply than maximum diet quality; the ecological significance
being that large herbivores have wider dietary tolerances than smaller
ones. The diet quality axis is an approximate guide, based on cell wall
percentages provided by Demment and van Soest (1985) for various
plant tissues, and on the proportional uses of different plant parts by
species within a size-structured browsing ruminant guild (du Toit,
1988).

support as well (Gross et al., 1995a,b). On the other hand, Pérez-Barbería
& Gordon (1999) found contradictory evidence from their field experiments on Soay sheep (Ovis aries). This highlights the problem of invoking
the Jarman–Bell principle within species in which inter-sexual differences in body size may not be large enough to result in measurable

differences in fibre digestibility (Conradt et al., 1999b). The ultimate test requires a species with extreme sexual size dimorphism, and because the application of the Jarman–Bell principle is not restricted to ruminants, the African elephant (*Loxodonta africana*) is a prime candidate (males reach ~6000 kg and females ~3000 kg; Owen-Smith, 1988).

Sex differences in elephant feeding ecology were studied in northern Botswana in the dry season (Stokke, 1999; Stokke & du Toit 2000). Adult males fed from a lower diversity of woody species than females did, but males used a greater diversity of plant parts. The implication is that males are less selective feeders, maximizing their intake rate by decimating the available browse (from leaves to roots) on the food plants they encounter, while females feed more selectively from a wider range of species to crop the most nutritious plant parts. Males also spent more than twice as much time feeding on each woody plant before moving on to the next one (males, 9 min; females, 3.6 min), which supports this deduction. Furthermore, adult males broke and bit off stems with greater diameters than those removed by females or subadult males, indicating that adult males satisfy their vast intake requirements at the expense of a comparative decline in feeding selectivity. Such differences in feeding ecology between age and sex classes within an elephant population are entirely consistent with predictions based on differences in body size, but have not yet been allowed for in management plans that relate levels of elephant impact on woodlands to total elephant counts. Adaptive management of elephant–woodland interactions could be refined through the recognition that male and female elephants represent ecologically distinct functional types of megaherbivore. For example, park managers would gain valuable insights through monitoring impacts on woody vegetation by an elephant population where the adult sex ratio is adjusted, while maintaining the overall population density at the same level as in a control area.

Across all large mammalian herbivores, any general effect of the Jarman–Bell principle on sex differences in ungulate feeding ecology should only be expected among species in which sexual size dimorphism is pronounced, and most of those occur in the upper region of the body size range (Loison et al., 1999). From modelling the allometry of food intake in grazing ruminants, Illius & Gordon (1987) suggested that a body mass of difference of >20% is required to cause measurable inter-sexual differences in digestive ability and, therefore, diet quality. While there is evidence that sexual segregation in ruminants may indeed begin at this 20% level (Ruckstuhl & Neuhaus, 2002), there is as yet no empirical evidence for such a threshold from inter-sexual

Table 3.1 *Evidence from published studies on free-living ruminants that indicate whether the quality of food ingested, or available in occupied habitats, differs between the sexes. Extracted from Appendix 1 of Mysterud (2000), where source references are cited.*

Species	Sexual size dimorphism[a] (%)	Food quality ingested		Food quality available in habitat	
		males < females	males ≥ females	males < females	males ≥ females
Capra ibex	52.6			✓	
Dama dama	42.0		✓		
Cervus nippon	37.2		✓		✓
Giraffa camelopardalis	36.4	✓			
Ovis dalli	36.3		✓		
Odocoileus hemionus	36.0		✓	✓	
Rangifer tarandus	35.0		✓		✓
Odocoileus virginianus	33.8	✓			✓
Cervus elaphus	32.8	✓		✓	
Rupicapra rupicapra	32.5			✓	
Ovis canadensis[b]	23.2–37.8		✓		✓
Alces alces	24.9	✓			
Kobus leche	23.6			✓	
Aepyceros melampus	23.2	✓			
Kobus ellipsiprymnus	23.1		✓		
Rupicapra pyrenaica	23.0	✓			
Capreolus capreolus	3.40				✓
All species		6	7	5	5

[a] Calculated as the difference in body mass between adult males and females, expressed as a percentage of the male's body mass, and ranked from highest to lowest difference. In all these species the male is larger than the female.
[b] *O. canadensis canadensis* and *O. canadensis nelsoni/mexicana*.

comparisons of diet quality. Using information on ruminant diet quality extracted from Mysterud's (2000) database, it can be seen that a meta-analysis based on previous autecological field studies yields no general pattern to associate sexual size dimorphism with the quality of food ingested by, or available to, each sex (Table 3.1). Furthermore, even when *Capreolus capreolus* is excluded due to its virtual absence of sexual size dimorphism, there is no statistical support for males being associated with lower food quality, whether in their diets or in their habitats ($\chi^2 = 0.752$, d.f. $= 1$, $P > 0.5$). This could be because field studies on sex differences in ruminant feeding ecology do not yet cover enough species that display enough of a range in sexual size dimorphism. In the case of browsers, increased sexual size dimorphism is associated with an increased likelihood of intersexual differences in diet composition and habitat use (Mysterud, 2000), and this may be due to the wider range in body size (and therefore presumably sexual size dimorphism) displayed among browsers compared to grazers (Cerling *et al.*, 2003; Clauss *et al.*, 2003). However, studies that attempt to test the effects of sexual size dimorphism on inter-sexual differences in diet quality will be inconclusive unless they are conducted in predator-free environments, where males and females have equal access to all available habitats (see Main & du Toit, Chapter 9). Also, and perhaps most importantly, they must be conducted during periods of high population density and/or seasonal resource limitation, which might only occur in critical time windows during late winter on temperate pastures (Clutton-Brock *et al.*, 1987a; Kie & Bowyer, 1999), or the dry season in tropical savannas (du Toit, 2003). That is when each sex is forced to accept food of lowest possible quality in relation to metabolic requirements, which declines more steeply with increasing body mass than does the maximum food quality that can be achieved within foraging constraints (Figure 3.1).

Horizontal segregation in grazers

It is a common observation that where males and females of a grazing ungulate species share the same pasture outside the breeding season, the sexes tend to be spatially segregated (reviewed by Ruckstuhl & Neuhaus, 2002). Because females are usually smaller than males, and bear the added nutritional demands of pregnancy and lactation, it is logical to suspect that the females will be more selective for grazing sites of high nutritional quality. But why would males voluntarily forgo nutritional benefits by avoiding the high quality sites used by females? One explanation was provided by the indirect competition hypothesis, and the rise and fall of this hypothesis (as it applies to grazers) will be described here in some detail because it has dominated the debate on sex differences in ungulate feeding ecology for the past two decades.

Studies on Scottish red deer (*Cervus elaphus*) provided evidence that the larger-bodied stags are disadvantaged in terms of meeting their intake requirements when feeding on close-cropped swards (Clutton-Brock *et al.*, 1982, 1987a). The proposed mechanism is that, when grass is very short (<5 cm), increasing intake can only be achieved by increasing bite rate and/or breadth of the incisor arcade. However, the allometric scaling of muzzle width (a linear dimension) has a much lower mass exponent (<0.33) than that for metabolic requirements (0.75), so when hinds and stags are feeding at maximum bite rate ($\sim$ 65 bites/min) and reducing the height of a shared sward, the stags will become intake limited before the hinds. This explanation (Clutton-Brock *et al.*, 1987a) for why red deer stags feed on taller swards than hinds (thereby ingesting more structural tissue, and therefore fibre), under conditions of high population density in winter, introduced the compelling hypothesis that sexual segregation in ungulates is driven by indirect (scramble) competition between the sexes. In effect, the hypothesis implies that by selecting patches of high quality grass and then grazing them short, the smaller-bodied females can competitively displace males from high quality resources. By virtue of their larger size the males can maintain themselves on lower quality grass of higher abundance, but if the standing crop of grass becomes critically low in the 'lean' season then the males would be expected to suffer a higher mortality than the females.

The indirect competition hypothesis has been widely discussed in the context of sexual segregation at the habitat scale (reviewed by Main *et al.*, 1996; see also Main & du Toit, Chapter 9), although it was initially proposed as an explanation for segregation at the patch scale (i.e. within habitats). Nevertheless, it has now been rejected in tests at both the habitat scale (Conradt *et al.*, 1999b) and the patch scale (Conradt *et al.*, 2001), using the same red deer population in the same study area (Isle of Rum, Scotland) in which Clutton-Brock *et al.* (1982, 1987a) found indirect evidence for its original formulation. We now find that despite its ecological elegance, and its support from mathematical modelling (Illius & Gordon, 1987), there is actually no empirical support for the hypothesis that spatial segregation is an outcome of indirect competition *within* grazing ungulate species. Indeed, in mountain sheep (*Ovis canadensis*), which display both sexual size dimorphism and sexual segregation, the males have been found to obtain diets of *higher* quality than the females (Bleich *et al.*, 1997). Furthermore, even if adult males in a sexually dimorphic and segregated grazing species do ingest consistently poorer diets than the females and subadults, as in bison (*Bos bison*) for example (Post *et al.*, 2001), it is difficult to determine if the dietary difference is the cause or the result, or even a relevant feature, of sexual segregation.

Notwithstanding the lack of evidence to support it so far, the indirect competition hypothesis remains in contention among the various explanations for spatial segregation among grazing ungulates that differ substantially in body size. When considering grazing interactions *between* species, it currently offers the best explanation for the 'grazing succession' phenomenon on the fertile plains of East Africa (Bell, 1971). There, larger-bodied grazing species keep moving away from smaller-bodied species that follow behind them, apparently because the smaller species are more efficient at selecting out high quality plant parts, thereby diluting the overall food quality of shared swards (Illius & Gordon, 1987; Murray & Illius, 2000).

Vertical segregation in browsers

One of the problems with tests of the indirect competition hypothesis to date is that inadequate attention has been given to the different mechanisms by which inter-sexual competition may occur in grazing and browsing ungulates respectively. The differential scaling of incisor arcade breadth and basal metabolism has little relevance to browsers, which do not feed off a two-dimensional grass carpet, but from a three-dimensional arrangement of branches, twigs and leaves. Consequently, as demonstrated by studies on sexual segregation in Alaskan moose (*Alces alces gigas*), the horizontal displacement of males from prime feeding sites by females should not be expected in a browsing species in the same way that it might in a grazing species (Miquelle *et al.*, 1992; Spaeth *et al.*, 2004). This does not preclude the possibility of indirect inter-sexual competition, however, but calls for closer consideration of how syntopic browsers of different size may interact through their dependence on the same food resources.

Vertical stratification of feeding within the woody canopy has been observed across species in various herbivorous mammal groups, ranging from small arboreal rodents (Barry *et al.*, 1984) to large browsing ungulates (du Toit, 1990). Furthermore, sex differences in feeding height have also been reported within species such as the grey bamboo lemur, *Hapalemur griseus* (Grassi, 2002), and the giraffe, *Giraffa camelopardalis* (Pellew, 1983; du Toit, 1990; Young & Isbell, 1991; Ginnett & Demment, 1999). Among browsers in which males are substantially larger than females, such as kudu (*Tragelaphus strepsiceros*), moose, giraffe and elephant, the males have access to foliage in the canopy that is above the reach of females and their young, so if inter-sexual competition occurs

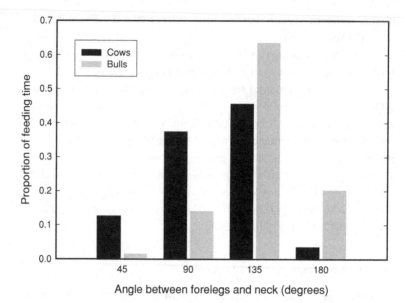

Figure 3.2 Distribution of feeding time across feeding postures adopted by giraffe cows (black bars) and bulls (grey bars) in the Kruger National Park, South Africa (du Toit, 1990). Bulls generally feed above the levels of cows, not only because they are taller, but also because they spend more time feeding with their necks inclined at steeper angles than cows.

it should cause vertical segregation of feeding (du Toit, 1995; Mysterud, 2000).

In the case of giraffes, bulls typically feed at a higher level above ground than cows, not only because they are taller but also because they frequently adopt a different feeding posture (du Toit, 1990). Although giraffe bulls and cows both feed most frequently with the neck inclined at an angle of 135° to the forelegs, the bulls differ from the cows in allocating a greater proportion of feeding time to standing with the neck and head extended vertically upwards, at 180° to the forelegs (Fig. 3.2). Two competing hypotheses have been suggested to explain this sex difference. Firstly, bulls could be displaced vertically as an outcome of indirect competition with cows and subadults, which may be more efficient at selectively cropping young shoots and stripping leaves from twigs (du Toit, 1990). Secondly, bulls could derive a vigilance advantage by holding their heads high during feeding, thereby concurrently improving their detection of intruding bulls and advertizing their size to male competitors while feeding (Young & Isbell, 1991).

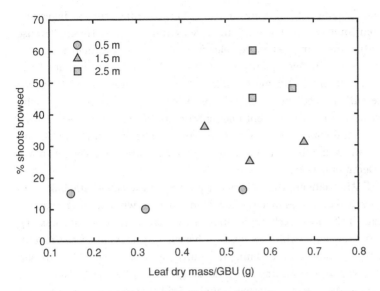

Figure 3.3 Browsing pressure (% shoots freshly browsed), mainly by giraffes, on *Acacia nigrescens* trees at three heights above ground (0.5 m, 1.5 m, 2.5 m) in three separate stands in the Kruger National Park, South Africa (Woolnough & du Toit, 2001). The plot shows that leaf biomass per standardized giraffe browse unit (GBU) increases with height up the canopy, and giraffe preferentially browse shoots at heights that yield a high leaf biomass ($p = 0.001$ for the overall relationship). Giraffe bulls should thus derive an additional bite-size advantage by feeding above 2.5 m, which is the level at which giraffe cows usually feed (du Toit, 1990).

As pointed out by Ginnett & Demment (1999), the indirect competition hypothesis requires that giraffe bulls feeding at 180° derive a feeding benefit that offsets costs associated with this posture (if there were no costs then cows should also use this posture as much as bulls). Feeding rates of giraffe bulls and cows are greatest (~26 bites/min) at intermediate neck angles, at feeding heights of ~3 m (Young & Isbell, 1991), probably because this allows free articulation of the head on the neck and thus greatest efficiency in manipulating and removing foliage from the tree. However, the profitability of each bite increases with height up the canopy, for at least two species of African savanna tree (*Acacia nigrescens* and *Boscia albitrunca*) that are staple food plants used by giraffes (Woolnough & du Toit, 2001). This is not due to any change in leaf chemistry, but an increase in leaf biomass on each standardized shoot available for browsing (Fig. 3.3). Giraffe bulls spend less time foraging and allocate more of their foraging time to ingestion, when compared with cows (Ginnett & Demment, 1997). Hence, because giraffe cows can feed more selectively it may be assumed that they remove the highest quality shoots from the height zone that allows them to maintain the most efficient feeding posture. This could induce the bulls to feed above this level on browse to which they have exclusive access (by virtue of both being taller and feeding at a steeper neck angle) and could explain why bulls choose taller patches of trees, and remain feeding in them for longer, when compared with cows (Ginnett & Demment, 1997).

Alternatively, the vigilance hypothesis, by which giraffe bulls feed at steep neck angles to derive a vigilance benefit while feeding, requires that: (i) the 180° neck angle improves vigilance; (ii) cows are less vigilant than bulls while foraging, otherwise cows should use the 180° neck angle as much as bulls; (iii) bulls interrupt feeding less to perform vigilance scanning when feeding at the 180° neck angle than when feeding at shallower neck angles. Evidence so far is contrary to all three requirements. Firstly, it seems unlikely that vigilance is improved when giraffes feed with the head and neck extended vertically upwards, because their eyes are directed skywards, and so giraffes feeding in this posture should actually incur a vigilance cost (du Toit, 1990). Secondly, Ginnett & Demment (1997) found no difference in vigilance between giraffe bulls and cows while foraging. Thirdly, a recent study has found that giraffe bulls interrupt their feeding significantly *more* to perform vigilance scans when foraging at neck angles of 45° and 180° than at 135° (Cameron & du Toit, 2005). The vigilance hypothesis is, therefore, unsupported for giraffes. Although the indirect competition hypothesis does have support as an explanation for sex differences in browsing height, further studies are required to test whether such differences vary predictably in response to variation in both browse abundance and browser density.

Moving the focus away from giraffes, intraspecific differences in browsing height were studied in a population of elephants occurring at high density (>4 animals/km) during the dry season in northern Botswana (Stokke & du Toit, 2000). No differences in feeding height were found between elephant bulls and cows, although this proved unsurprising because the sexes were spatially segregated in the dry season (Stokke & du Toit, 2002). Of interest, however, was that when elephant cows were feeding in family units together with their offspring and other related subadults and juveniles, they fed (on average) at higher levels in the canopy than when feeding alone, separate from

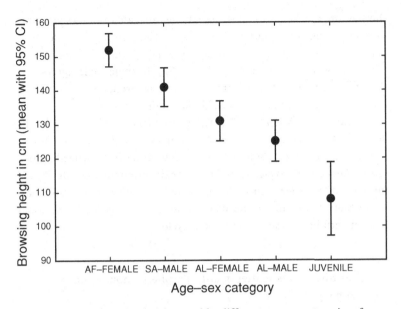

Figure 3.4 Browsing heights used by different age–sex categories of elephant in Chobe National Park, Botswana (redrawn from Stokke & du Toit, 2000). The categories are: adult females in family groups (AF–FEMALE); subadult males in family groups (SA–MALE); adult lone females (AL–FEMALE); adult lone males (AL–MALE); juveniles in family groups (JUVENILE).

their herds (Fig. 3.4). This could be an outcome of either indirect competition between elephant size classes within mixed herds, or active avoidance of competition between adults and juveniles within kin groups. Either way, this elephant example adds weight to the previously discussed evidence from giraffe studies that, in large browsing herbivores, intraspecific competition for food could be reduced by the segregation of sex/age/size classes along a feeding height axis. Hence, among sexually size-dimorphic browsers, males are partially released from the costs of inter-sexual competition, which may enable them to achieve larger body sizes (relative to females) than is the case for grazers.

DIFFERENCES IN DIETARY SELECTION CRITERIA

Other factors operating additionally to, or independently from, those relating to the allometric scaling of gut capacity vs. metabolic requirements can cause sex differences in dietary selection within ungulate species. Such factors relate to differences in life history, which influence the ways in which adult males and females allocate resources and time to maximizing their reproductive success. The priority for males is to gain and retain access to reproductive females, which involves diverting time away from foraging and investing in male–male aggression. On the other hand, females invest heavily in producing and rearing offspring, which imposes a nutritional drain especially during lactation (Clutton-Brock *et al.*, 1982). The consequence is that, even in the absence of sexual size dimorphism, adult male and female ungulates occupying the same habitat are expected to differ in their feeding patterns. Males should typically maximize their instantaneous rate of food intake to offset their reduced foraging time, while females should maximize their diet quality in terms of protein gains relative to energy, at least at certain stages of the seasonal cycle.

Sex differences in diet quality, despite monomorphism in body size

An example of a large monomorphic browsing herbivore is the black rhinoceros (*Diceros bicornis*), with males and females both typically weighing 1000 kg as adults (Owen-Smith, 1988). In the absence of intense human predation, which is an evolutionarily recent phenomenon, black rhino populations have balanced sex ratios due to mortality being roughly equal between males and females (Berger & Cunningham, 1995). This differs from markedly dimorphic savanna browsers such as giraffe (du Toit, 1990) and kudu (Owen-Smith, 1993) with population sex ratios strongly biased towards females, apparently due to the multiple costs to males of maintaining their larger body sizes. Black rhinos, therefore, provide an opportunity to test for sex differences in feeding ecology that can be ascribed to factors other than sexual size dimorphism, such as sex differences in reproductive physiology and social behaviour (black rhino bulls are solitary but cows are usually accompanied by a calf).

Experienced game scouts are able to sex black rhinos remotely by examining the size of twig fragments in their dung (personal observation). Being perissodactyls, black rhinos use their upper and lower incisor rows to shear twigs and thin branches from their woody food plants, and the game scouts claim that black rhino bulls process twigs of visibly larger diameter than cows. This was tested in Pilanesberg National Park, South Africa, where free-ranging black rhinos of known identity (11 bulls; 15 cows) were followed on foot and their dung was sampled during wet and dry seasons (Deere, 2001). During both seasons

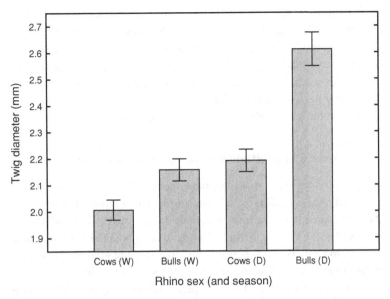

Figure 3.5   Diameters of woody twigs sampled from the dung of black rhino cows and bulls in Pilanesberg National Park, South Africa, during wet (W) and dry (D) seasons (Deere, 2001). Error bars indicate 99% confidence intervals.

the bulls did indeed process twigs of greater diameter than the cows, and in both sexes the diameter of ingested twigs increased in the dry season (Fig. 3.5). An increase in the mean diameter of ingested twigs indicates either that larger bites were taken, or that rebrowsing occurred at sites from which the thinner twigs had previously been browsed off. If larger bites were taken then entire shoots would have been processed, including thin terminal twigs, and so the variance in twig diameters would have been higher for the bulls than for the cows, but this was not the case (compare the 99% confidence intervals in Fig. 3.5). It would appear, therefore, that to meet their requirements for gestation and lactation the black rhino cows were feeding selectively on thinner shoots (minimizing lignin intake), and the bulls were taking the thicker shoots that were left. At the time of this study the Pilanesberg black rhino population was exhibiting density-dependent limitation (Hrabar & du Toit, in press), so it appears that to meet their intake requirements the bulls were forced to feed on 'left-over' browse. Because black rhinos are monomorphic, the bulls do not have the option of feeding at higher levels in the canopy than the cows, so inter-sexual scramble competition for browse resources may become a management issue as black rhino populations recover in small, enclosed reserves.

Because most large mammalian herbivore species are sexually size dimorphic, there has been a dearth of field-based research on differences in foraging ecology that can be ascribed to differences in sex per se, rather than allometry. For example, size-independent differences in the specific nutritional requirements of adult males and pregnant, lactating, and non-reproductive females, and links between these requirements and foraging ecology, are fertile topics for further studies.

Learned diet selection: feedback to social segregation?

Controlled experimental studies have shown that learning from peers and dominant members of a social group has a strong influence on the development of feeding behaviour in ungulates (Thorhallsdottir et al., 1990). The developmental 'window' during which dietary learning appears to be most active is the period surrounding weaning, although young animals continue refining their diet selection criteria into adulthood (Provenza & Balph, 1988; Launchbaugh et al., 2001). In ungulate species that segregate into same-sex social groupings outside the rut (see Conradt, 1998a; Ruckstuhl & Neuhaus, 2002), the implication is that young females have a foraging advantage over young males, especially if bachelor groups occupy different habitats than nursery groups (see Chapter 9 for examples). Consequently, on the basis of feeding efficiency, it is expected that females will continue to use feeding areas and foods that their mothers exposed them to, and the stimuli that cause young males to leave their natal herds must override the nutritional benefits of remaining. My point is that the benefits of maintaining learned feeding patterns should reinforce phyllopatry in females, so the question of why sexual segregation occurs in a certain species should be refined to the more specific question of why non-breeding males segregate themselves from females. Furthermore, social learning in bachelor herds should reinforce sexual segregation by virtue of young males conforming with social models, those being higher ranking males that have adapted their feeding patterns to the vegetation occurring in male feeding areas.

SUMMARY

To identify general patterns among sex differences in the foraging ecology of large herbivores requires the compilation of data from multiple independent studies. But, as demonstrated by Table 3.1, such patterns are elusive, at least as far as food quality is concerned. The use of meta-analysis in ecology is fraught with methodological problems and increased meta-analytical rigour involves imposing stricter criteria by which studies may be included, with the cost of reduced sample size (Gates, 2002). Consequently, the evidence is still too limited to allow conclusive and widely substantiated statements to be made about the mechanisms underlying sex differences in the foraging ecology of large mammalian herbivores. What follows, therefore, is a summary of theoretically plausible expectations arising from research conducted so far.

As a general principle in large herbivore feeding ecology, the adult females of a species tend to feed more selectively (for food patches, plant species and plant parts) than the males when both sexes occur in the same habitat (Clutton-Brock et al., 1982). Adult males maximize their instantaneous rate of intake of acceptable forage and thereby minimize the overall time that feeding activities take away from social interactions related to dominance hierarchies, territoriality, mate acquisition and guarding, etc. Females have to transfer nutrients to their young and they therefore maximize the nutritional quality of the food they ingest, to the extent allowed by environmental constraints (predation risk, weather exposure, etc.). This sex difference in feeding pattern is expected from differences in reproductive strategy, as well as differences in body size within sexually size-dimorphic species. When males are substantially (>20%) larger than females the Jarman–Bell principle predicts that allometric factors will additionally allow males to tolerate diets of lower quality than females. This prediction has been extended to include the possibility that females can actually displace males from high quality sites through indirect scramble competition. The mechanism should differ between grazing and browsing guilds, with the expected displacement of males occurring horizontally in grazers (to other feeding patches) and vertically in browsers (to higher levels in the canopy). Although the indirect competition hypothesis has been rejected for grazers on empirical grounds, there is increasing evidence in its support from the larger members of the browsing guild. Nevertheless, the hypothesis applies to differences in feeding ecology between categories of animals that differ only in body size, rather than sex. Body size differences are inadequate to explain all sex differences in diet, because even in sexually monomorphic large herbivore species, or those in which sexual size-dimorphism is low (<20%), the differences between male and female reproductive strategies impose differences in nutritional demands, which (at least in some cases) translate into sex differences in diet quality.

Research to date on the feeding ecology of large mammalian her-bivores has demonstrated that males and females of the same species can, in general, be expected to differ in the composition of their diets. A unifying hypothesis is elusive, and is likely to remain so until more data are available from studies that (i) uncouple body size differences from sex differences, and (ii) focus on both sexes when they are using the same habitats during the same 'lean' season. For all practical pur-poses, however, there is ample evidence that the males and females of large polygynous species represent functionally distinct categories of consumers, and traditional population-based approaches to manage-ment and conservation should be adapted to recognize this.

4

Sexual segregation in seals

INTRODUCTION

Seals have a worldwide distribution ranging from the high Arctic, through the tropics, to the coast of the Antarctic continent. They occupy the higher positions in many of the world's marine food webs and are increasingly being used as model species to monitor the health of oceans worldwide (Jouventin & Weimerskirch, 1990a). Historically, knowledge of their biology was restricted to periods when they could be observed on land or ice but even this was limited to the more accessible species. Advances in animal-borne recording devices, such as satellite transmitters and time–depth recorders, have lead to a rapid increase in the understanding of seal behaviour (Boyd, 1993b). A few decades ago, any attempt to investigate sexual segregation would have been extremely limited in its scope. Even today, knowledge is skewed towards the more heavily studied species and we can only consider segregation in terms of large spatial scales or distinct behaviours. Information that is often taken for granted for terrestrial organisms, such as foraging locations and basic diet, is simply not known for many species of seal.

The Pinnipedia is a diverse suborder, represented by some 33–37 extant species spanning three families (Berta, 2002). This diversity, coupled with their global distribution, is reflected in a large range of life history strategies. They show the greatest range of sexual size dimorphism of any higher vertebrate group (Ralls & Mesnick, 2002). In some species males can be ten times heavier than females (Fig. 4.1), whereas females are slightly heavier in others (Table 4.1).

Sexual Segregation in Vertebrates: Ecology of the Two Sexes, eds. K. E. Ruckstuhl and P. Neuhaus. Published by Cambridge University Press. © Cambridge University Press 2005.

Figure 4.1   A southern elephant seal male attempting to copulate with a small female, highlighting the extreme sexual dimorphism observed in this species. Note the enlarged proboscis (nose) of the male and the scarring, from fights, around his neck. Photographer: Callan Duck (British Antarctic Survey).

Pinnipeds have three main lactation strategies. Phocid seals (true seals) generally act as capital breeders (Jonsson, 1997): fasting ashore whilst suckling their young with milk produced via the mobilization of fat reserves. Their suckling periods are short, ranging from four days in hooded seals, *Cystophora cristata*, to two-and-a-half months in Baikal seals, *Phoca sibirica* (Bowen *et al.*, 1985; King, 1983). For the rest of the year (apart from the moult in some species) females are not confined spatially in their foraging and can exploit profitable feeding areas away from the breeding beaches. In otariids (eared seals) lactation can last for over a year and mothers act as income breeders: alternating suckling their young ashore with foraging at sea to maintain their milk supply. Pups do not forage with the mother and, as a result, the mother's foraging range is restricted. However, as they fast ashore whilst mating, and do not subsequently invest in the pup, male otariids could be viewed as capital breeders akin to phocids. In the only existing member of the Odobenidae, the walrus, *Odobenus rosmarus*, calves are suckled for at least a year. However, mothers are not restricted in their foraging range as calves are able to swim immediately after birth and can be suckled on land, ice and in the water (Kastelein, 2002).

Table 4.1 *Breeding habitat, mating system, degree of sexual dimorphism and the nature of segregation in seal species. Sexual dimorphism is shown as the difference in mass of the two sexes expressed as a percentage of the female mass using maximum values (Ruckstuhl and Neuhaus, 2002). Values in bold indicate where males are larger.*

| Species | Breeding habitat | Mating system | Sexual dimorphism | | | Nature of sexual segregation |
			% diff	♂ mass (kg)	♀ mass (kg)	
Monachinae						
Hawaiian monk seal *Monachus schauinslandi*	Islands (aquatic mating)	Promiscuity	16%	230	273	Slight difference in the relative composition of diet, may segregate by sex when foraging (Goodman-Lowe, 1998; Stewart *et al.*, 1998)
Mediterranean monk seal *Monachus monachus*	Islands, remote caves / beaches (aquatic mating)	Promiscuity	5%	315	300	Unknown
Northern elephant seal *Mirounga angustirostris*	Islands (terrestrial mating)	Extreme polygyny	177%	2000–2500	500–900	Differences in foraging location, diving behaviour, suggestion of diet differences (Le Boeuf *et al.*, 2000)
Southern elephant seal *Mirounga leonina*	Islands (terrestrial mating)	Extreme polygyny	288%	2000–3500	500–900	Differences in foraging location and diving behaviour (Slip *et al.*, 1994)
Weddell seal *Leptonychotes weddelli*	Fast-ice, Antarctic (aquatic mating)	Moderate polygyny	0%	400	400	No evidence of segregation
Ross seal *Ommatophoca rossi*	Pack-ice, Antarctic (aquatic mating?)	Unknown	6%	129–216	159–204	Unknown

(cont.)

Table 4.1 (*cont.*)

Species	Breeding habitat	Mating system	Sexual dimorphism			Nature of sexual segregation
			% diff	♂ mass (kg)	♀ mass (kg)	
Crabeater seal *Lobodon carcinophagus*	Pack-ice, Antarctic (mating on ice)	Serial monogamy	1%	224	227	No evidence of segregation
Leopard seal *Hydrurga leptonyx*	Pack-ice, Antarctic (aquatic mating?)	Unknown	23%	200–455	225–591	Unknown diet and foraging, equal mix of sexes at South Georgia in winter (Jessopp et al., 2004)
Phocinae						
Hooded seal *Cystophora cristata*	Pack-ice, Arctic (mating occurs on both ice and in water)	Serial monogamy	**24%**	250–435	180–350	Males have a higher trophic position (Lesage et al., 2001)
Bearded seal *Erignathus barbatus*	Pack-ice, Arctic (aquatic mating)	Serial monogamy?	0%	275–340	275–340	No evidence (Antonelis et al., 1994; Hjelset et al., 1999)
Harp seal *Pagophilus groenlandicus*	Pack-ice, Arctic (mating occurs on both ice and in water)	Promiscuity	0%	130	130	No evidence, sexes have the same trophic position (Lesage et al., 2001)
Ribbon seal *Phoca fasciata*	Pack-ice, Arctic	Promiscuity	0%	70–100	70–100	No evidence of segregation
Caspian seal *Phoca caspica*	Fast-ice, Arctic	Serial monogamy	0%	50–60	50–60	Slight separation in timing of moult (King, 1983)
Baikal seal *Phoca sibirica*	Fast-ice, Arctic (aquatic mating?)	Promiscuity	0%	63–70	63–70	No evidence, though females have long lactation and pups are confined to birth lairs (King, 1983)

Species	Breeding habitat (mating)	Mating system	%			Notes
Ringed seal *Phoca hispida*	Fast-ice, Arctic (aquatic mating)	Serial monogamy	0%	68	68	No evidence, though females have long lactation and pups are confined to birth lairs (King, 1983)
Largha seal *Phoca largha*	Pack-ice, Arctic	Serial monogamy?	30%	85–150	65–115	Unknown
Harbour seal *Phoca vitulina*	Islands, remote coasts or Arctic pack-ice (aquatic mating)	Serial monogamy	20%	55–170	45–142	No difference in trophic position (Lesage et al., 2001). Males have longer foraging trips and a greater foraging range (Thompson et al., 1998), difference in haulout patterns at moult and breeding times (Thompson et al., 1989, 1997)
Grey seal *Halichoerus grypus*	Islands, pack-ice or fast-ice (mating occurs on ice, land and in water)	Polygyny	56%	170–400	103–256	Males have a higher trophic position (Lesage et al., 2001). Males take larger older prey (Hauksson & Bogason, 1997), sexes have different energy storage strategies and diving behaviour (Beck et al., 2003a, 2003b)
Otariidae California sea lion *Zalophus californianus*	Islands (terrestrial mating)	Polygyny	254%	275–390	91–110	Males migrate farther north females remain on breeding grounds (Bigg, 1973)

(cont.)

Table 4.1 (cont.)

Species	Breeding habitat	Mating system	Sexual dimorphism			Nature of sexual segregation
			% diff	♂ mass (kg)	♀ mass (kg)	
Galapagos sea lion *Zalophus californianus wollebaeki*	Islands (terrestrial mating)	Polygyny	354%	363	80	Unknown
Steller sea lion *Eumetopias jubatus*	Islands (terrestrial mating)	Polygyny	115%	566	263	Males migrate farther north, females remain on breeding grounds (King, 1983). Males have a higher trophic position (Hobson et al., 1997)
Southern sea lion *Otaria byronia*	Islands, remote coasts (terrestrial mating)	Moderate polygyny	100%	300	150	Males travel farther on foraging trips, dietary differences (Campagna et al., 2001)
Australian sea lion *Neophoca cinerea*	Islands (terrestrial mating)	Moderate polygyny	172%	250–300	70–110	Separation in timing of moult (Ling, 2002)
New Zealand sea lion *Phocarctos hookeri*	Islands, remote coast (terrestrial mating)	Extreme polygyny	181%	300–450	160	Males have a greater range at sea (Gales, 2002)
Northern fur seal *Callorhinus ursinus*	Islands (terrestrial mating)	Extreme polygyny	440%	180–270	30–50	Males migrate to different foraging areas (Kajimura, 1985; Bigg, 1990)
Guadalupe fur seal *Arctocephalus townsendi*	Islands, remote caves (terrestrial mating)	Polygyny	280%	190	150	Unknown
Juan Fernandez fur seal *Arctocephalus philippii*	Islands (terrestrial mating)	Moderate polygyny	180%	140	50	Suggestion of diet differences in relative composition of diet (Acuna & Francis, 1995)

Species	Habitat	Mating system	%			Notes
Galapagos fur seal *Arctocephalus galapagoensis*	Islands (terrestrial mating)	Polygyny	150%	60–75	30	Unknown
South American fur seal *Arctocephalus australis*	Islands, remote coasts (terrestrial mating)	Polygyny	77%	160	60–90	Unknown
New Zealand fur seal *Arctocephalus forsteri*	Islands, remote coasts (terrestrial mating)	Polygyny	270%	120–185	25–50	Unknown
Antarctic fur seal *Arctocephalus gazella*	Islands (terrestrial mating)	Polygyny	282%	90–210	25–55	Differences in foraging location, diving behaviour, suggestion of relative diet composition differences (see text)
Sub-Antarctic fur seal *Arctocephalus tropicalis*	Islands (terrestrial mating)	Polygyny	200%	165	55	Males distributed at sea whilst females constrained by provisioning pup (King, 1983)
South African fur seal *Arctocephalus pusillus*	Islands (terrestrial mating)	Polygyny	350%	200–360	40–80	Males spend more time at sea and have longer foraging trips (King, 1983)
Australian fur seal *Arctocephalus doriferus*	Islands, remote coasts (terrestrial mating)	Polygyny	227%	220–360	36–110	Males distributed at sea whilst females constrained by provisioning pup (King, 1983)
Odobenidae						
Walrus *Odobenus rosmarus*	Pack-ice, Arctic (aquatic mating)	Polygyny	50%	800–900	500–600	Differences in diet, migration (King, 1983)

? = suggested or based on very low sample size

Unknown: little known about species or only data on females available

No evidence of segregation: no distinction made between sexes in major studies

In addition to the different lactation strategies, seals show seasonal movements ranging from general dispersal within a limited area to migrations over thousands of kilometres. Some pinnipeds also show a high degree of intraspecific variation in their behavioural ecology. For example, the timing of breeding and the length of lactation can vary significantly between populations of harbour seals, *Phoca vitulina*, throughout their large latitudinal range (Riedman, 1990).

Because of their diversity, the sexes segregate in many different ways and to varying extents within pinnipeds. This, and the lack of data for many species, means that it is impossible to generalize the entire group. Therefore, in order to illustrate the range of sexual segregation in seals the following sections focus on three species with contrasting life-histories: two phocid, capital breeders, the monomorphic Weddell seal, *Leptonychotes weddelli*, and the extremely size dimorphic northern elephant seal, *Mirounga angustirostris*, and an otariid income breeder, the Antarctic fur seal, *Arctocephalus gazella*. In this chapter, I will examine how the disparate breeding systems of these species place different demands and constraints on the sexes, and lead to different degrees and modes of segregation.

WEDDELL SEALS

Mating systems

Weddell seals live within the Antarctic fast-ice, accessing the surface through naturally occurring cracks or holes they scrape out themselves (Fig. 4.2). Because of their docile behaviour towards humans, they are one of the most widely studied seal species despite their remote distribution. Mating occurs aquatically and males hold underwater territories that are associated with breathing holes near pupping sites (Testa, 1994). Male territorial fights also occur aquatically and the nature of these fights differs from that of land-based breeding species where size and strength are paramount. In aquatic fights, males attempt to bite their opponents by chasing and spinning around one another so that agility and speed appear to be as, or more, important than size.

During the breeding season, male activity can be classified into two main behaviours: basking and territorial (Kaufman *et al.*, 1975). Basking individuals spend the majority of their time on the pack-ice and seldom enter the water. These individuals are normally found away from the pupping sites but still within the general breeding area. Territorial individuals are rarely seen on the pack-ice, spending periods of up

Figure 4.2 A dispersed group of Weddell seal mothers and their pups. The seals are hauled out on fast-ice, attached to land, and access to the water is limited to breathing holes and tidal cracks. Photographer: Doug Allan (British Antarctic Survey).

to seven days in the water. Underwater tracking has shown that males can defend territories up to 6.5 km^2 (Siniff *et al.*, 1977). Males displaced from their territories adopt basking behaviour and are often covered in numerous cuts as a result of conflicts with other males (Smith, 1966).

Aquatic mating is very difficult to observe and it is thought that Weddell seals are polygynous, though at a low level (Stirling, 1969). Females haul out on pack-ice to pup during spring, the exact timing depending on the location. They form groups associated with breathing holes but the individuals within them are dispersed to avoid aggressive encounters with their neighbours (Fig. 4.2). Pups can enter the water as early as eight days old and suckling mothers make frequent trips into the water (Tedman & Bryden, 1979). It was originally thought that mothers did not feed during this time but there is some evidence that they will take prey if it is available (Sato *et al.*, 2002). Lactation continues for six to seven weeks, and the mother's milk is gradually replaced by a diet of small crustaceans as the pup develops (King, 1983).

Sex differences

There is little detectable segregation of males and females in Weddell seals. On the pack-ice it may appear that females are spatially separated from males during the breeding period. However, when the sub-surface distribution is included there is actually very little spatial separation (Siniff *et al.*, 1977).

The aquatic nature of mating and fighting negates the need for large male size. The stresses of fasting are also reduced for both sexes as they are thought to feed opportunistically during the breeding season and the moult. Because of these reduced selective pressures, males and females attain the same body size (400 kg). Although they may feed during the breeding season, females act primarily as capital breeders and are required to provision their pups from their body reserves. However, at this time males are also spatially and temporally restricted in their foraging and must expend increased amounts of energy through competitive interactions for mates. In a number of phocid seals the reproductive cost to males, through competition and fasting, has been shown to be very high (Kovacs *et al.*, 1996; Reilly & Fedak, 1991; Walker & Bowen, 1993). In a comparative study of male and female elephant seals the reproductive costs of the two sexes was equal (Deutsch *et al.*, 1990). Although the male elephant seal fast is much longer than in Weddell seals, these data highlight the fact that the reproductive effort of the sexes, when both act as capital breeders, is not as different as it initially appears. Therefore, with an equivalent body size the predicted metabolic requirements of male and female Weddell seals will be more or less comparable.

Direct observation of feeding is usually not possible in marine predators and their diet must be inferred through alternative techniques (Pierce & Boyle, 1991). In seals, the most common of these is the use of prey remains resistant to digestion recovered from either stomachs or faecal material (scats). Scat analysis has a number of biases but these are the same for both sexes so the technique is still useful for a direct comparison between males and females. Another method, stable isotope analysis, uses the isotope ratios of elements, such as carbon and nitrogen, to estimate the trophic level at which predators are operating. The technique uses the assumption that heavier isotopes are selectively retained as the element is passed up the food chain. There is no reported sex difference in the diet of Weddell seals based on isotope ratios in their scats (Burns *et al.*, 1998). No significant difference was found between the carbon and nitrogen isotope ratios of males and females, suggesting they forage at the same trophic level (Burns *et al.*, 1998).

The monomorphism of the sexes and the subsequent similarity of their energy budgets, at least outside the immediate breeding season, is in turn reflected in their indistinguishable diet, distribution and predicted diving capabilities. As a result in most studies of Weddell seal behaviour (apart from those directly related to breeding) researchers make no distinction between males and females, and data are usually pooled.

ELEPHANT SEALS

Mating system

Northern elephant seals show huge sexual size dimorphism, with males having up to ten times the body mass and nearly one and a half times the body length of females (Deutsch et al., 1994). Dominant males arrive at rookeries, situated on islands along the coast of California and Mexico, in early December (King, 1983). Elephant seals exhibit harem defence polygyny, where males establish a dominance hierarchy using their immense size and strength to challenge and battle each other for access to breeding females (Le Boeuf, 1991). In fights, males rear up, chest to chest, rocking back and forth until an opportunity occurs for one of them to deliver a downward blow with an open mouth, tearing the opponent's skin with strong canine teeth. Males have a thickened shield of skin on their chests to protect them so that wounds, which often look horrific, are usually only superficial and even severe ones heal quickly (Deutsch et al., 1994). The enlarged male proboscis is used to make a deep roar that is thought to help males assess one another's size. A challenger can sometimes displace a harem bull by simply posturing and roaring (King, 1983). A dominant harem bull stays close to a group of females and defends them from lower members of the hierarchy found towards the periphery. A group of around 40 females can be controlled by a single male (known as a beach master) but as the group size increases, and defence becomes more difficult, lower ranking males can gain access (Le Boeuf, 1971).

Consequently, relatively few dominant males are responsible for most of the mating on a beach, and there is great variance in male reproductive success. However, genetic paternity data show that reproductive success for some alpha males was lower than that expected from observed matings (Hoelzel et al., 1999). In a study of 91 male pups, only 19 reached breeding age (Le Boeuf & Reiter, 1988). Of these survivors 3 inseminated a total of 281 females, a further 5 mated 69 times and the remaining 11 failed to mate. The most successful males were ones that gained high ranks in the dominance hierarchies.

Although males become sexually mature at five years they do not achieve high rank until they have attained their full size, when at least eight years old. Even successful males, with a life span of up to 14 years, usually only maintain their dominance for a year or two (Clinton, 1994).

Sex differences

Size and growth

A hypothesis is that the intense competition between males for mates provides a strong selective force favouring their continued growth and delayed maturity. Consequently, between the ages of three and five, when similar aged females are pupping, young males invest resources in increased growth rather than in reproductive effort (Deutsch et al., 1994). In comparison, competitive social interactions are much less important to the reproductive success of females. Their investment pattern needs to optimize the opposing effects of investment in the pup and maternal fitness. Hence, females have a smaller body size, younger age of first breeding (3–5 yrs) and a longer lifespan (up to 18 yrs) than males (Sydeman & Nur, 1994).

Annual cycle and migrations

The annual cycle of elephant seals comprises two periods of intensive foraging interspersed by periods of fasting (Le Boeuf & Laws, 1994). Both sexes require an enormous capacity to store fat and a need to accumulate energy during their long trips to sea. After weaning their pups, females spend 70 days at sea before returning to land to moult. Following one month ashore between March and June, the females depart to sea for eight months. Because of delayed implantation of the blastocyst, this period coincides with the duration of active gestation (Boyd et al., 1996).

Males spend four months at sea after the breeding season before they begin their moult, between June and August. After a month ashore, they return to sea for a further four months prior to the breeding season. Males are therefore still foraging at sea when females are ashore moulting and vise versa. Because of their greater size, males are estimated to require up to three times the amount of energy needed by females (Le Boeuf et al., 1993). Although the energy requirement of adult males exceeds those of adult females, they spend less time at sea foraging (eight compared to ten months: Le Boeuf et al., 2000).

As well as the slight temporal separation in the foraging of the two sexes there is a strong degree of spatial separation (Stewart & Delong, 1994). Adult males and females show strong differences in their distribution at sea and behaviour, during both of their annual foraging trips. Males have been reported to travel from Año Nuevo, California, to foraging areas 4000 km north on and around the continental shelf break of North America (Le Boeuf *et al.*, 2000). The journey to these areas is direct and rapid, suggesting that there is a strong drive for males to reach their destination quickly. Their diving behaviour within feeding areas suggests they target prey near the ocean bottom, either by moving slowly over the seabed or using a sit-and-wait approach (Le Boeuf *et al.*, 2000). In contrast, females from the same colony do not proceed directly to any one particular feeding site (Le Boeuf *et al.*, 2000). Instead, they follow a range of trajectories travelling in the open ocean over deep water. Females forage intensively during periods of slow, or no, horizontal movement and appear to be feeding on pelagic prey within the deep scattering layer. When moving between these temporary foraging areas, females change direction often, in active search of prey utilizing both transit and pelagic foraging dives (Le Boeuf *et al.*, 2000).

Diet

Little is known of the diet of northern elephant seals throughout their distributional range as investigations have been confined to stomach content analyses of animals at haul-out sites. This method reflects only prey consumed during the last few days and is therefore not representative of the diet for such a wide-ranging animal. From such analysis, it appears that females mainly forage on pelagic cephalopods and Pacific hake, *Merluccius productus*, in the open ocean (Antonelis *et al.*, 1994). Stomach contents data from males suggest that they also consume such pelagic species. However, the large spatial separation between the males' main foraging areas and the haul-out sites (4000 km) means that their stomach contents will simply reflect food taken on their return journey. Diving data from the feeding grounds suggest that, in contrast to the stomach contents collected on shore, males may actually focus their foraging on unidentified benthic animals (Le Boeuf *et al.*, 2000).

One explanation for the differences in foraging areas between males and females is that the shelf-break areas exploited by males have a higher risk of predation than the open ocean (Le Boeuf *et al.*, 2000). Because of their greater metabolic requirements, males may exploit coastal areas with a high-energy return at the cost of greater exposure to predators such as white sharks, *Carcharodon carcharias*, and killer whales, *Orcinus orca*. In contrast, females are able to forage efficiently on smaller, or less predictable, food resources in the open ocean that are closer to the breeding areas and where there is less risk of predation. Indeed, shark inflicted wounds are five times more likely to occur on males than females at breeding sites, despite, there being twice as many females present (Le Boeuf & Crocker, 1996).

To sum up, the competition for mates between male elephant seals means that they are considerably larger than their female counterparts. This dimorphism creates a disparity in the energy budgets of the two sexes that leads to segregation of their distributions, diving behaviours and possibly their diets.

ANTARCTIC FUR SEALS

Mating system

Antarctic fur seals also show a high degree of sexual dimorphism (Fig. 4.3). Males can grow up to 200 kg whereas the largest females are around 50 kg (Laws, 1993). However, unlike elephant seal mothers, female fur seals do not fast during lactation. Instead, they obtain resources for milk production by interspersing periods suckling ashore with foraging at sea (Doidge *et al.*, 1986). Females give birth in December, within a couple of days of their arrival at the breeding beaches. They stay ashore for the first week suckling their pups using body reserves and thereafter commence foraging trips. A pattern of two to ten days at sea followed by one to four days ashore continues for around four months until the pup is weaned (Staniland & Boyd, 2003). Pups remain ashore during the suckling period. Therefore, throughout the summer breeding season, fur seal mothers are confined to a limited foraging area close to the breeding beach.

Males arrive on the breeding beaches from October through to late December. Rather than defending females, fur seals defend an area of beach (resource-defence polygyny) with boundaries that are vigorously defended from neighbouring males (McCann, 1980). Territories are not equal in quality. Prime locations, such as those just above the high water mark, will attract more females than others on the periphery of the beach. On heavily populated beaches, territories may extend to only a few metres. In order to challenge a male for a territory, a contender must first run the gauntlet of the fiercely guarded territories between himself and his opponent.

Figure 4.3 Antarctic fur seals on a crowded breeding beach. In the centre, the large male and his thickened neck contrasts with the more slender and smaller females around him. Photographer: T. D. Williams (British Antarctic Survey).

Territorial conflicts can be intense (McCann, 1980) and infected injuries, sustained during disputes, have been identified as a major cause of death (Baker & McCann, 1989). Males fight by pushing chest to chest, attempting to bite and slash each other with their enlarged canines. As their chests are protected by a thick mane of guard hairs, males attempt to attack the exposed skin around the top of the fore-flippers. Size is important in these fights, as larger males are able to push their opponents into neighbouring territories causing their occupants to join the defence. However, speed and agility are as important as size because skilled fighters can avoid lunges of their opponents and make retaliatory strikes. Once a male has a position on the beach he must fast, as returning to sea to feed would mean the loss of the territory and a subsequent fight to regain it. The ability to resist starvation by building up fat reserves, whilst not the only factor, increases a male's mating success through increased length of tenure (Arnould & Duck, 1997).

Sex differences

Size dimorphism

Selection pressures, in the form of increased mating opportunities, drive males to be larger in order to compete for high quality territories and, through increased fasting abilities, to retain them for longer. However, the advantages of size are modified by the need for agility and speed. Therefore, whilst there is still a large sexual dimorphism in Antarctic fur seals, it is slightly reduced compared to elephant seals.

Summer movements

Because of their large size, and associated problems in experimental handling, there is limited knowledge of the feeding and at sea behaviour of male Antarctic fur seals. Males begin to leave the breeding beaches at the end of December once most of the females have been mated. Of three males tracked from South Georgia all headed south and one reached Signy Island in the South Orkneys (Boyd et al., 1998). The build-up of male numbers on Signy, and surrounding areas, suggests that this southwards migration is common to males breeding at South Georgia. Because females are restricted to the breeding areas, there is a large separation of the sexes during the summer.

As female fur seals must feed throughout their lactation, breeding beaches need to be located in close proximity to foraging areas that are optimal for females. The spatial separation of the sexes may reflect either that South Georgia is a sub-optimal site for males or that foraging by females causes increased competition for resources. Males, having no constraints on their movements, may be able to forage more efficiently in regions where there are no females. The South Orkneys support large numbers of males during the summer, however, less than six pups are born there each year (Boyd, 1993a). This suggests that the local conditions, whilst ideal for males, are not favourable for females during the lactation period.

Winter movements

During the winter months females disperse at sea. Some have been recorded as far north as the River Plate region of South America and others to the south, in the region of the Antarctic ice edge (Boyd et al., 2002). It is thought that females remain at sea for extended periods during the winter months as they are rarely observed on land and are often infested with goose barnacles, Lepas australis, upon arrival at breeding beaches (Bonner, 1968; Arnbom & Lundberg, 1995).

Males have not been satellite tracked during the winter but land-based observations at South Georgia indicate that a small proportion of the population regularly haul out there. Adult males reappear on the breeding beaches around May and June, after the females have left, although not in the numbers associated with peak breeding. During these months males appear to alternate periods resting ashore with foraging trips to sea (British Antarctic Survey, unpublished data).

Diet

During the summer, the diet of lactating females at South Georgia is dominated by krill, *Euphausia superba*, with small amounts of fish (Reid & Arnould, 1996). The males at the South Orkneys also appear to consume mostly krill but fish prey is also common in their scats. In areas close to colonies, significant numbers of penguins are also taken (Daneri & Coria, 1992).

The winter diet of females is unknown. However, as Antarctic krill do not occur north of the polar front, females foraging on the Patagonian shelf will probably be consuming other items that are abundant in this region, e.g. lobster krill, *Munida gregaria*, squid and fish. The winter diet at South Georgia, when the population is dominated by males, shows a large increase in the amount of fish taken. Whilst krill is still important, the number of scats containing fish rises by around 25% in winter (Reid, 1995).

Diving behaviour

Even when males and females are located in the same area (for around a month of the breeding season) males fast. Immediately before and after their haul-out males actively feed and there is potential for an overlap with the foraging areas of females. During this time, females forage in both oceanic and shallow shelf waters exploiting the surface pelagic layers with a median dive depth of between 35 and 40 m (Staniland & Boyd, 2003). Because of their larger body size, male fur seals are predicted to have increased diving capabilities compared to females (Boyd & Croxall, 1996). The limited studies of male diving behaviour suggest that they do not exploit the same region of the water column as females and instead target prey at depths greater than 40 m in the bottom of the surface mixed layer (Boyd et al., 1998). Therefore, even though males and females may have a broadly similar diet, at least for part of the year, their foraging areas do not overlap.

Male fur seals are larger than females, in part because the intense competition for territories, although this is partly moderated by the need for agility and speed. In addition, the male breeding strategy could be considered a form of capital investment. Large size, through increased fasting abilities, will therefore allow males to remain on the breeding beaches for longer. As well as different energy requirements, the sexes have different constraints placed upon their foraging. As a result, male and female Antarctic fur seals are segregated in their year-round spatial distributions, their diving behaviour, and there is evidence of a difference in the relative compositions of their diets.

THE INFLUENCE OF MATING SYSTEMS ON SEXUAL SEGREGATION

Sexual segregation in seals ranges from little or no separation (e.g. Weddell seals) to large habitat and foraging differences (e.g. elephant seals). The nature and degree of separation appears to be most strongly influenced by the sexual size dimorphism of species which in turn is influenced by their mating systems. Pinnipeds must haul out to give birth. To avoid land-based predators they breed on isolated areas such as islands, remote beaches and ice. Seal mating systems are thought to have evolved in response to their breeding habitats and the way in which these influence the behaviour of females (Stirling, 1983). For example, how males can compete for mates will be shaped by the proximity of females to one another, female site fidelity and the synchronicity of oestrous. These factors are in turn influenced by the habitat stability, access to water and competition for space.

Land breeders

Because of limited suitable habitat for species that breed on land, females are usually clumped together. Large aggregations of females allow males to compete for, and mate with, a large number of females. This extreme polygyny means there is intense competition between males and a large variation in their breeding success. Because of the stability of the breeding habitat, site fidelity is common in land breeders and the timing of reproduction is usually consistent between years. The predictability of breeding location and timing increases the competition between males by allowing them to establish territories or hierarchies before the arrival of females. The different selective pressures acting on the sexes in land breeding systems leads to large size differences. Males have a larger body size than females and develop secondary sexual characteristics such as enlarged canines, protected neck regions and modified organs for display (e.g. the enlarged proboscis of male elephant seals: Fig. 4.1).

Larger animals have higher absolute energy requirements; for example, male southern elephant seals, *Mirounga leonina*, require twice as much energy as females (Boyd *et al.*, 1994). Bigger, stronger males will also have different predation risks, prey handling dynamics and increased diving capabilities compared with females. These factors mean that, in effect, the sexes should be treated as separate species. Given these differences it therefore seems highly unlikely that males and females displaying large size dimorphism would be able to feed optimally in the same areas or even on the same food. The examples of elephant seals and Antarctic fur seals confirm this expectation and Table 4.1 shows other examples of sexual segregation in dimorphic species.

In sexually dimorphic species, differences between males and females are apparent during their development. Delayed sexual maturity and increased investment in growth is apparent in male elephant seals and Antarctic fur seals (Riedman, 1990). Even as pups there are differences: male Antarctic fur seals are born heavier and grow faster than females (Goldsworthy, 1995). Male pups appear to direct more energy towards lean tissue growth, whereas females tend to accumulate greater fat reserves. As a result, males have a lower mass specific metabolic rate than females (Arnould *et al.*, 2001; Lunn & Arnould, 1997). There are also behavioural differences; male pups are more likely to be found on the beach interacting with their peers whereas females are more commonly found in the relative safety of the surrounding tussock (British Antarctic Survey, unpublished data).

Ice breeders

Ice as a breeding substrate can be divided into two main groups: pack-ice that floats and offers unrestricted access to the sea, and fast-ice which is attached to land and from which the sea is only accessible through breathing holes and natural fissures. Both ice types often provide a ready access to food near breeding sites, and some security from predation.

Pack-ice provides an extensive habitat for breeding and females are usually widely distributed over it. Males therefore cannot defend more than one female at a time. The unstable nature of pack-ice means that the breeding season is usually very short, sometimes only a few days. The result of these factors is that the extreme polygyny seen in many land-based breeders is not possible. Monogamy or very low-level polygyny, usually described as serial monogamy, is more common. It is not clear how males gain access to females in this system but it is believed that the majority of competitive interactions and copulations take place in the water.

Although fast-ice appears to be similar to pack-ice, offering abundant breeding space, the usable habitat is restricted because females can only haul out in regions that are accessible from the water (i.e. near cracks, fissures or dug-out breathing holes). Species that breed on fast-ice do not form large groups as only a limited number of individuals can share a breathing hole. In the Arctic, the presence of land-based predators also discourages the formation of groups (Riedman, 1990). Therefore, only monogamy or low-level polygyny is observed in fast-ice breeding seals and the sexes are usually similar in size. Even in the example of Weddell seals, where males compete for territories, the aquatic nature of fights and copulations has removed the pressure for large male body size (Riedman, 1990).

Ice breeding seal species (and other aquatically mating species) are characterized by monomorphism or slightly larger females. Individuals therefore have equivalent metabolic requirements (apart from possibly lactation), diving capabilities and predation risks that result in little detectable sexual segregation. Males are often able to feed opportunistically during the breeding season. As a result, their length of tenure (i.e. time at the breeding site) will not be as dependent on their body reserves and the selective pressure for large male size is again diminished. There are, however, exceptions to these generalizations, notably the sexually dimorphic, ice breeding walrus and the hooded seal (see Stirling, 1983).

Parental investment in offspring

The dichotomy of parental care and investment in the offspring can also influence the segregation of the sexes. The need for female otariid seals to feed during their protracted lactation means that, for this period, their foraging range is restricted. Males who do not provision offspring are therefore free to move to productive feeding areas away from the breeding females e.g. northern fur seals, *Callorhinus ursinus* (Table 4.1).

There are also temporal differences in the energy demands of the sexes. For example, females require increased resources during the active gestation period. The delayed implantation used by females means that, at a time when adult males are on a maintenance diet, females are required to increase their energy consumption. Although the additional resources required to sustain gestation are small. Grey seals, *Halichoerus grypus*, show significant sex differences in their seasonal patterns of total body energy. Females accumulate body energy stores earlier and carry a relatively higher level than males. Males appear only to gain mass in the three months immediately prior to the breeding season (Beck *et al.*, 2003a).

Very few studies have directly examined sexual segregation in seals and Table 4.1 highlights the lack of published information for many species. In some cases, such as the Ross seal, *Ommatophoca rossi*, this is because very little is known about the animal. In others, such as many of the fur seal and sea lion species, knowledge is restricted to lactating females because of problems associated with handling the larger and more aggressive males.

Implications of sexual segregation

Seals attract a great deal of interest because of their interactions with man through direct culling, competition with fisheries or disturbance in breeding areas. Several species, or populations, are in decline or under threat; indeed two species (*Monachus tropicalis* and *Zalophus californianus japonicus*) are listed as possibly extinct (Reijinders *et al.*, 1993). Because of the economic implications, interactions of seal populations with commercial fisheries receive particular interest. In certain species, the nature and magnitude of these interactions will be heavily influenced by the observed segregation of the sexes, especially the separation of foraging areas and differences in diet. Where it does occur this segregation must be identified and each sex treated separately in order to fully quantify any potential effect.

One example is the small-scale management units currently being developed in the Southern Ocean by the Commission for the Conservation of Antarctic Marine Living Resources (CCAMLR). These management zones are being proposed to minimize the effect of developing fisheries on areas where there are large populations of higher predators (Constable & Nicol, 2002). Clearly, the design of these zones will need to account for any potential sexual segregation in the numerous species of seal distributed within the Antarctic.

5

Sexual differences in foraging behaviour and diets: a case study of wandering albatrosses

OVERVIEW

Albatrosses and petrels (Procellariiformes) are a group of pelagic seabird species that exhibit a wide range in body mass and some degree of sexual dimorphism (Warham 1990; Croxall 1995). Within this order, the K-selected, single-egg clutch, monogamous, biennial breeder, wandering albatross *Diomedea exulans* (Fig. 5.1), which breeds on various sub-Antarctic islands (Fig. 5.2), is the most sexually dimorphic of any albatross species; although structurally similar to females at all ages, males are approximately 20% heavier and larger than females, and have a whiter plumage (Tickell, 1968; Weimerskirch *et al.*, 1989). Moreover, when comparing the morphometric characteristics of female and male wandering albatrosses, 11 (out of 12) parameters were significantly different ($P < 0.001$) (Shaffer *et al.*, 2001).

As body size, wingspan and flight performance in wandering albatrosses are known to be positively related (i.e. males have longer wingspan and higher wing loading than females) (Shaffer *et al.*, 2001), sexual size dimorphism may have a functional influence on the capacity to transport food (particularly important whilst breeding) and at-sea distribution of wandering albatrosses. With the recent development of small satellite tracking devices and additional instruments (e.g. activity recorders, stomach probes, GPS devices; Prince & Francis, 1984; Prince & Walton, 1984; Jouventin & Weimerskirch, 1990b; Wilson *et al.*, 1992; Weimerskirch *et al.*, 2002), a detailed characterization of foraging patterns of male and female wandering albatrosses can be obtained. Furthermore, as wandering albatrosses are accomplished opportunistic

Sexual Segregation in Vertebrates: Ecology of the Two Sexes, eds. K. E. Ruckstuhl and P. Neuhaus.
Published by Cambridge University Press. © Cambridge University Press 2005.

(a)

(b)

Figure 5.1 Wandering albatross. (a) A male wandering albatross, *Diomedea exulans*, incubating, and (b) a couple of wandering albatrosses (male left, female right) during the chick-rearing period, at Bird Island, South Georgia.

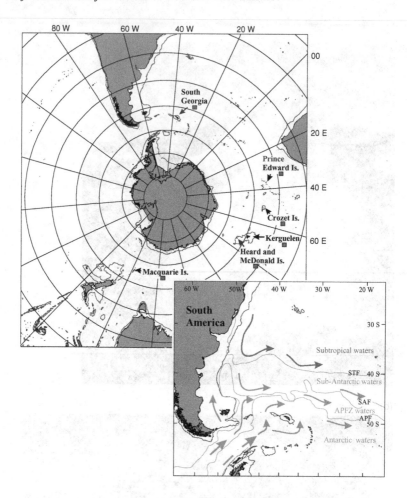

■ Breeding sites of wandering albatross *Diomedea exulans*

Figure 5.2   Breeding islands of wandering albatrosses around the
Southern Ocean and the various water masses close to South Georgia
(Legend: APF – Antarctic Polar Front; APFZ – Antarctic Polar Frontal
Zone; SAF – Sub-Antarctic Front; STF – Subtropical Front).

foragers, highly attracted to fishing vessels (Brothers, 1990; Croxall
et al., 1998; Weimerskirch & Jouventin, 1998), it can also permit a critical
evaluation of the effects of sex-specific albatross interactions with longline fisheries, whereby wandering albatrosses are attracted to the bait
and get caught by the longline hooks and drown (Brothers, 1990; Gales,
1993; Croxall *et al.*, 1998). Largely as a result of the population decrease
caused by by-catch in longline fisheries, wandering albatrosses are

currently classified as vulnerable according to the Red List criteria of the International Union for the Conservation of Nature (IUCN) (Croxall & Gales, 1998; Gales, 1998; BirdLife International, 2000). It is, therefore, imperative to quantify how much wandering albatrosses sexually segregate in terms of foraging, and how much their foraging is associated to longline fisheries.

Here we describe research into sexual differences in the foraging behaviour of wandering albatrosses at South Georgia (54°S 38°W), where 27% of the world's population of wandering albatrosses breed (Croxall et al., 1998). Male wandering albatrosses arrive at South Georgia in late November, two weeks earlier than females (Tickell, 1968). Wandering albatrosses breed biennially if successful, taking about one year to fledge a chick, which is reared through the Antarctic winter (Tickell, 1968). Their diet are mainly cephalopods and fish (Xavier et al., 2003a,b; Xavier et al., 2004). Whilst rearing chicks at South Georgia (March until December), wandering albatrosses forage in many different marine habitats, including Antarctic, sub-Antarctic, subtropical and tropical waters and there is some evidence that females forage further north than males (Prince et al., 1998; Xavier et al., 2004). Their foraging trips are long (in comparison with other seabirds), typically lasting 7–13 days but trips of 57 days have been recorded (Prince et al., 1999; Berrow & Croxall, 2001). Foraging in such a wide range of habitats potentially provides a particular opportunity to evaluate sexual differences in terms of foraging preferences and allows consideration of how this might relate to prey choice and to the location of longline fisheries.

The aims of this chapter are:

(a)   to test if male and female wandering albatrosses from South Georgia segregate into different habitats of the foraging range during their chick-rearing period;

(b)   to review sexual segregation in wandering albatrosses in relation to foraging preferences through their life-cycle and evaluate implications for diets and interactions with longline fisheries;

(c)   to suggest a possible explanation on the origins of sexual spatial segregation and sexual dimorphism in wandering albatrosses.

ASSESSING SEXUAL FORAGING DIFFERENCES

Breeding wandering albatrosses were tracked using satellite transmitters at Bird Island, South Georgia during the chick-rearing period between May and August of 1999 and 2000.

The deployment procedure followed Xavier et al. (2004). The transmitters provided: the geographic position of the seabird at a particular time, obtained from signals detected by orbiting satellites, when the bird was in contact with salt-water (i.e. the time the bird spent on the sea surface; devices have a salt-water switch that when immersed in sea-water initiates an electrolytic process); and the foraging trip duration, based on the difference between colony departure and return times. The oceanic regime and foraging tracks in relation to sea surface temperature followed Xavier et al. (2004), where albatross foraging tracks were overlayed on the sea surface temperature (SST) images and the timing of the geographic position given by satellite uplinks was used to determine the time spent over each water mass. Kernel analysis was used to characterize spatial distribution following Wood et al. (2000) and Hooge & Eichenlaub (1997) in order to identify the preferred areas used by wandering albatrosses. As overall foraging distributions were similar between years, these were combined in order to evaluate differences between the sexes (e.g. males, trip durations for both years were combined). A foraging trip is defined as the time taken from the departure of the albatrosses from the colony until their return. Food samples were collected after a single foraging trip and dietary analysis was performed (see Xavier et al. (2004) for details). A randomization test was used for sex differences in diet species composition. Diet diversity was quantified using Shannon–Wiener index following Xavier et al. (2003a).

Foraging efficiency was measured as the proportionate daily mass gain whilst foraging, by weighing albatrosses before and after a foraging trip, following González-Solís et al. (2000a). Where appropriate, both years were combined when performing comparative statistical analysis.

Sex differences in wandering albatross foraging strategies through their life-cycle

During their life-cycle, wandering albatrosses forage from Antarctic to tropical waters in the Southern hemisphere (Tickell, 2000, this study). Even when breeding, they forage in these water masses, but are limited in terms of longitude to areas close to their breeding islands (Tables 5.1 and 5.2).

Differences in foraging patterns between the sexes are reported in all wandering albatross breeding sites and are evident throughout the life-cycle of wandering albatrosses (Tables 5.1 and 5.2). During the incubation period (between January and February), when both male and female alternate duties, differences between the sexes in both foraging and food provisioning strategy were found; females had longer trips, foraged further north, and visited their nest less often than males

Table 5.1 *Foraging trip durations (average ± SE), time spent in water masses and main diets of wandering albatrosses breeding at South Georgia. Statistics compared males and females with both years data combined (APFZ – Antarctic Polar Frontal Zone waters).*

	Males		Females		Overall		
	1999 (n = 9)	2000 (n = 10)	1999 (n = 9)	2000 (n = 10)	Males (n = 19)	Females (n = 19)	Statistics
Mass (kg)	10.7 ± 0.4	10.4 ± 0.3	8.8 ± 0.1	8.6 ± 0.3	10.5 ± 0.2	8.7 ± 0.2	P < 0.01
Trip duration (mean ± SE)	9.3 ± 2	5.4 ± 2	18.1 ± 4	7.8 ± 2	7.2 ± 1	12.7 ± 2	P = 0.08
Distance from colony (range; km)	78–1443	16–907	580–1502	102–1345	16–1443	102–1502	P < 0.01
Foraging efficiency (g/day foraging ± SE)	102.0 ± 50	348.1 ± 93	1.9 ± 23	99.1 ± 42	231.5 ± 79	53.1 ± 35	P = 0.09
Mean distance from colony (km)	827 ± 145	391 ± 109	1016 ± 96	732 ± 164	571 ± 103	874 ± 99	P = 0.06
Water masses (%):							
Antarctic	29.3	48.3	12.7	35.4	36.8	20.0	P = 0.95
APFZ	16.0	10.5	23.2	15.8	13.8	20.8	P = 0.07
Sub-Antarctic	32.9	27.0	43.0	33.8	30.6	40.1	P = 0.06
Subtropical	21.8	14.2	19.8	12.4	18.8	17.4	P = 0.40
Tropical	0.0	0.0	1.2	2.5	0.0	1.6	
Food sample (g)	1022 ± 160	1025 ± 164	517 ± 128	679 ± 142	1024 ± 112	602 ± 95	P < 0.01

(cont.)

Table 5.1 (cont.)

	Males		Females		Overall		Statistics
	1999 (n = 9)	2000 (n = 10)	1999 (n = 9)	2000 (n = 10)	Males (n = 19)	Females (n = 19)	
Estimated prey size (mm)	1096 ± 357	1680 ± 397	218 ± 18	640 ± 239	1380 ± 110	409 ± 267	$P < 0.01$
Main diet (% by mass ± SE):							
Fish	73.6 ± 13	90.5 ± 8	32.7 ± 13	78.2 ± 10	82.5 ± 8	56.6 ± 9	$P = 0.01$
Bathylagus sp.	0.0	0.0	12.4	0.0	0.0	2.2	
Dissostichus eleginoides	68.3	70.1	0.0	56.3	76.6	36.6	
Macrourus holotrachys	3.85	0	0	13.4	2.1	8.7	
Cephalopods	17.4 ± 11	8.5 ± 8	66.7 ± 13	14.0 ± 8	12.7 ± 7	39.0 ± 10	$P = 0.02$
Kondakovia longimana	0.0	7.7	24.6	2.1	9.0	17.9	
Moroteuthis knipovitchi	3.9	0.4	10.8	0.0	1.1	4.2	
Others	8.9 ± 8.8	0.9 ± 1	0.6 ± 1	7.8 ± 6	4.7 ± 4	4.3 ± 3	$P = 0.79$
Number of species present	12	10	16	18	17	27	$P = 0.14$
Number of prey per sample (mean ± SE)	2.6 ± 0.6	1.6 ± 0.3	7.2 ± 2.5	4.7 ± 1.0	2.1 ± 0.3	5.9 ± 1.3	$P = 0.02$
Shannon–Weaver index	0.48	0.57	0.93	0.75	0.64	0.98	$P = 0.60$

Table 5.2 *Some differences between sexes in the wandering albatross at their main breeding sites.*

Study site	South Georgia		Prince Edward Islands		Crozet Islands	
Breeding pairs (annual)	2178		3071		1734	
Sex	Males	Females	Males	Females	Males	Females
Mass (kg; mean and range)	9.8 (8.2–12.4)	7.7 (6.7–10.2)	9.3 (8.2–11.3)	7.8 (6.4–9.0)	9.4	7.8
Main foraging habitat	Antarctic	Sub-Antarctic	Sub-Antarctic	Subtropical	Antarctic/Sub-Antarctic	Sub-Antarctic–Tropical
Main diet (% mass):						
Fish	83	57	Cephalopods (59); fish (37)		32	20
Cephalopods	13	39	(both sexes combined)		68	80
Sources	1, 2 and 3		2, 4, 5 and 6		7 and 8	

sources: 1. Present study; 2. Gales (1998); 3. Tickell (1968); 4. Cooper et al. (1992); 5. Nel et al. (2002); 6. Tickell (2000); 7. Weimerskirch et al. (1997a); 8. Weimerskirch (1998).

(Weimerskirch *et al.*, 1993; Weimerskirch, 1995; Weimerskirch *et al.*, 1997b; Croxall *et al.*, 1999; Nel *et al.*, 2002).

During the brood-guard (between March and April when parents still alternate attendance and foraging, and when range is particularly reduced) sex-specific foraging behaviour differences also occurred, with females foraging further away from the colony, off the shelf, travelling greater distances and with greater foraging effort than males (Salamolard & Weimerskirch, 1993; Weimerskirch *et al.*, 1993; Croxall & Prince, 1996; Xavier *et al.*, 2003c). This might be related to the competitive exclusion of females by larger males (Weimerskirch *et al.*, 1993). If most males forage at the shelf and shelf edge of their breeding islands, intraspecific competition could be high and access to a patchy resource could be difficult for the females. This is consistent with what is found at South Georgia during the brood-guard period (Xavier *et al.*, 2003c), where despite the apparent similarities in overall diet (both fed primarily on fish (40–49% by mass for both sexes) and cephalopods (25–35% by mass)) males fed mainly on the Patagonian toothfish *Dissostichus eleginoides* (50% of diet by mass; squid was represented only by *Batoteuthis skolops*) whereas females fed on a wider variety of different fish (*Antimora rostrata*, *Lionirius filicauda* and *Macrourus holotrachys*) and squid (*Chiroteuthis* sp., *Galiteuthis glacialis*, *Kondakovia longimana* and *Histioteuthis* B).

During the chick-rearing period (from April to December), females also foraged further north than males. At South Georgia, when the distance between each known albatross location (from satellite tracking data) and the breeding colony were measured, females were foraging significantly further north than males (Table 5.1). Male wandering albatrosses ($n = 19$) concentrated their foraging area close to South Georgia and foraged predominantly in Antarctic waters (37% of total foraging time; Fig. 5.3; Table 5.1). Females ($n = 19$) foraged mainly in sub-Antarctic waters (40% of total foraging time), particularly at the edge of the Patagonian shelf, but also concentrated close to South Georgia and in an area close to 50°S 45°W (Fig. 5.4; Table 5.1).

Despite the highest density of foraging of both sexes overlapping greatly (reflecting similar foraging concentrations close to South Georgia), females foraged in areas further north (Figs. 5.3 and 5.4). The level of foraging overlap of both sexes in Antarctic waters could be a reflection of birds having to return to their breeding colony (in this case South Georgia) and, consequently, over-represent the importance of Antarctic waters as a foraging area. This is particularly evident for

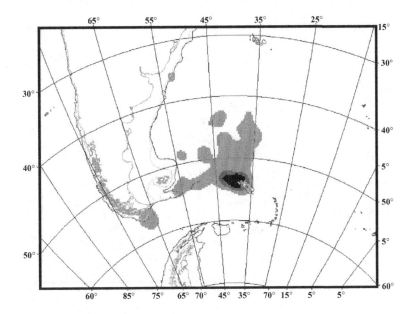

Location densities

50% of locations

75% of locations

95% of locations

Figure 5.3 The distribution of satellite tracked male wandering albatrosses breeding at South Georgia between May and July, using density contours resulting from kernel analysis.

females that spend most of their time foraging in sub-Antarctic waters (Table 5.1), have significantly longer trips than the males (Prince *et al.*, 1998; Prince *et al.*, 1999, this study; Berrow *et al.*, 2000) and also spend a greater proportion of their time foraging in oceanic waters (Xavier *et al.*, 2004).

The foraging behaviour of South Georgia wandering albatrosses during the chick-rearing period is broadly similar to that recorded for wandering albatrosses breeding elsewhere. At Crozet Islands, the males generally foraged in Antarctic waters and on shelves of Crozet and Kerguelen Islands with females foraging more in subtropical oceanic waters (Weimerskirch & Jouventin, 1987; Weimerskirch *et al.*, 1993).

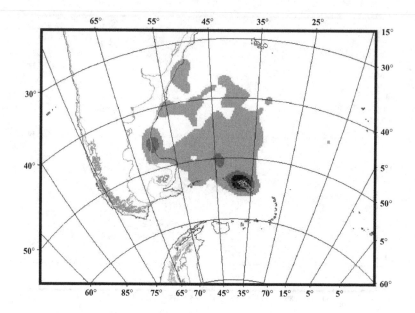

Location densities

█ 50% of locations

▓ 75% of locations

░ 95% of locations

Figure 5.4 The distribution of satellite tracked female wandering
albatrosses breeding at South Georgia between May and July, using
density contours resulting from kernel analysis.

Similarly, at the Prince Edward Islands, the females spent more time
foraging in warmer waters than the males (Cooper *et al.*, 1993; Nel
et al., 2002).

Outside the breeding season, wandering albatrosses from South
Georgia, and probably elsewhere, circumnavigate the Southern Ocean,
visiting the Indian and Pacific Ocean waters (usually overwintering in
Australian waters) before returning to their breeding colonies in the
following year (Nicholls *et al.*, 1997; Croxall, 1998; Croxall *et al.* (2005);
Tickell, 2000). There is evidence to suggest that, even between breeding
attempts, the sexes still segregate: females flying more into tropical and
subtropical waters and males into Antarctic and sub-Antarctic waters
(Weimerskirch & Wilson, 2000; Nel *et al.*, 2002).

CONSEQUENCES OF SEXUAL SEGREGATION
IN WANDERING ALBATROSSES

Diet differences between the sexes

Exploiting different, although overlapping, foraging areas when breeding may lead to a reduction in inter-sexual food competition and consequently produce sexual differences in the diets, particularly of the species taken. At South Georgia, the females consumed a higher diversity of cephalopods and fish, and of smaller size, than the males (see Table 5.1). Indeed, the females also caught a significantly wider range of species than the males; out of the total prey species found in both males and females (31 species), 14 were only found in female samples against 4 found only in male diets (t-test; $P = 0.01$; Tables 5.3 and 5.4). Thirteen prey species were found in both diets. In terms of species, *D. eleginoides*, macrourids, and the squids *K. longimana* and *M. knipovitchi* dominate in both sexes (Xavier *et al.*, 2003a; this study); the diet of male wandering albatrosses from South Georgia was dominated by fish (range: 74–91% by mass), particularly by *D. eleginoides*, which contributed 68–70% of the diet by mass (Table 5.1 and Table 5.3). Cephalopods, mainly the onychoteuthids *K. longimana* and *M. knipovitchi*, only represented 9–17% (by mass) of their diet (Table 5.1). The female diet varied greatly between years; cephalopods predominated in 1999 and fish in 2000 (Table 5.1). The *K. longimana* was the most important prey in 1999, whereas in 2000, as in male diets, *D. eleginoides* was the most important prey (Table 5.1).

Therefore, at South Georgia, both sexes fed primarily on fish and cephalopods (Prince & Morgan, 1987; this study). Elsewhere, however, cephalopods were the main component of the diet of both sexes. At Crozet Islands, cephalopods were the main prey in both female and male wandering albatrosses, with females consuming *K. longimana* and males eating predominately the onychoteuthid *Moroteuthis ingens* and *D. eleginoides*, feeding fish to their chicks more often than the females (Weimerskirch *et al.*, 1997b; Table 5.2). At Prince Edward Islands, cephalopods are also the main prey (59% by mass; both sexes combined), with *K. longimana* as the main prey (Cooper *et al.*, 1992).

Foraging differences between the sexes and interactions with fisheries

Sexual spatial segregation in wandering albatrosses is expressed by a more northerly distribution of females (i.e. males tend to stay in

Table 5.3 *Frequency of occurrence (FOO), number of individuals (n) and mass (m) of female wandering albatrosses (data from 1999 and 2000 combined).*

Family	Cephalopods	FOO	%	n	%	m	%
	Cephalopods						**39.0**
Alluroteuthidae	*Alluroteuthis antarcticus*	4	21.1	10	9.1	2945	4.0
Chiroteuthidae	*Chiroteuthis* sp.	2	10.5	3	2.7	474	0.6
Cranchiidae	*Galiteuthis glacialis*	7	36.8	8	7.3	1079	1.5
	Taonius sp.	2	10.5	3	2.7	913	1.2
Gonatidae	*Gonatus antarcticus*	2	10.5	3	2.7	504	0.7
Histioteuthidae	*Histioteuthis atlantica*	2	10.5	7	6.4	1816	2.5
	Histioteuthis macrohista	1	5.3	1	0.9	51	0.1
	Histioteuthis eltaninae	4	21.1	13	11.8	872	1.2
	Histioteuthis miranda	1	5.3	1	0.9	253	0.3
Ommastrephidae	*Illex argentinus*	3	15.8	20	18.2	5241	7.1
	Todarodes sp.*	1	5.3	1	0.9	1133	1.5
Onychoteuthidae	*Kondakovia longimana*	3	15.8	3	2.7	9652	13.0
	Moroteuthis knipovitchi	3	15.8	6	5.5	3943	5.3
	Unknown	1	5.3	1	0.9	0	0.0
	Fish						**56.6**
Anopteridae	*Anopterus pharao*	1	5.3	1	0.9	0	0.0
Bathylagidae	*Bathylagus* sp.	1	5.3	1	0.9	670	2.2
Clupeidae	*Sardinops? sagax*	1	5.3	7	6.4	1522	4.9
Channichthyidae	*Chaenocephalus aceratus*	1	5.3	1	0.9	441	1.4
Macrouridae	*Lionurius filicauda*	2	10.5	2	1.8	70	0.2
	*Macrourus holotrachys**	3	15.8	3	2.7	2699	8.7
	Macrourus sp.*	1	5.3	1	0.9	381	1.2
	Ventrifossa nasuta	1	5.3	1	0.9	19	0.1
Myctophidae	*Diaphus* sp.	1	5.3	1	0.9	3	0.0
Muraenolepididae	*Muraenolepsis microps*	1	5.3	1	0.9	270	0.9
Nototheniidae	*Dissostichus eleginoides**	2	10.5	4	3.6	11337	36.6
	Gobionotothen gibberifrons	1	5.3	1	0.9	103	0.3
	Others						**4.3**
Amphipoda	*Themisto gaudichaudii*	1	5.3	1	0.9	1	0.0
	Carrion (whale, seal)	1	5.3	1	0.9	562	3.6
	Scyphozoa (jellyfish)	1	5.3	1	0.9	104	0.7
	Nematodes	1	5.3	2	1.8	2	0.0
	Chilli peppers	1	5.3	1	0.9	4	0.0

* probably caught as offal/discards from longline fisheries.

Table 5.4 *Frequency of occurrence (FOO), number of individuals (n) and mass (m) of male wandering albatrosses (data from 1999 and 2000 combined).*

		FOO	%	n	%	m	% overall
Family	Cephalopods						12.7
Alluroteuthidae	*Alluroteuthis antarcticus*	1	5.3	1	2.6	415	1.3
Cranchiidae	*Galiteuthis glacialis*	3	15.8	3	7.7	317	1.0
Gonatidae	*Gonatus antarcticus*	3	15.8	3	7.7	1034	3.2
Histioteuthidae	*Histioteuthis atlantica*	1	5.3	1	2.6	253	0.8
	Histioteuthis eltaninae	2	10.5	4	10.3	311	1.0
Ommastrephidae	*Illex argentinus*	1	5.3	1	2.6	236	0.7
Onychoteuthidae	*Kondakovia longimana*	2	10.5	2	5.1	673	2.1
	Moroteuthis knipovitchi	2	10.5	2	5.1	850	2.6
	Unknown	0	0.0	1	2.6	0	0.0
	Fish						82.5
Channichthyidae	*Chaenocephalus aceratus*	1	5.3	1	2.6	724	2.3
	Champsocephalus gunnari	1	5.3	1	2.6	0	0.0
	Pseudochaenichthys georgianus	1	5.3	1	2.6	1172	3.7
Nototheniidae	*Dissostichus eleginoides**	6	31.6	7	17.9	24517	76.6
	Gobionotothen gibberifrons	1	5.3	1	2.6	255	0.8
Macrouridae	*Lionurius filicauda*	1	5.3	1	2.6	27	0.1
	*Macrourus holotrachys**	1	5.3	1	2.6	680	2.1
Moridae	*Antimora rostrata**	2	10.5	2	5.1	1690	5.3
	Others						4.7
Natantia	*Pasiphaea scotiae*	1	5.3	1	2.6	3	0.0
	Carrion (whale, seal)	2	10.5	2	5.1	602	4.7
	Fishing hook*	1	5.3	1	2.6	1	0.0
	Nematodes	1	5.3	2	5.1	2	0.0

* probably caught as offal/discards from longline fisheries.

Antarctic waters around their breeding islands and females tend to forage in more oceanic, sub-Antarctic/subtropical, waters, Tables 5.1 and 5.2) and one of the possible consequences is in terms of foraging efficiency, when females tended to gain less weight while foraging (mean gain: 53 g/day trip) than males (232 g/day trip) (Table 5.1). The different nutrient requirements for females (e.g. to restore their calcium levels after producing the egg during breeding periods) or sexual differences in terms of moulting (almost exclusively during the inter-breeding periods) may contribute to maintain these foraging differences; males always replace more feathers than females, who thus have more older secondaries and consequently influence, at least temporarily, their foraging abilities (Weimerskirch, 1991; Tickell, 2000).

Another consequence of females foraging more northerly than males is the different levels of interactions of wandering albatrosses with longline fishing vessels, and to a lesser extent trawl fisheries, which are thought to be a major cause of incidental mortality of wandering albatrosses (Brothers, 1990; Croxall & Prince, 1996; Weimerskirch, 1998; Nel et al., 2002; Xavier et al., 2004). Longline fisheries on the Southern Ocean are concentrated at various scales and areas: locally, where wandering albatrosses breed (around island shelves, particularly represented by the *Dissostichus eleginoides* fisheries), in oceanic waters (represented by the Southern bluefin tuna fisheries) and on continental shelves (such as the Patagonian Shelf; represented by the Patagonian toothfish and kingclip, *Genypterus blacodes*, fisheries) (Tuck et al., 2003; Xavier et al., 2004). In Antarctic waters, on the shelf and shelf slope of South Georgia and Crozet Islands, males interact more with the local longline fishery (wandering albatrosses breeding at Crozet islands might also interact with nearby island longline fisheries such as Kerguelen). In addition, as females have a more northerly distribution they interact with other longline fisheries, increasing their probabilities of incidental mortality. Females breeding at South Georgia forage mainly in pelagic sub-Antarctic waters, where a Southern bluefin tuna longline fishery has been operating since 1951 (Tuck et al., 2003; Xavier et al., 2004), and in the shelf-slope of the Patagonian shelf, where a Patagonian toothfish fishery has operated since 1988. Females at Crozet Islands interact with the Southern bluefin tuna longline fishery in sub-Antarctic, subtropical and tropical waters (Weimerskirch, 1998). Bearing in mind that the oceanic fisheries are larger than the local ones, female wandering albatrosses, therefore, might interact with more longline fisheries throughout the Southern hemisphere (including the Southern Ocean), leading to a higher mortality in females caused by fisheries.

Sex-biased mortality and implications for population structure

Interactions with fisheries, particularly longliners, have been reflected in survival rates of adult wandering albatrosses. At South Georgia and at Crozet Islands, adult females had a lower survival rate (92.4–94.2%)

than males (93.2–95.2%) (Croxall *et al.*, 1990; Weimerskirch *et al.*, 1997a; Croxall *et al.*, 1998). Although a 1–2% difference in survival rate between sexes may appear to be small, it has important repercussions for the sex ratio and breeding structure of the population, particularly if occurring over a long period of time. Weimerskirch *et al.* (1997a) provided a good example: if 1600 breeding birds at Crozet Islands are considered, of which, in 1969, 800 were males and 800 were females, a difference of 1–2% in survival rate would lead, in 1985, to a population having 580 males and 420 females. Consequently, only 420 breeding pairs could be formed. This would represent another source of pressure on wandering albatross populations, many of which are currently in decline, arising from sex-biased survival rates.

Sexual size dimorphism in relation to sexual segregation in wandering albatrosses

One of the aspects of sexual dimorphism in wandering albatrosses at South Georgia is that the males were considerably heavier (i.e. 18% heavier) than the females, weighing 10.5 kg on average compared to an average female weight of 8.7 kg (Tables 5.1 and 5.2). Furthermore, we can recognize that sexual dimorphism in wandering albatrosses occurs in every breeding site (Tables 5.1 and 5.2; Tickell, 2000).

Sexual dimorphism in wandering albatrosses may be caused by a combination of various factors. These may include different parental roles during their reproductive life-cycle (i.e. breeding role specialization), avoidance of food competition within and between species and/or sexual selection (e.g. female preference for larger males). The relative importance of these factors is unknown. Female wandering albatrosses work harder (e.g. undertake more trips; possible need to gather reserves to produce an egg) than males at certain stages of the breeding cycle (Weimerskirch, 1995; Berrow *et al.*, 2000; Tickell, 2000), but overall, wandering albatrosses display little differentiation in reproductive sex roles in comparison with other seabird species (Coulson, 2002). Inter-sexual food competition within wandering albatrosses might be an important factor, especially in seasons when food may be scarce, which may lead to competitive exclusion of females by larger males (Weimerskirch *et al.*, 1993, see earlier). Food competition between species might not contribute greatly to sexual dimorphism as both sexes of wandering albatrosses are still larger than any other seabird species and therefore capable of out-competing other seabird species when fighting for food (Harper, 1987). Sexual selection is well illustrated in other seabirds, such as penguins (Davis & Speirs, 1990; Clarke et al., 1998). For the monogamous wandering albatrosses, mating with more experienced, bigger partners would be especially advantageous because foraging skills increase with age and/or experience (Weimerskirch, 1992). However, they seem to show significant age-assortative pairing, with wandering albatrosses mating with partners of similar age and experience (Jouventin et al., 1999). If females chose the eldest (i.e. most experienced) males, they would mate with those individuals that are more likely to die in subsequent years. Searching for a new partner could be costly in terms of breeding performance and therefore choosing experienced males of similar ages may represent the best solution (Bried & Jouventin, 2002). Wandering albatrosses have a unique courtship mating behaviour, named dance display, which includes a full extension of the wings and vocalizations (Pickering, 1989; Pickering & Berrow, 2002). The main purpose of this behaviour is to find a partner and it is mainly performed in very spacious areas where females can evaluate the size and performance of each male (Pickering & Berrow, 2002). As larger male albatrosses can potentially fast for longer (e.g. larger males have more energy reserves; Tickell, 2000) sexual selection might be related to the time spent by males ashore; females prefer males that spend more time on shore (Pickering, 1989; Croxall, 1991). By doing this, females maximize the chances of survival of their chick (i.e. by decreasing male desertion). Therefore, in an ideal situation, we suggest that a female wandering albatross would probably choose the larger, most experienced male of similar age during these display rituals. To date, however, no studies have focused on whether females do avoid young males, given they have a real choice of a range of aged mates, nor have studies focused on how the size of male wandering albatrosses might affect female preferences, particularly when mating.

These factors mentioned earlier, as potential contributions to sexual dimorphism, might also apply to sexual spatial segregation. Some authors hypothesized that sexual segregation in birds may arise from despotic exclusion of females from favoured areas by dominant (i.e. male) conspecifics (social dominance hypothesis) or by sex-specific habitat preferences (specialization hypothesis) (Selander, 1966; Greenberg, 1986). Regarding wandering albatrosses, the first hypothesis might apply, as inter-sexual food competition could occur with bigger males out-competing smaller females. The second hypothesis, however, cannot apply; differences in feeding mechanisms are non-existent (i.e. although smaller in females, the male and female bills are identical in shape and structure, Tickell, 1968) and both sexes feed on the same prey types (i.e. mainly fish and cephalopods).

In summary, pronounced sexual size dimorphism is relatively rare in seabirds. In this analysis of the most sexually dimorphic albatross species, we reviewed the sexual segregation in terms of foraging and in terms of diet through the life-cycle in wandering albatrosses. Female wandering albatrosses are smaller than males in all breeding sites, forage further north and feed on some different prey species than males. These foraging differences might be affected by differences in female nutrient requirements and/or the different demands of moulting. We also suggest that female selection for male size may play an important role in the development and maintenance of sexual dimorphism in this species. This issue has been relatively neglected in studies of seabirds and much work is needed to evaluate critically which factors affect, and maintain, sexual dimorphism and sexual segregation, particularly those that influence mating preferences. In addition, we require more detailed information on the ecological and demographic consequences of sexual segregation on seabird species, such as sex-biased incidental mortality with fisheries, and their effects on the survival rates, population structure and breeding population size.

ACKNOWLEDGEMENTS

The authors thank Andy Wood for producing the kernel analysis maps, Peter Rothery for statistical discussions, Nathan Cunningham for providing multi-variate analysis software advice, Rob Pople (BirdLife International) for seabird conservation issue discussions, and Catherine Smith, Richard Phillips and Paulo Catry for helpful discussions relevant to this chapter. This research was partly supported by the Ministry of Science and Higher Education, Portugal, the British Antarctic Survey and University of Cambridge.

JACOB GONZÁLEZ-SOLÍS AND JOHN P. CROXALL

6

Differences in foraging behaviour and feeding ecology in giant petrels

OVERVIEW

The two sibling species of giant petrels (northern *Macronectes halli* and the southern *M. giganteus*), the dominant scavengers of the sub-Antarctic and Antarctic environment, are one of the best examples of sexual segregation in avian foraging and feeding ecology. During breeding, males and females feed mainly on penguin and seal carrion, but females also feed extensively on marine prey such as cephalopods, fish and krill. Sexual differences in diet are reflected not only in analyses of regurgitations, but also in the isotopic composition of carbon and nitrogen, as well as in their heavy metal burdens. Direct observations and tracking of pelagic movements showed that males of both species usually forage closer to the breeding grounds, exploiting carcasses on beaches, whereas females show more pelagic habits. In consequence, foraging effort, foraging efficiency, predictability of resources exploited, optimum foraging time and activity budgets differ between sexes. During winter, however, studies on activity and pelagic movements suggest more similar feeding habits between sexes. Overall, differences in foraging and feeding ecology are probably related to the substantial sexual size dimorphism; males of both species are 16–35% heavier and have disproportionately larger bills than females. The importance of size in contest competition to access carrion may explain their large body size as well as the competitive exclusion of females from coastal habitats, reducing intraspecific competition for food. Ultimately, a differential exploitation of carrion probably increased sexual dimorphism

Sexual Segregation in Vertebrates: Ecology of the Two Sexes, eds. K. E. Ruckstuhl and P. Neuhaus. Published by Cambridge University Press. © Cambridge University Press 2005.

not only in body size but also in the feeding structures such as bill size. The apparently greater profitability of the feeding strategies of males compared to females suggests that giant petrels reflect an evolutionary process that benefits each sex differently.

SEGREGATION IN SEABIRDS

Males and females may vary considerably in the way they exploit food resources and differences are usually associated with diet. Most studies on seabirds have focused on to this aspect of the trophic niche, given the difficulty to study birds in a pelagic environment (Shealer, 2002). However, niche segregation is not restricted to dietary divergences but can also relate to foraging locations and activity, and consequently foraging strategies may also differ. These issues can now be addressed in detail, even for seabirds, by using a wide array of lightweight recording instruments, such as satellite transmitters, geolocators (GLS) and activity loggers. In general, most studies on trophic segregation reported sexual differences restricted to prey size, microhabitat use, or small differences in foraging locations (e.g. Lewis *et al.*, 2002 and references therein). Here, we report one of the most dramatic examples of trophic segregation and sexual size dimorphism in birds, in giant petrels. We: (i) review evidence that directly or indirectly demonstrates sexual differences in diet, foraging areas and foraging strategies; (ii) document the nature of the remarkable sexual dimorphism in body size as well as disproportionalities in bill and wing length with respect to overall body size; (iii) discuss whether the covariation of sexual segregation and sexual dimorphism is mainly driven by sexual selection or by ecological factors.

GIANT PETRELS

Northern (*Macronectes halli*) and southern (*M. giganteus*) giant petrels are closely related seabird species, so similar morphologically that until recently they were considered conspecific (Bourne & Warham, 1966). Both species are large (4 to 6 kg) and are the dominant avian scavengers in sub-Antarctic and Antarctic waters (Hunter, 1985). Inter-sexual differences in size are greater than interspecific differences, with males being between 16 and 35% heavier than females at different localities (Table 6.1). Both species lay a single egg and incubate for 60 days (González-Solís, 2004b). Laying dates of the two species never overlap, northern giant petrels laying about 40 days earlier than southern giant petrels in the four sub-Antarctic archipelagos where both species breed sympatrically (Hunter, 1987).

Table 6.1 *Percentage difference* * *of weight, wing, tarsus and culmen measurements in northern and southern giant petrels at different localities (only those traits with at least five measurements available for each sex on live birds were included).*

Species	Locality	Weight	Wing	Tarsus	Culmen	Reference
NGP	Crozets	27.50	5.30		13.70	Voisin, 1968
	Crozets	17.09	9.87	14.73	15.99	Johnstone, 1974 unpublished**
	Chatham Is.	25.66	5.84			Johnstone, unpublished**
	Macquarie Is.	28.90	6.89		14.00	Johnstone, 1977 and unpublished**
	South Georgia	27.30	6.30		14.40	Hunter, 1984
	South Georgia	22.90	4.90	10.6	14.90	González-Solís, 2004a
	Mean	**24.89**	**6.87**	**12.67**	**14.60**	
SGP	Frazier Is.		5.69	9.38	11.66	Voisin & Bester, 1981
	Macquarie Is.	20.13	4.67	8.63	12.73	Johnstone, 1977
	Crozets	23.99	5.38		13.74	Voisin, 1968
	Crozets	23.78	8.21	10.51	14.27	Johnstone, 1974 unpublished**
	South Georgia	34.48	5.92	10.41	15.37	González-Solís et al., 2000a
	South Georgia	31.06	5.63		14.67	Hunter, 1984
	Signy Is.	24.80	6.45	10.03	14.83	Conroy, 1972
	King George Is.	28.08	5.89	9.65	15.05	Peter et al., 1988
	Patagonia	27.75		11.82	12.55	F. Quintana & S. Copello, unpublished
	Mean	**26.76**	**5.98**	**10.06**	**13.87**	

* Storer's index $= 100 \,((X - Y)/((X + Y)\, 0.5))$, where X are male measurements and Y female measurements.

** Provided by E. J. Woehler, Australian Antarctic Division.

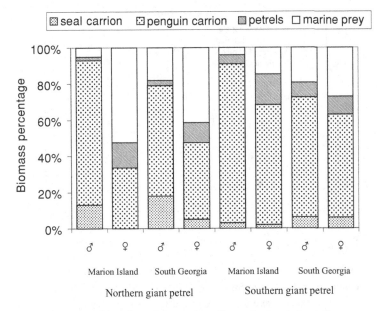

Figure 6.1   Diet of northern and southern giant petrels at the
sub-Antarctic archipelagos of Marion Island and South Georgia. Data
from Hunter (1983) and Hunter & Brooke (1992).

SEXUAL SEPARATION IN DIET

Early studies on the feeding ecology of giant petrels did not distinguish between male and female diet (Mougin, 1968; Mougin, 1975;
Conroy, 1972; Johnstone, 1977; Voisin, 1991), and suggested that penguins formed most of the nestling's diet at various sites. Recently, more
detailed studies found a noticeable sex-related differences in diet of
both species at South Georgia and Marion Island (Fig. 6.1, Hunter,
1983; Hunter & Brooke, 1992). Males fed almost exclusively on penguin and fur seal carrion, whereas females fed extensively on marine
prey as well. These differences are probably related to the greater size of
males, which can monopolize carrion close to the breeding grounds (see
later). Fish, squid and crustaceans formed from 41 to 53% by biomass
of the diet of female northern giant petrel, whereas corresponding values were only 5 to 18% in males. In southern giant petrels, differences
between females and males were more modest, the proportion of the
marine prey in the diet being 15–27% and 4–19%, respectively (Fig. 6.1).
Northern giant petrel males seem particularly specialized in feeding
on seal carrion, which comprises 13–18% of the biomass of the diet.
In contrast, the proportion of seal carrion is 0–5% in female northern

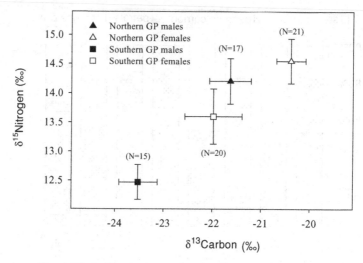

Figure 6.2   Stable nitrogen and carbon isotope values (mean ± confidence intervals at 95%) in the blood of northern and southern giant petrel males and females sampled in November 1998 at Bird Island (South Georgia).

giant petrels but 2–6% in both sexes of southern giant petrels. This striking difference in seal carrion consumption has also been confirmed by direct observation at fur and elephant seal carcasses, which are nearly exclusively attended by male northern giant petrels (>90%, Hunter, 1983; Cooper et al., 2001).

STABLE ISOTOPES AS DIETARY TRACERS

Sexual segregation in diet can also be investigated by analysis of nitrogen and carbon stable isotopes, as dietary tracers (Hobson et al., 1994; Forero & Hobson, 2003). Generally, a higher ratio of the heavier isotope with respect to the lighter ($\delta^{15}N$) in the tissues of consumers indicates higher trophic position. Likewise, in marine ecosystems, consumers in pelagic habitats should have lower levels of $\delta^{13}C$ compared to those relying on benthic food webs. Values of $\delta^{15}N$ in blood of giant petrels sampled during the incubation period were greater than the values found in the blood of other Antarctic petrels (Hodum & Hobson, 2000), as would be expected from a scavenger operating at an upper trophic level. However, in both species females showed greater $\delta^{15}N$ and $\delta^{13}C$ values than males (Fig. 6.2), which presumably relates to the greater consumption of marine prey by females as opposed to the more carrion-based diet of males (Fig. 6.1). These results suggest that carrion is a resource of lower trophic level and more indirectly related to pelagic environments than fish and squid. Indeed, at South Georgia penguins and seals feed primarily on krill (Reid & Arnould, 1996; Croxall et al., 1997), which shows low isotopic signatures of both nitrogen and carbon (Hodum & Hobson, 2000). Alternatively, higher isotopic values in females may arise from their greater exploitation of waters north of the Antarctic polar front compared to males during the breeding season (see later). Food webs in Antarctic ecosystems are comprised of few trophic steps (Rau et al., 1992; Hodum & Hobson, 2000). Therefore, isotopic signatures of predators and preys are relatively low compared to those in the neighbouring Patagonian waters (Forero et al., 2002). Isotopic analyses on the preys are needed to disentangle the reasons of the detected differences in the isotopic signature of males and females.

SEXUAL DIFFERENCES IN FORAGING HABITATS
AND FEEDING GROUNDS

Spatial segregation at sea

The first studies on giant petrels suggested that northern giant petrels showed more coastal habits than southern giant petrels (Voisin, 1968; Johnstone, 1977). Likewise, from studies on diet and observation at carcasses, some habitat segregation between sexes was suspected. However, this could not be properly addressed until the emergence of satellite telemetry. Giant petrels were among the first birds to be tracked by satellite (Parmelee et al., 1985; Strikwerda et al., 1986), but intensive studies on their foraging ecology are more recent (González-Solís et al., 2000a,b).

At South Georgia, satellite tracking of northern giant petrels during the incubation period revealed that most males foraged mainly on nearby coasts whereas females foraged at sea. Male southern giant petrels foraged closer to the South Georgian coast than females, though males did not show such a strong association with the coastal habitat as northern giant petrel males (Fig. 6.3). As a consequence, the distance covered in each foraging trip was greater for females than for males in both species (Fig. 6.4). Similar studies of southern giant petrels during the breeding season in Patagonia and Palmer Station (Antarctic Peninsula) also showed a similar sexual segregation in foraging areas (Quintana & Dell'Arciprete, 2002; Patterson & Fraser, 2003). In addition to satellite transmitters, some birds at South Georgia were also fitted

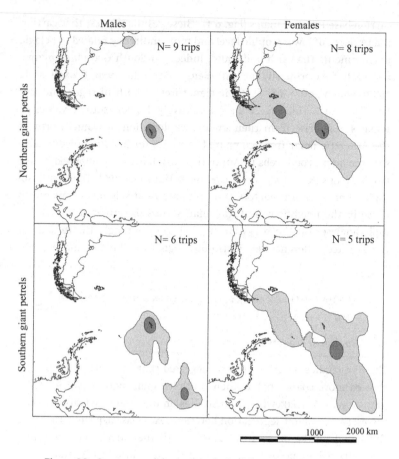

Males   Females

Northern giant petrels

N= 9 trips   N= 8 trips

Southern giant petrels

N= 6 trips   N= 5 trips

0   1000   2000 km

Figure 6.3   Summer activity ranges derived from kernel analyses on the validated locations from satellite tracking for males and females of northern and southern giant petrels breeding at Bird Island (South Georgia). Each device was deployed for a single foraging trip during the incubation period. Activity ranges encompassed 95% (general range) and 50% (core area) of locations as indicated by light and dark grey respectively. Data from González-Solís et al. (2000b).

with activity loggers on the leg, which recorded wet (on sea surface) and dry periods (flying or on land). Results from these loggers showed that coastal trips were associated with longer dry periods, particularly at night, suggesting that coastal trips were likely related to periods on land scavenging on beaches (Table 6.2, González-Solís et al., 2002a).

Though giant petrels show a remarkable sexual segregation in diet and foraging areas during the breeding period, foraging ecology

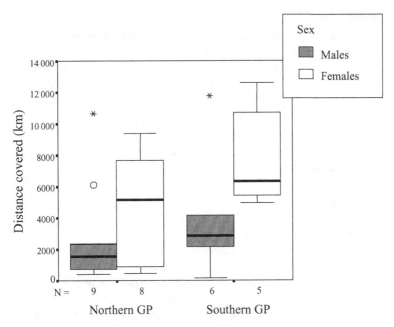

Figure 6.4 Estimated distances covered during foraging trips by males and females of the northern and southern giant petrels. Differences between sexes were significant only for southern giant petrels (NGP, Mann–Whitney U test, $Z = -0.8$, $P = 0.44$; SGP, $Z = 2.0$, $P < 0.05$). Asterisks are extreme values; circles are outlier values; the length of the box is the inter-quartile range; the length of the bar is the range, excluding extreme and outlier values. Data from González-Solís *et al.* (2000a).

during the wintering period could not be studied until light level geolocation devices for birds were developed (Global Location System; Afanasyev, 2003). Geolocators are less accurate than satellite transmitters (average error about 186 ± 114 km), but accuracy is less important when studying species that can travel vast distances in few days (Phillips *et al.*, 2003). These devices permit birds to be tracked on year-round deployments, thus it is possible to obtain information on the main foraging areas during winter (e.g. Weimerskirch & Wilson, 2000). Our studies show that northern giant petrel males and females spend the winter around South Georgia and the Falkland Islands. Males use the Falkland Islands region, the Patagonian coast, and shelf break to a greater extent than females (Fig. 6.5). Southern giant petrel males and females also showed differences between foraging areas. Males were mainly observed

Table 6.2 *Main differences in foraging strategies between male and female
northern giant petrels during the incubation period. Data derived from
González-Solís et al., 2000b and 2002a.*

	Females	Males	Sample size ♀/♂	Test	Probability
Foraging habitat (% of birds)	At sea, 100	On coast, 78	8/9 satellite tracked trips	Fisher's	$P < 0.05$
Correlation of the mass gain while foraging with the duration of the trip (r^2)	0.37	0.10	35/33 foraging trips	Pearson	$P < 0.001/$ $P = 0.08$
Time on sea surface (%)	41.4	19.3*	2/4 satellite tracked trips		
Core activity range (km^2)**	151.4	17.3	8/9 satellite tracked trips		
Median trip duration (days)	5	3	35/33 foraging trips	M–W U	$P = 0.11$
Flight speed (m/s)	8.2	4	8/9 satellite tracked trips	M–W U	$P < 0.05$
Maximum foraging range (km)	1222	180	8/9 satellite tracked trips	M–W U	$P = 0.32$
Daily distance covered (km)	389	173	8/9 satellite tracked trips	M–W U	$P = 0.11$
Daily mass gain at sea (% of body mass)	1.26	2.46	35/33 foraging trips	M–W U	$P < 0.05$
Mass loss average in 30 days	0.4 kg	No loss	35/33 foraging trips	Pearson	$P < 0.001/$ $P = 0.88$

* Only coastal trips considered.
** Area derived from kernel analyses encompassing 50% of locations.

around South Georgia and the neighbouring South Sandwich Islands,
whereas females used the Patagonian shelf to some extent (Fig. 6.5).

Association with nocturnal fisheries and land-based activity

Because geolocators record light level over the day and night, they can
be used to assess the association of giant petrels with nocturnal fisheries. If a bird approaches a ship using lights at night, the logger records
a spike of light at that time. This can be recorded, as small spikes for a

Males   Females

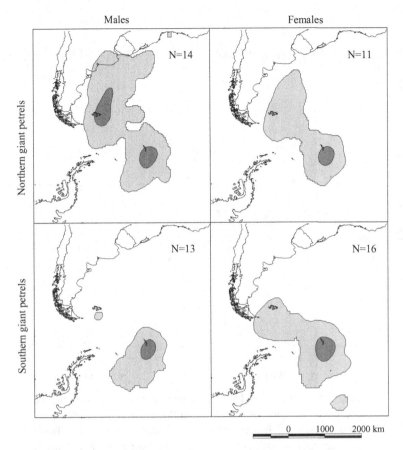

Figure 6.5   Winter activity ranges in 2000 derived from kernel
analyses on the validated locations from light level geolocators (GLS)
for male and female northern and southern giant petrels from Bird
Island population (South Georgia). Data from spring to autumn
equinox are included for each bird. Activity ranges encompassed 95%
(general range) and 50% (core area) of locations as indicated by light
and dark grey respectively.

short time period up to the saturation of the photocell (as in daylight
conditions), for several hours. On the Patagonian shelf break there is
an important squid fishery that attracts squid to the surface by using
powerful lights that can be detected by orbiting satellites (Rodhouse
et al., 2001). Average light levels from the geolocators indicate that male
northern giant petrels associate with nocturnal fisheries only in winter, in contrast to females which associate with them all year round
(Fig. 6.6). This result again supports the more coastal habits of males

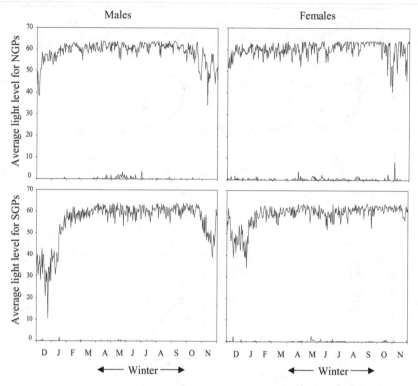

Figure 6.6 Average light levels recorded by the geolocators during day (top line) and night (bottom line) from December 1999 to November 2000 for male and female northern and southern giant petrels (light level saturates at 64). Light drops during daylight indicate that bird is on land, whereas light spikes at night suggest association with nocturnal fisheries (see text).

during the breeding season, but shows that both sexes have similar pelagic habits during winter. Southern giant petrels do not use the Patagonian shelf as much as northern giant petrels do (Fig. 6.5), as suggested by the light spikes which were only detected occasionally (Fig. 6.6). Nevertheless, more light events were recorded for females, in accordance with their greater use of the Patagonian shelf compared to males (Figs. 6.3 and 6.5).

Geolocation tags can also be used to evaluate land-based activity of giant petrels during the day. When on land, birds sit on the device which causes a sharp decrease in the light intensity recorded by the geolocator. During the breeding season, data from males of both species showed an average light level that was less than females, confirming that scavenging habits of males are associated with beach areas. In contrast, drops in light intensity in winter were less frequent and light levels were similar for both sexes and species (Fig. 6.6), suggesting comparable pelagic habits outside the breeding season for both sexes and species.

Metal contamination

Differences in foraging areas and diet between sexes and species should be reflected in the levels of metal contamination, since metal exposure in birds is thought to be mainly through food ingestion (Furness, 1993). Although South Georgia can be considered a relatively pristine area, the Patagonian shelf is polluted by discharges from riverine transport, offshore oil operations and high shipping activity. The Patagonian shelf is far more exploited by northern than by southern giant petrels breeding at South Georgia (Figs. 6.3 and 6.5); thus the former species shows higher levels of metal contamination in the blood than the latter species (Fig. 6.7). Moreover, blood cadmium levels were found in significantly higher concentrations in females compared to males in both giant petrel species (Fig. 6.7; González-Solís et al., 2002b). This is consistent with differences in diet between the sexes, since krill and particularly cephalopods generally contain high levels of cadmium (Honda et al., 1987; Gerpe et al., 2000). Mercury levels in the body feathers of both giant petrel species were also found to be higher in females compared to males at South Georgia (Becker et al., 2002). Consistent with results of the stable isotope analyses (Fig. 6.2), females seem to feed at a higher trophic level than males, thus higher concentrations of mercury contamination can be expected in females.

SEXUAL DIFFERENCES IN FORAGING STRATEGIES

In birds, sexual divergence in prey size or microhabitat exploited can emerge even when males and females search the same macrohabitat (e.g. Radford & Plessis, 2003). However in giant petrels, differential exploitation of coastal or pelagic habitats implies a fundamental difference in the distance, spatial pattern and predictability of the resources exploited, thus imposing different foraging strategies on each sex. These differences have been studied in detail for northern giant petrels during the incubation period (González-Solís et al., 2000a). Since most males search for carrion on the nearby beaches of South Georgia and most

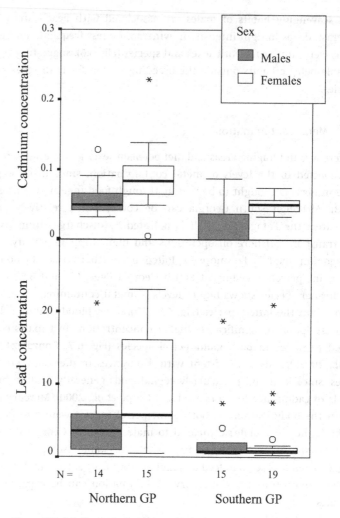

Figure 6.7 Median concentrations (μg/g dry weight) of cadmium and
lead in the blood of incubating giant petrels, according to sex and
species. NGP = northern giant petrel; SGP = southern giant petrel.
Asterisks are extreme values; circles are outlier values; the length of
the box is the inter-quartile range; the length of the bar is the range,
excluding extreme and outlier values. Data from González-Solís et al.
(2002b).

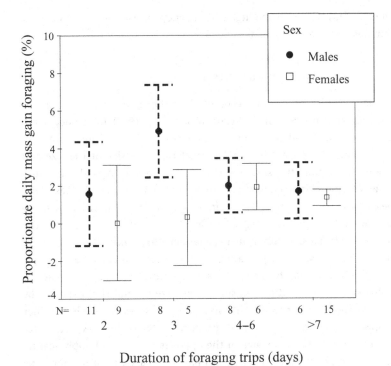

Figure 6.8 Efficiency, as indicated by the proportionate daily mass gain while foraging, in relation to the duration of the foraging trip for male and female northern giant petrels. Values are means + 95% confidence interval. Data from González-Solís *et al.* (2000b).

females engage in pelagic trips, foraging effort is lower for males compared to females. That is, the distance covered, flight speed and duration of foraging trips are substantially greater for females than for males (Fig. 6.4, Table 6.2). Carrion is an unpredictable resource compared to marine prey, as indicated by the lower correlation of mass gain while foraging with the duration of the trip (González-Solís *et al.*, 2000a). However, carrion seems to be a more profitable resource than marine prey, since foraging efficiency, defined as the proportionate daily mass gain while foraging, is significantly greater for males than for females (Table 6.2). In males, efficiency decreases with the duration of the foraging trip, being greatest for foraging trips of three-day durations. In contrast, efficiency increases with the duration of the foraging trip in females; reaching its highest value in foraging trips that last four to six days (Fig. 6.8). Not surprisingly, median values of foraging trip duration (Table 6.2) coincide with the optimum trip duration, as indicated by the trip duration with the highest efficiency for each sex, i.e. 5 days for females and 3 days for males (Fig. 6.8).

SEXUAL SIZE DIMORPHISM

Sexual selection is usually recognized as the main force driving sexual dimorphism or sex-linked traits (Andersson, 1994). In essence, large body size can be advantageous in contest over mates, thus increasing the mating success of large males. Likewise, body size or the size and colour of certain traits is often used by one sex, typically females, to assess individual quality and select partners, which increases the bearer's mating success. However, ecological factors, such as intersexual food competition, may also affect the size of a certain trait or the degree of sexual size dimorphism (Selander, 1972; Shine, 1989a; Blondel et al., 2002). If sexual segregation on feeding resources is favoured by a difference in size between the sexes, natural selection may act to enhance sexual size dimorphism or trait size beyond that expected by the existing difference in body size between sexes. To study whether males or females have traits disproportionate to their expected size with respect to the body size of the opposite sex, we can apply scaling to morphometric data. This analysis has been carried out for northern giant petrels breeding at South Georgia (González-Solís, 2004a). Tarsus, wing and bill length were determined to be isometric traits within each sex with respect to body mass. However, whereas isometry could also explain differences in tarsus size between males and females, this was not the case for bill and wing length. The bills of males were longer than expected from an isometric relationship with female body mass. In contrast, wings of females were longer than expected with respect to male body mass (Fig. 6.9). For example, according to the isometric regression models obtained for each sex, a female should be 11% heavier than a male to achieve the average bill size of a male; a female with an average body mass of a male should have a wing length 3% longer than a male. A greater size for bill length in males and for wing length in females than those expected by the isometric predictions is seen in the dimorphism index for each trait in both giant petrel species (Table 6.1). That is, the average percentage difference for tarsus in both giant petrel species was about 10%. However, for wing length it was only 5.5% whereas bill length was about 13–14%. These values suggest directional selection pressures in the past for bill and wing length in relation to their functional role in contest competition and flight performance.

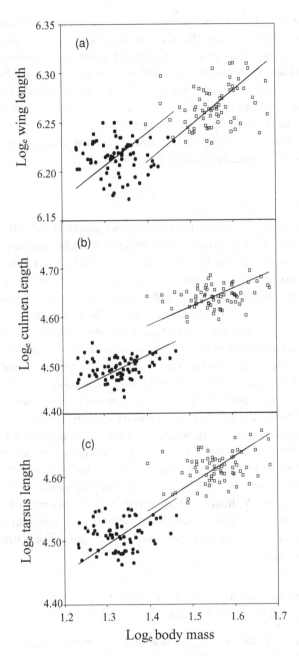

Figure 6.9 Scaling of wing, culmen and tarsus length (cm) in relation
to body mass (kg) of 71 male (open squares) and 77 female (solid dots)
northern giant petrels. All variables are log$_e$ transformed and lines
are plotted for each sex by reduced major axis method. None of the
slopes deviated significantly from the expected value under isometry
(1/3) in either sex. Data from González-Solís (2004a).

Larger bills in males could make them more efficient for contest competition when competing for access to a carcass, as well as for dismembering carrion. Longer wings in females probably correspond to lower wing loadings compared to males; particularly since wing loading increases more steeply with increasing body size (Pennycuick, 1987). In such cases, females could be more efficient at foraging over longer distances than males. Indeed, recent studies suggest that inter-sexual differences in flight performance in albatrosses may mediate spatial segregation while foraging (Phillips et al., 2004).

SEX RATIO

If males were more numerous than females, we would expect bill size to be important in contest competition for females. Differences in maximum body mass attained by male and female chicks are similar to those of adults in both giant petrel species (Hunter, 1984; Cooper et al., 2001). Moreover, in both species the fledging period is about 4–5% longer for male than for female chicks (Hunter, 1987). The larger size of male chicks requires a larger energy investment from adults during the chick rearing period (Anderson et al., 1993; Weimerskirch et al., 2000; Daunt et al., 2001). Indeed, total food intake for a male chick was estimated to be 23% and 37% greater than for a female chick, in northern and southern giant petrels, respectively (Hunter, 1987). In such circumstances, a male-biased nestling mortality can be expected (e.g. Clutton-Brock et al., 1985; Griffiths, 1992). However, studies at South Georgia indicate that sex ratio at fledging does not depart from parity over four consecutive years for either of the two species (Table 6.3). At Marion Island, there appears to be a bias towards higher female fledgling success in both giant petrel species (Williams, 1982). Nevertheless, the operational sex ratio in giant petrels is likely to be close to parity (or female-biased) suggesting that contest competition over females is not particularly strong (they are monogamous).

EVOLUTION OF BODY SIZE, SEXUAL SIZE DIMORPHISM AND SEXUAL SEGREGATION

Given the disproportionately large size of giant petrels compared to the rest of the species in its family, it is likely that a process of directional selection favouring an increase in body size has occurred in the past. Larger body size in giant petrels probably evolved from an

Table 6.3 *Fledging sex ratios (males:females) for northern and southern giant petrels from 2000 to 2003 at Bird Island, South Georgia. None of them was significantly different from parity in any year, neither when years were pooled (all $\chi^2 < 1.47$, d.f. = 1, all P > 0.22). Sex ratio was determined by culmen length, after a graphical examination of the bimodal distribution; culmen length lower than 9 mm in northern and lower than 90 mm in southern giant petrel fledglings correspond to females and higher than this value to males. Data for northern GPs from González-Solís, 2004a.*

Year	Northern GPs (*n*)	Southern GPs (*n*)
2000	1.06 (70)	0.85 (74)
2001	1.23 (78)	0.92 (50)
2002	1.00 (134)	1.02 (105)
2003	0.94 (70)	1.39 (55)

evolutionary 'arms race' through contest competition, as modelled for diploid species from a game theory background (Maynard Smith & Brown, 1986). Whether this contest competition was over females or over carrion is unknown. However, the covariation of a disproportionately large size with a massive bill, apparently stronger than any other procellariiform, suggests that both are mainly related to the exploitation of carrion. Nevertheless, the competitive advantage for getting food in males conferred by a larger body and bill size may in turn be used by females to choose their partners. The unpredictability of a resource such as carrion may also have played an important role in the size increase. Increasing body size enhances fasting endurance because the storage of energy reserves increases with body size faster than does metabolic costs (Millar & Hickling, 1992; Warham, 1990).

Ever since Darwin, there has been intense discussion amongst ecologists on the reasons for the trophic niche divergence between the sexes, found in many species. Giant petrels show a noticeable sexual size dimorphism as well as sexual segregation in their trophic ecology. Whether trophic segregation arises from sexual dimorphism in body size or results from intraspecific competition for food is difficult to assess. In the former case, resource partitioning is mediated by sexual dimorphism, this being attributed to sexual selection or to differences in the reproductive roles of males and females (Hedrick & Temeles, 1989). In the latter case, an ecological cause is invoked as a mechanism reducing inter-sexual competition for food and sexual dimorphism is considered to evolve from the niche segregation between sexes (Shine, 1989a).

Larger size of males confers greater contest competition for access to carcasses and greater fasting endurance for exploiting unpredictable resources, allowing the monopolization of carrion close to the breeding grounds. However, for the more pelagic females, a larger body size could be energetically costly due to the increase in flight costs. In such circumstances, small body size in females may be more advantageous, as suggested by the better condition of chicks reared by smaller females (González-Solís, 2004a). Indeed, females may have experienced an indirect increase of size through a genetic correlation of body size between the sexes beyond their optimum size. Genetic correlation between the sexes in body size is common in several taxa and seems particularly strong in birds (Andersson, 1994).

If selection operates primarily on dietary divergence, trophic structures are likely to be modified in addition to the sexual size dimorphism. A modification of the trophic structures between the sexes is considered the best example of an ecological causation of sexual segregation (e.g. Selander, 1972; Shine, 1989a). This seems to be the case for the bill of male giant petrels being longer than it would be expected from sexual size dimorphism alone (Fig. 6.9).

It is not clear how niche segregation between sexes can originate from purely ecological factors. Its origin is therefore usually attributed to sexual selection. At present, sexual selection is likely to be weak in giant petrels given their monogamy, similarity in sex roles, sex ratio parity and the important parental investment of males. Once sexual size dimorphism evolved, however, males may have excluded females from access to carrion closer to the breeding grounds, thus forcing females to use alternative resources. Ultimately, exploiting different resources can promote sexual size dimorphism and shape morphological traits of males and females differently.

ACKNOWLEDGEMENTS

We thank the field team at Bird Island for invaluable support; Chris Hill, Dafydd Roberts, Ben Phalan and Chris Green for help in the field and for collecting sex ratio data; Vsevolod Afanasyev and Dirk Briggs for their assistance in providing geolocators for this study; Joan Carles Abella for help in stable isotope analyses and Geir Sonerud and Scott Shaffer for their helpful comments. This research was funded by the Training and Mobility Program of the European Commission

(ERBFMBICT983030), the British Antarctic Survey, the Dirección General de Investigación Científica del Ministerio de Educación y Ciencia Español (PF960037747430), the Universitat de Barcelona, the Generalitat de Catalunya, the Ministerio de Ciencia y Tecnología (Ramon y Cajal, 2001) and fondos FEDER. Flavio Quintana, Sofia Copello, Eric J. Woehler, Markus Ritz and Hans-Ulrich Peter kindly provided information on giant petrel measurements.

Part IV Predator avoidance and reproductive strategies

7

Predation risk as a driving factor for size assortative shoaling and its implications for sexual segregation in fish

OVERVIEW

Fish shoals have been fundamental in developing our understanding of the evolution of sociality (see Krause & Ruxton, 2002). They have been used in a range of laboratory and field studies, and have been particularly useful in investigating the adaptive significance of the phenotypic assortment of social groups (see Krause et al., 2000a for a review). In particular investigations on fish shoals have elucidated the mechanisms underlying assortment by body size (Krause et al., 2000b; Croft et al., 2003). However, in contrast to mammals (mainly ungulates) fish have largely been neglected from the literature on sexual segregation (see Sims et al., 2001a; Croft et al., 2004 for exceptions).

Sexual segregation is often associated with sex differences in body size (see Ruckstuhl & Neuhaus, 2000, 2002 for a review in ungulates – see also Chapter 10). In such cases, segregation by body size will automatically result in segregation by sex. Body size differences, however, are not restricted to sexual dimorphism but often found within the sexes as well. Thus, by understanding the mechanisms underlying body size assortment we may better understand a more general phenomenon that includes sexual segregation.

Group living is an important anti-predator strategy (Hamilton, 1971; Krause & Ruxton, 2002), consequently predation risk is a strong selective force influencing the composition of social groups. Initially, we examine the anti-predator benefits of the phenotypic and behavioural assortment of social groups in fish, paying particular attention to assortment by body length. We then describe the observed patterns of group

Sexual Segregation in Vertebrates: Ecology of the Two Sexes, eds. K. E. Ruckstuhl and P. Neuhaus.
Published by Cambridge University Press. © Cambridge University Press 2005.

assortment in natural fish populations and discuss the underling mechanisms. Finally, we discuss whether sexual segregation may result from similar mechanisms.

WHY SHOULD PREDATION SELECT FOR THE ASSORTMENT OF SOCIAL GROUPS?

The decision to join a group represents a behavioural trade-off, whereby the benefits of grouping, including reduced predation risk and increased foraging efficiency, are traded off against the costs, mainly increased competition for limited resources (reviewed in Magurran, 1990; Pitcher & Parrish, 1993; Krause & Ruxton, 2002). The antipredator benefits of group living have been well documented and include increased vigilance, encounter dilution, predator confusion, predator swamping, communal defence, predator learning and selfish herd effects (see Krause & Ruxton, 2002 for a review). Simply associating with a group of random individuals can confer certain antipredator benefits. For example, as group size increases the overall vigilance of the group may be enhanced, irrespective of the phenotypic composition of the group. In contrast, predator confusion is dependent on the phenotypic assortment of individuals within a group, with odd individuals suffering an increased risk of predation due to the 'oddity effect' (Landeau & Terborgh, 1986; Theodorakis, 1989). Predator confusion results from the simultaneous movement of phenotypically identical individuals, making it difficult for a predator to focus on a particular target, reducing the predator's attack-to-kill ratio (Ohguchi, 1978; Landeau & Terborgh, 1986). This process is thought to be an important selective force leading to the homogeneity of group composition in free-ranging fish shoals, particularly by body size (Ranta et al., 1994). Although sexual segregation is often associated with sexual dimorphism in body size (Ruckstuhl & Neuhaus, 2000; 2002), the oddity effect has received little attention as an adaptive benefit of sexual segregation.

The confusion effect may also select for associations between individuals with similar behaviours, with behaviourally 'odd' individuals potentially being more vulnerable to predation due to the 'oddity effect'. Furthermore, when behaviour is plastic (e.g. swimming speed) the 'oddity effect' may select for synchrony in behaviour between individuals. For example, in an investigation on swarms of *Daphnia pulex* (zooplankton), Jensen et al. (1998) observed that swimming speed was more uniform in groups under threat of predation than in control groups where swimming speed was unsynchronized. In fish, the optimal foraging rate is size dependent (Hjelm & Persson, 2001) and may be dependent on swimming speed (Weihs & Webb, 1983), which in turn is body length dependent (Beamish, 1978). Individuals in a group of others of dissimilar body length may be forced to travel and forage at sub-optimal speeds, to minimize behavioural oddity. This can incur an energetic cost and potentially contribute to the selection for group assortment. Furthermore, the selective pressure to conform to the majority of a group, both phenotypically and behaviourally, may be dependent on the motivational state of the individual (Reebs & Saulnier, 1997), which may be dependent on body size. For example, small individuals may take greater risks due to proportionally greater metabolic demands (Krause *et al.*, 1998).

Behavioural oddity may be reduced through associations between familiar individuals, with the behaviour of familiars being more predictable. For example, in an experiment on fathead minnows, *Pimephales promelas*, Chivers *et al.* (1995) found that in response to a predation threat from the northern pike, *Esox lucius*, minnows in groups that were familiar showed greater shoal cohesion in comparison to unfamiliar groups. An increase in shoal cohesion may result in increased anti-predator success. Whilst associations between familiar individuals may decrease behavioural oddity, there may be other benefits for associating with familiars, including increased foraging efficiency, reduced aggression and the potential for co-operative interactions (see Griffiths, 2003; Ward & Hart, 2003 for reviews).

In support of the confusion effect as an adaptive benefit of group assortment, there is empirical evidence that when a predator detects and attacks a group, prey are not selected at random, and that the presence of odd individuals in a group increases the predator's capture success rate. For example, in an experiment on silvery minnows, *Hybognathus nuchalis*, Landeau and Terborgh (1986) found that odd individuals in a group were particularly vulnerable to predation when being predated upon by largemouth bass, *Micropterus salmoides*. Furthermore, when shoals contained a small number of odd individuals the success rate of the predator was greatly increased (Fig. 7.1). The presence of odd individuals in a group allows a predator to fixate on single prey items, overcoming the confusion effect and increasing the predator's attack-to-kill ratio. In support of body length assortment being important in minimizing phenotypic oddity, Theodorakis (1989) found that when bass were presented with shoals of minnows that were dominated by one size category, the minority size was taken more often than would be predicted by chance alone. The oddity effect is not limited to fish shoals but has been described in other taxa as well. For example, Kruuk

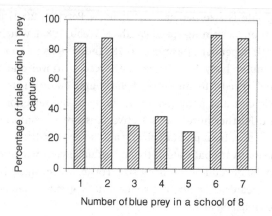

Figure 7.1   Percentage of five-minute trials that ended with prey capture, as a function of the number of odd prey in a school of eight individuals. Odd individuals were created by dyeing the normally brown fish blue. (Redrawn from Landeau & Terborgh, 1986)

(1972) reported that wildebeest that had their horns painted white were exceptionally vulnerable to predation by lions.

Although these empirical investigations provide support for the oddity effect selecting for homogeneous groups, there may be other benefits to group assortment. For example, Ranta et al. (1994) used a simulation model to explore the factors promoting phenotypic assortment of fish shoals. The model considered both foraging and predator avoidance as the key factors in generating phenotypic assortment. In fish, body size is often positively correlated with foraging ability (Ranta & Lindström, 1990; Krause, 1994). In the model it was assumed that an individual's ability to find and share food was a function of its body size, with larger fish being more effective in finding food, and attaining proportionally more food after being located than smaller fish. It was further assumed that the ability of a group to find food was a function of the group size. The model predicted the phenotypic assortment of shoals when the differences in food finding and food sharing between individuals were large and dependent on body size. It is likely that these two mechanisms (foraging and predation risk) are both active simultaneously within a population (Ranta et al., 1994).

WHAT DO WE OBSERVE IN THE FIELD?

In teleost fish growth is indeterminate, being a function of the age of a fish, its ability to attain resources and the environmental characteristics

(Wootton, 1998). As a result, in most wild fish populations there can be huge variability in body size both within and between cohorts (Helfman et al., 1997). In the face of such phenotypic variability, the oddity effect predicts that free-ranging shoals will be assorted by body length.

Accordingly, a large number of investigations (across a range of fish species) have reported free-ranging shoals to be assorted by body length (see Krause et al., 2000a for a review). Furthermore, there is limited evidence that free ranging groups of individuals are composed of familiar individuals. For example, Griffiths and Magurran (1998) found that female guppies, Poecilia reticulata, preferred to shoal with familiar conspecifics (members of a shoal in which they were captured).

WHAT ARE THE MECHANISMS UNDERLYING THE ASSORTMENT OF FISH SHOALS?

Given that most animals that segregate by sex are sexually dimorphic in body size, we may further our understanding of the potential mechanisms underlying sexual segregation by examining group assortment, particularly by body size. Various mechanisms have been proposed to explain group assortment in shoaling fish, which can be divided into two categories (active and passive).

Active mechanisms

Firstly, during shoal encounters fish may make active choices with regards to joining, leaving or staying with a shoal based on the composition of the shoal. Much of the evidence for active choice as a mechanism underlying shoal assortment comes from laboratory investigations, whereby test fish are given a binary choice between two shoals of conspecifics of different body length (with one shoal being size matched to the test fish). Such investigations have demonstrated active association preferences for a number of phenotypic characters including: size (see Krause et al., 2000a for a review), species (Wolf, 1985; Krause & Godin, 1994; Magurran et al., 1994), parasite load (Krause & Godin, 1996a) and colour (McRobert & Bradner, 1998), with individuals preferring to shoal with others that are phenotypically similar. Furthermore, using similar methods, active preference for familiar over non-familiar individuals has been demonstrated in a number of fish species (see Griffiths, 2003; Ward & Hart, 2003 for reviews).

More recent investigations have attempted to quantify the role of active choice in determining the phenotypic composition of open

groups (groups where individuals are free to leave and join). For
example, in an investigation on the banded killifish, Krause *et al.* (2000b)
followed individual shoals in a wild population. The outcome of shoal
encounters (to join or not to join) was found to be dependent on the
body length differences between the shoals, with shoals of equal body
lengths more likely to join, thus providing support for active choice
being an important mechanism underlying the body length assortment
of free ranging groups in this species.

Passive mechanisms

The second group of mechanisms can be classified as passive. For
example, a positive relationship between body length and speed of locomotion (Blaxter & Holliday, 1969) has been proposed as a mechanism
for creating assortment by size in a number of taxa (krill: Watkins *et al.*,
1992; African ungulates: Gueron *et al.*, 1996; see Couzin & Krause, 2003
for a review). Alternatively, phenotypic assortment of groups could arise
by body-length-specific habitat preferences, which will limit the opportunities for social interactions to occur between individuals of different
size classes, and may lead to the passive phenotypic assortment of social
groups (see Croft *et al.*, 2003).

IMPLICATIONS FOR SEXUAL SEGREGATION IN FISH

In species that are sexually dimorphic in body size, the same selective
pressures that select for body length assortment due to the 'oddity
effect' (Theodorakis, 1989) may select for sexual segregation. The extent
of sexual dimorphism may be closely linked to the mating system. For
example, in polygamous species males have a much larger reproductive potential than females, leading to sexual selection (Darwin, 1871).
This can take the form of intra-sexual selection, predominantly through
male–male competition for access to females, or inter-sexual selection,
predominately through female choice for high quality mates. Sexual
selection may result in sexual dimorphism in body size (Andersson,
1994), as seen in a number of fish families including; Centrarchidae,
Chaenopsidae, Cottidae, Cyprinidae, Gasterosteidae, Gobiidae, Labridae,
Percidae, Pomacentridae, Poeciliidae, Salmonidae and Tripterygiidae
(Kodric-Brown, 1990). However, the patterns of sexual dimorphism in
fish are complex, with some of the most extreme examples of sexual dimorphism coming from monogamous species (in particular deep-
sea fishes of the genus *Photocorynus* – see Magurran and Garcia, 2000).

Interestingly, patterns of sexual dimorphism in fishes are often a reversal of that seen in other taxa (e.g. ungulates, where intra-sexual competition results in males being larger than females (Huntingford & Turner, 1987)), with female fecundity in fish often covarying with body size, selecting for larger females, which often outgrow males (see Magurran and Garcia, 2000 for a review).

Furthermore, in fish, sexual selection (particularly inter-sexual selection) may result in sexual dimorphism in nuptial colouration, which in turn may create phenotypic oddity in mixed sex groups. Sexual dimorphism in nuptial colouration has been reported in a number of fish families including: Cyprinidae, Cyprinodontidae, Gasterosteidae, Hexagrammidae, Percidae and Poeciliidae (Kodric-Brown, 1990). The extent of sexual dimorphism in nuptial colouration may depend on the seasonality of breeding. For example, some poeciliids have a more or less continuous breeding season, therefore male nuptial colouration is expressed throughout the year (e.g. guppies see Houde, 1997). In contrast, in species with distinct breeding seasons, for example, three-spine sticklebacks, *Gasterosteus aculeatus*, male nuptial colouration may only be expressed during the breeding season (Wootton, 1976). This may partly be because the signal is energetically costly to produce (Frischknecht, 1993), however, it may also function to reduce conspicuousness and phenotypic oddity outside the breeding season. A comparative study investigating the link between sexual dimorphism, mating system and sexual segregation is needed in this context.

Parr (1931) first noted that species showing the highest shoaling tendency have the lowest levels of sexual dimorphism. Hobson (1968) related this trend to the 'oddity effect' and suggested that the selection against oddity is so strong that is has suppressed the evolution of sexual dimorphism in pelagic shoaling fish species in marine environments. Similarly, in freshwater environments, strongly shoaling species such as characins (e.g. *Paracheirodon innesi, Paracheirodon axelrodi, Hemigrammus erythrozonus*) and some cyprinid species appear to display little if any sexual dimorphism (e.g. *Brachydanio rerio, Puntius tetrazona, Balantiocheilos melanopterus*). In contrast, sexual dimorphism can be quite pronounced in weakly shoaling species. For example, in guppies (a relatively weakly shoaling species (Croft *et al.*, 2003)) males are smaller than females and brightly coloured, whereas females are cryptic (see Houde, 1997). Furthermore, there is some evidence of sexual segregation in the species (see later). Despite this apparent trend for an increase in sexual dimorphism with a decrease in shoaling tendency, there are many non-shoaling species that display little if any sexual dimorphism (e.g. *Epalzeorhyncus bicolor*, *E. kalopterus*). A comparative study exploring the potential link between sexual dimorphism, shoaling tendency and sexual segregation in closely related species would be very beneficial.

Some evidence for a relationship between shoaling tendency and sexual dimorphism can be gleaned from studies of guppy populations that have evolved under different predation pressures. In low predation areas where the shoaling tendency is weak (Seghers, 1974) (and the oddity effect less important), female choice has led to the evolution of more colourful males (Endler, 1980). In contrast, in high predation sites, males are less colourful (presumably at least partly due to the oddity effect, but conspicuousness outside shoals may play a role as well) but mature earlier because of their short lifespan due to greater predation risk (Reznick, 1996). Early maturation of males means that size differences are more pronounced between the sexes in high predation sites (Reznick, 1996) whereas colour differences are more pronounced in low predation populations (Endler, 1980). This shows how complex the effects and interactions between predation pressure and female mate choice can be.

In addition to selecting for phenotypic assortment, the 'oddity effect' may select for behavioural assortment and synchronization of behaviours (see earlier). In fishes, the nature of the mating system may lead to sex differences in behaviour. For example, in a review of the literature Magurran and Garcia (2000) reported sex differences in foraging behaviour, aggression and anti-predator behaviour (including shoaling tendency, predator detection and anti-predator responses). Such behavioural differences may select for sexual segregation, with individuals that are in the minority in mixed sexed groups potentially suffering increased risk of predation due to their behavioural oddity. Furthermore, in species that are sexually dimorphic in body length, costs due to synchrony of behaviour between individuals with different optimum speeds of locomotion and foraging patterns, may contribute to the evolution of sexual segregation (Conradt, 1998a; Ruckstuhl, 1999). Future work examining the effect of behavioural oddity, predation risk and the cost of synchronization of behaviour is eagerly anticipated.

Finally, in species that are sexually dimorphic in body length, the mechanisms underlying sexual segregation may be indistinguishable from the mechanisms underlying body length segregation. For example, in poeciliids (e.g. guppies) where body length differences can be very pronounced (with females being considerably larger than males (see Houde, 1997)) much of the observed sex segregation can be explained

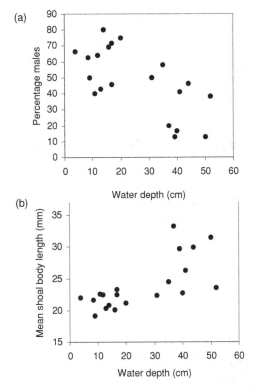

Figure 7.2 (a) The relationship between the sex composition of a shoal (expressed as the percentage of males) and water depth ($N = 20$) in the Arima River, Northern Mountain Range of Trinidad (Pearson's Correlation: $r_{19} = -0.66$, $P = 0.001$) (redrawn from Croft *et al.*, 2004). (b) The relationship between mean body length of fish in a shoal and water depth ($N = 20$) in the Arima River (Pearson's Correlation: $r_{19} = 0.71$, $P < 0.001$). (Redrawn from Croft *et al.*, 2004.)

by body length segregation. For example, Croft *et al.* (2004) observed that female guppies occupied areas of deeper water further away from the riverbank than males (see Fig. 7.2a). Much of this variation in habitat use between the sexes could be explained by differences in body length, with shoals in deeper water having a larger mean body length (see Fig. 7.2b). Furthermore, this trend was not just a product of sexual segregation, with body length differences in habitat use being present in both males and females independently (Fig. 7.3). Thus, the same body-length-related mechanism that leads to segregation between adults and juveniles (see Croft *et al.*, 2003) can be invoked to explain sexual segregation.

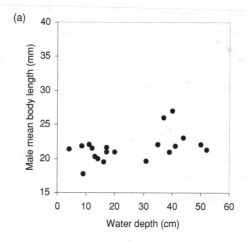

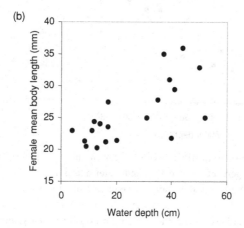

Figure 7.3   (a) The relationship between mean body length of males in a shoal and water depth ($N = 20$) in the Arima River (Pearson's Correlation: $r_{19} = 0.44$, $P = 0.05$) (redrawn from Croft *et al.*, 2004). (b) The relationship between mean body length of females in a shoal and water depth ($N = 20$) in the Arima River (Pearson's Correlation: $r_{19} = 0.69$, $P = 0.001$). (Redrawn from Croft *et al.*, 2004.)

METHODOLOGY FOR INVESTIGATING SEXUAL SEGREGATION IN SHOALING FISHES

Investigation of the degree of phenotypic assortment of social groups (e.g. by sex or body length) requires the choice of an appropriate null model with which to compare the observed data. Outside the breeding season a valid null model may be a random distribution of males and females across the observed shoal sizes. The most appropriate way of doing this is by comparing the observed degree of assortment with that from a computational null model in which shoal composition is randomized. The details of the randomization will depend on which features of the wild population should be preserved. For example, the degree of sexual assortedness of a shoal can be measured by determining the sex ratio. For comparison, all the individuals in the population are pooled into one data set. From this pooled data set, individuals are selected at random and assigned to the observed shoal sizes. This procedure is repeated many (typically 1000) times to generate random simulations with which the observed shoals can be compared. Each simulation generates a sex ratio for the shoal of interest, and the position of the observed sex ratio in the ranked sex ratios from the simulations can be used to determine whether we can reject the null hypothesis that the shoal is composed of a random set of individuals (see Croft *et al.*, 2003).

In contrast, during the breeding season, the distribution of the sexes is unlikely to be independent. In fact, males are likely to map onto the existing female distribution in an ideal free manner (resulting in an even distribution of males relative to females) with each male maximizing its own mating opportunities. Thus, in species that have a continuous breeding season (e.g. guppies) an even distribution might provide a more appropriate null model. However, caution is required when applying an ideal free distribution in this context, as it assumes that all competition interactions between individuals are equal and this is unlikely to be the case in many mating systems. Nevertheless, an even distribution may be a good first approximation of the distribution of males between shoals.

CONCLUSIONS AND OUTLOOK

As a taxonomic group, fish are especially suitable for investigations of group living. In particular, small freshwater species that form shoals in shallow littoral habitats (e.g. minnows, banded killifish, sticklebacks) have proved most valuable in investigating the adaptive significance of, and the mechanisms underlying, group living. In such species, entire groups of free-ranging individuals can easily be captured, allowing a direct assessment of group composition. Furthermore, such species can be captured in large numbers and maintained under laboratory conditions, allowing manipulative experiments to be conducted.

Most of the work on sexual segregation has been done on ungulates. In comparison, fish provide a number of interesting options and opportunities. For example, in guppies, the existence of multiple populations of the same species that have evolved under different predation pressures (see Seghers, 1973) provide a scenario in which some of the existing hypotheses regarding sexual segregation could be critically tested using the comparative approach. Comparing populations that occur in the presence and absence of predators would elucidate the importance of sex differences in predation risk as a factor underlying sexual segregation. Provisional observations suggest that predation risk is an important factor driving sexual segregation in guppies (Croft et al., 2004). The opportunity for experimental manipulation combined with replication should be a particularly attractive feature in this context. Another promising aspect is to unravel the importance of body length generating sexual segregation, by looking at segregation within sex and between the sexes as a function of body length.

8

Differences in habitat selection and reproductive strategies of male and female sharks

OVERVIEW

Segregation of the sexes within a species is a widespread behavioural phenomenon in both terrestrial and aquatic animals. In the marine realm, sexual segregation is exhibited by many taxa including whales, seals, seabirds and fish. Of the latter group, sharks may be particularly appropriate model animals to test theories on the mechanisms underlying sexual segregation, because sexual segregation is a general characteristic of shark populations, with both sexually dimorphic and monomorphic species being well represented among the approximately 400 extant species (Springer, 1967; Compagno, 1999). The reproductive modes of sharks are diverse ranging from egg-laying (oviparity) to placental live-bearing (viviparity) (Wourms & Demski, 1993). Among sexually dimorphic, viviparous shark species it is generally the female that is larger than the male, whilst in some oviparous species males are larger than females. Sexually monomorphic species also occur. Therefore, sharks possess a number of characteristics that make them an interesting alternative to terrestrial animal models for investigating the causes of sexual segregation.

In this chapter the prevalence and nature of sexual segregation in sharks is described and the relationship with reproductive modes is explored. Hypotheses suggested to account for sexual segregation in sharks are examined with respect to new field and laboratory behaviour studies of males and females of a monomorphic species, the lesser spotted dogfish (*Scyliorhinus canicula*). The chapter concludes by drawing

Sexual Segregation in Vertebrates: Ecology of the Two Sexes, eds. K. E. Ruckstuhl and P. Neuhaus.
Published by Cambridge University Press. © Cambridge University Press 2005.

together the main points from all shark studies to date, and suggests future directions for research in this area.

SEXUAL SEGREGATION IN SHARKS

Sharks and their relatives are members of an ancient lineage having first appeared in the late Silurian period about 415 million years ago. The cartilaginous fishes, class Chondrichthyes, to which sharks belong, are the oldest surviving group of jawed vertebrates (Wourms & Demski, 1993). Extant sharks are a large and diverse group numbering about 400 species. They form an important component of marine ecosystems as macropredators and scavengers on other fish and invertebrates, although some large species feed opportunistically on whales, whilst others consume only zooplankton (Sims & Quayle, 1998).

Sharks, skates and rays generally have larger brains than other ectothermic vertebrates. The relative size and structural complexity of sharks' brain-mass to body-mass ratios overlap the range for mammals and birds (Northcutt, 1977; Demski & Northcutt, 1996), suggesting they may be capable of complex behaviours, for example, social systems with dominance hierarchies and segregation by age and sex (Myrberg & Gruber, 1974; Klimley, 1987). Segregation by age is thought to be a universal feature of shark populations (Springer, 1967). Size-assorted schools of active shark species may be maintained by the different swimming speeds that can be sustained by different-sized individuals (Wardle, 1993). In addition to swimming capability, segregation by age in sharks may be common because of the increased risk of cannibalism and depredation of juveniles and subadults by mature individuals (Snelson et al., 1984; Morrissey & Gruber, 1993; Ebert, 2002). This probably accounts for the fact that gravid females of some shark species undertake long-distance migrations to sheltered nursery grounds to give birth away from adult sharks (Feldheim et al., 2002).

Similarly, segregation by sex is considered a general characteristic of shark populations (Springer, 1967; Klimley, 1987) and is also common among bony fishes (e.g. Becker, 1988; Robichaud & Rose, 2003; Chapter 7). The first substantive evidence for sexual segregation in sharks came from studies made during the early part of the twentieth century (e.g. Ford, 1921; Hickling, 1930). Observations of fishery landings showed significant bias in the sex ratio of fish caught by longlines and trawls, indicating that same-sex individuals aggregate, or that different sexes may occur preferentially in particular places. One species that attracted particular attention in this context was the spurdog (*Squalus acanthias*), a pelagic-demersal shark that forms large schools, which are targeted by commercial fishers (Compagno, 1984). This species also shows pronounced sexual size dimorphism with females reaching up to 1 m in length at maturity compared to a maximum of 0.72 m in males (Compagno, 1984). Ford (1921) collected data on the number of spurdog landed at Plymouth, England, and found that 92% of those captured in November were mature females. Over the following year, records showed that the sex ratio of landings varied widely, with four categories of schools evident: large females that were mostly gravid, exclusively mature males, immature females, and immature males and females in equal number. Ford (1921) concluded that inequality in sex composition of the schools was largely due to the tendency of individual *S. acanthias* to school with others of similar size and sex. This type of sexual segregation was termed 'behavioural' (Backus *et al.*, 1956).

Investigations of the lesser spotted dogfish (*Scyliorhinus canicula*) by Ford (1921) also described what was called 'geographical' sexual segregation (Backus *et al.*, 1956). Several thousand specimens of this benthic catshark examined at Plymouth, England, showed that males apparently dominated catches during winter (65% of numbers caught) whereas females marginally predominated in summer (58%). These changes over time were interpreted as being the result of same-sex individuals clustering more often in preferred habitat, rather than in other available habitats. Over the next 40 years, numerous studies similarly documented unequal sex ratios in fishery and fishery-independent catches of sharks. Geographical segregation was shown to be present in oceanic whitetip sharks (*Carcharhinus longimanus*) in the Gulf of Mexico (Backus *et al.*, 1956) and in school sharks (*Galeorhinus galeus*) off southern Australia (Olsen, 1954). Furthermore, landings of *G. galeus* off California showed that not only did catch composition vary by area with respect to the ratio of sexes present, but also with depth, with females occurring in shallower water than males (Ripley, 1946).

By the end of the 1960s, there was a burgeoning literature of observations of sexual segregation in sharks (Bullis, 1967; Springer, 1967). Further studies in the 1970s and 1980s expanded the number of species for which sexual segregation was observed or suspected (e.g. *Sphyrna tiburo*, Myrberg & Gruber, 1974; *Prionace glauca*, Stevens, 1976, Pratt, 1979; *Carcharias taurus*, Gilmore *et al.*, 1983; *Sphyrna lewini*, Klimley, 1985; *Carcharhinus amblyrhynchos*, McKibben & Nelson, 1986). A review of Compagno's (1984) catalogue of biological data on 340 species makes reference to sexual segregation for 38 species (Table 8.1). Despite this, by 1987 the causes of sexual segregation had not been formally investigated in any species, although differences in swimming capabilities,

Table 8.1 The 38 shark species for which sexual segregation has been documented (Compagno, 1984; Richardson et al., 2000; Pratt & Carrier, 2001). Very many of the other 400 or so shark species have not been studied in sufficient detail. The degree of sexual dimorphism is given as the relative (percentage) difference in minimum total length at sexual maturity between female and male, where positive values indicate larger females and negative values larger males. The reproductive modes are denoted by ovoviviparous (O), oviparous (Ov) and viviparous (V). Information on whether species were most often solitary and/or found in groups or aggregations (termed social) was taken from Compagno (1984).

Order	Family	Species	Sexual dimorphism (% difference of female)	Reproductive mode	Social (Sc) or solitary (So)
Squaliformes	Squalidae	Centroscyllium fabricii	–	O	–
		Centroscymnus plunketi	29.0	O	Sc
		Somniosus microcephalus	–	O	So
		Somniosus pacificus	–	O	So
		Squalus acanthias	18.6	O	Sc
		Squalus megalops	32.5	O	Sc
		Squalus mitsukurii	10.8	O	–
Heterodontiformes	Heterodontidae	Heterodontus portusjacksoni	14.3	Ov	Sc/So
Orectolobiformes	Ginglymostomatidae	Ginglymostoma cirratum	2.2	O	Sc/So
Lamniformes	Odontaspidae	Carcharias taurus	0	O	Sc
	Alopiidae	Alopias vulpinus	17.9	O	So
	Cetorhinidae	Cetorhinus maximus	62.0	O	So
	Lamnidae	Carcharodon carcharias	33.3	O	So
		Lamna nasus	−30.6	O	So

Carcharhiniformes	Scyliorhinidae	Galeus eastmani	16.1	Ov	–
		Halaelurus lineatus	-4.2	Ov?	–
		Holohalaelurus punctatus	-17.2	Ov	–
		Holohalaelurus regani	-24.0	Ov	–
		Scyliorhinus canicula	4.0	Ov	Sc/So
	Proscyllidae	Eridacnis radcliffei	-16.7	O	–
	Triakidae	Galeorhinus galeus	8.3	O	So
		Mustelus higmani	11.6	V	–
		Mustelus lenticulatus	1.3	O	Sc
	Carcharhinidae	Carcharhinus amblyrhynchos	-6.1	V	Sc/So
		Carcharhinus brevipinna	6.9	V	Sc/So
		Carcharhinus dussumieri	7.7	V	–
		Carcharhinus leucas	14.6	V	So
		Carcharhinus limbatus	-11.1	V	Sc/So
		Carcharhinus longimanus	2.9	V	So
		Carcharhinus macloti	10.1	V	–
		Carcharhinus melanopterus	5.5	V	Sc/So
		Carcharhinus obscurus	-8.2	V	–
		Carcharhinus plumbeus	9.9	V	So
		Negaprion brevirostris	6.7	V	So
		Prionace glauca	21.4	V	So
	Sphyrnidae	Sphyrna lewini	51.4	V	Sc
		Sphyrna mokarran	6.8	V	So
		Sphyrna tiburo	61.5	V	–

dietary preferences, absence of aggression between similar-sized sharks, or migration of gravid females to nursery grounds, were all forwarded as possible explanations (Springer, 1967; Klimley, 1987).

REPRODUCTIVE MODES AND SEXUAL DIMORPHISM

One suggested explanation for the cause of sexual segregation in mammals is the effects of sexual body-size dimorphism (Ruckstuhl & Neuhaus, 2000). Differences in size at maturity between males and females of the same species occurs widely in sharks and appears related to the mode of reproduction.

Sharks reproduce by internal fertilization via the paired intromittent organs of the male. Fecundity is low because females produce a few large, heavily yolked eggs, with embryonic development taking a relatively long time, such that offspring are generally large and well developed at parturition (Wourms & Demski, 1993). Oviparity is the primitive form of reproductive mode. However, whilst oviparity is the primitive form, it is not the most common reproductive mode as the majority of sharks (~70%) are live-bearers (Wourms & Demski, 1993). Within these, about five different types of viviparity occur (Dulvy & Reynolds, 1997). The numbers of young produced by live-bearing species varies, but generally most species (~80%) give birth to between 2 and 16 offspring. Generally speaking, larger live-bearing sharks have larger litter sizes, or greater relative offspring biomass, compared with smaller individuals. The planktivorous whale shark (*Rhiniodon typus*) for example, reaches over 12 m in length and also produces the highest number of embryos, around 300 (Joung *et al.*, 1996). The second largest species, however, the basking shark (*Cetorhinus maximus*) reaches about 11 m but produces only 6 offspring, although these are large (2 m long) individuals (Sund, 1943).

The adoption of live-bearing reproductive modes where gestation times can be as long as 22 months (Wourms & Demski, 1993) appears to have a profound effect on the body size of female sharks. Analysis of the difference in female to male size at maturity in 159 species for which there are sufficient data, reveals female live-bearing sharks to be between 10 and 16% longer than males of the same species (Fig. 8.1). The differences observed for ovoviviparous and viviparous modes are significantly higher than those observed in egg-laying species (~1%), which suggests that the evolution of larger female body size within a species arises from a live-bearing reproductive mode and the resulting impact of female body size on fecundity. Differences in size are likely to lead to

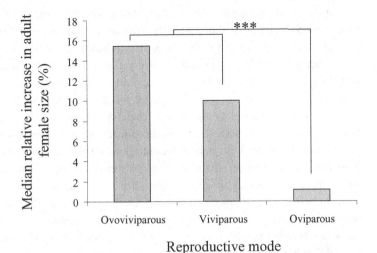

Figure 8.1   Sexual size dimorphism in 159 species of shark in relation to reproductive mode. Ovoviviparous and viviparous are live-bearing modes, whereas oviparous refers to egg-laying species. Significant differences examined using a Kruskal–Wallis rank test with multiple comparisons (***, $P < 0.001$).

differences in rates of energy intake and expenditure because larger animals have higher absolute energy requirements but lower mass-specific costs than smaller individuals (Schmidt-Nielsen, 1972; Gillooly et al., 2001). Different energy requirements may necessitate the selection of different forage conditions and/or habitat. The fact that the majority of shark species have females that are larger than males may mean that different energetic requirements could be a prime determinant of sexual segregation in sharks.

HYPOTHESES ABOUT SEXUAL SEGREGATION

The first study to propose a formal hypothesis to explain sexual segregation in a shark species (the scalloped hammerhead, Sphyrna lewini) focused on the apparent effects of sexual size dimorphism on behaviour and ecology (Klimley, 1987). Fishery data on the size, sex, reproductive state and stomach contents of S. lewini captured in bottom longlines or gill-nets in the Gulf of California were used to examine the possible underlying causes of sexual segregation. Klimley (1987) found that subadult females were captured more often in deeper water than similar-sized males, implying they moved offshore at a smaller size.

This observation was supported by stereophotographic records of free-swimming *S. lewini* at offshore sites showing an increased percentage of larger females compared to males. Although based on small sample sizes, dietary analysis suggested a divergence between the sexes, with females feeding more on epipelagic and mesopelagic prey than males. There was also marginal support for the biomass of stomach contents of females being greater than that of males. This led Klimley (1987) to speculate that greater consumption of energy-rich pelagic prey in an offshore habitat would allow females to grow faster than similar-aged males remaining inshore. Female *S. lewini* are no different from many shark species in that they mature at a larger size than males (Fig. 8.1) (Compagno, 1984). Therefore, Klimley (1987) proposed that female *S. lewini* segregate from males by moving to an offshore habitat that confers increased growth rates such that they reach maturity at a larger size than similar aged males, which is necessary to support large embryos. It was suggested that this strategy would act to match the reproductive lifetime of females with that of males.

Using the existing theoretical framework, developed largely from terrestrial mammal studies, to examine whether the factors underlying sexual segregation in sharks are consistent with those found for mammals may provide much needed comparative data useful for understanding the evolution of sexual segregation. In broad terms Klimley's (1987) hypothesis for sexual segregation in scalloped hammerhead sharks falls within the forage selection hypothesis (also called sexual dimorphism – body size hypothesis). This was proposed as a proximate mechanism for sexual segregation in some species of ungulates (Main *et al.*, 1996). Briefly, this states that physiological factors related to nutrition are largely responsible for sexual segregation, because each sex satisfies their different physiological requirements. It predicts that sexes segregate because sexual body size differences lead to differences in food selection arising from different energy requirements (Ruckstuhl & Neuhaus, 2000). For scalloped hammerheads off California, it was argued that sexual segregation by habitat was driven by the different physiological requirements between the sexes. However, there are several competing hypotheses that could explain Klimley's (1987) data equally well. Firstly, female scalloped hammerheads may move offshore prior to becoming sexually mature due to ecological factors (if, for example, predation risk differs between inshore and offshore areas). Sexual segregation arising from predation risk would then be consistent with the reproductive strategy hypothesis (predation risk hypothesis), which states that reproductive success of females is determined by offspring survival, and behaviours that reduce the risk of predation will be favoured through selection pressures (Main *et al.*, 1996). In this hypothesis, females are thought to choose habitats that are firstly safe from predators, but only secondly choose habitat by food availability, whereas males seek habitats with high food availability (Ruckstuhl & Neuhaus, 2000). This hypothesis cannot be discounted, because the difference in predation risk for female scalloped hammerheads between inshore and offshore habitats was not quantified by Klimley (1987). Secondly, sexual differences in activity budgets and movement rates (activity budget hypothesis) may be key factors of sexual segregation (Ruckstuhl, 1998). Here, sexes may become separated over time, even if initially in the same group, because activity budgets may differ significantly. Another possibility is that female scalloped hammerheads move offshore due to social factors. One aspect of the social factors hypothesis states that sexual segregation may be maintained by aggression of one sex towards the other (Main *et al.*, 1996). Therefore, subadult female scalloped hammerheads studied by Klimley (1987) could also leave inshore areas to avoid aggressive behaviour of males.

It remains unclear what causes sexual segregation in scalloped hammerheads because of equally valid and non-mutually exclusive competing hypotheses that were not evaluated. Therefore, systematic investigation of a number of factors and hypotheses within a single study of a single species provides a useful approach for examining which factors cause sexual segregation in sharks. It is possible that some elements of existing hypotheses developed from mammal studies apply to sharks, but there may also be important differences.

A NEW APPROACH

To remove the effects of body-size dimorphism on the behaviour of the sexes it seems profitable to study monomorphic species. Among sharks, these are principally the egg-laying species such as catsharks (Scyliorhinidae) and horn sharks (Heterodontidae) (Table 8.1, Fig. 8.1). However, it has been argued that segregation appears to be less well defined in these species compared to live-bearers because of the similarity in male to female sizes that result in similar nutritional requirements and, hence, much weaker sexual segregation as the forage selection (sexual dimorphism – body size) hypothesis suggests (Klimley, 1987). Critically though, what was not considered in the latter author's discussion of monomorphic species was the role of sampling scale. Trawls, longlines and gill-nets have been used to capture sharks in most studies where sexual segregation has been described. Because egg-laying shark species are mostly small-bodied, their dispersal distances and home ranges are smaller than those of large-bodied species (McLoughlin & O'Gower, 1971; Gruber et al., 1988; Bowman et al., 2002). The fishery-based sampling techniques are, therefore, imprecise tools for document-ing small animal movement and distribution patterns because deploy-ments tend to last several hours or days and so integrate any fine-scale changes that may occur. This suggests that studies relying on fishery catches are less likely to find well defined segregation in smaller bodied species, simply due to the mismatch in the scale of sampling compared to a species' distribution patterns.

Studies on the behaviour of free-ranging monomorphic sharks, where movements and associations between individuals are measured directly, presents an appropriate system for testing hypotheses about the causes of sexual segregation. Here, sampling techniques can be matched readily to the same spatio-temporal scale as their movements and home ranges without the confounding effect of sex differences in body size.

STUDIES ON FREE-RANGING MONOMORPHIC SPECIES

There have been very few studies on sexual segregation of free-ranging monomorphic sharks. This is principally because of the inherent dif-ficulty of tracking fish movements and interpreting activity patterns in relation to marine habitats that are equally difficult to sample ade-quately. Recently though, causes of sexual segregation in monomor-phic sharks have begun to be investigated in detail using three main hypotheses outlined earlier as a framework. Taking the forage selec-tion hypothesis as a starting point leads to a testable prediction that (i) for species lacking sexual size dimorphism, segregation between the sexes should not occur because nutritional requirements are the same for males and females. The predation risk hypothesis predicts that (ii) males exploit areas where prey is most abundant while females reduce the risk of predation by increasing offspring security. Finally, one aspect of the social factors hypothesis indicates that (iii) sexual segregation is the product of aggression between males and females.

Behaviour of oviparous dogfish

The lesser spotted dogfish (Scyliorhinus canicula) has been used as a model species in recent studies that employ a multifaceted approach to unravel the causes of sexual segregation. The dogfish is a noctur-nal predator and scavenger and is an appropriate model because it is

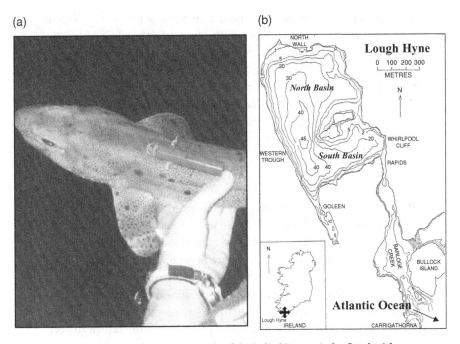

Figure 8.2 (a) An adult male dogfish, *Scyliorhinus canicula*, fitted with a temperature-sensitive acoustic transmitter. (b) Map showing the dimensions and bathymetry of Lough Hyne and its location in southwest Ireland (inset). Contour lines are in metres. After Minchin (1987).

sexually monomorphic, is large enough to be tracked in the wild by attaching electronic tags, but is also small enough to enable behavioural experimentation under controlled laboratory conditions. The recent research uses acoustic transmitters to track the movements of free-ranging male and female dogfish (Fig. 8.2a), together with simultaneous surveys of population abundance and distribution (Sims *et al.*, 2001a). Direct observations of male–female interactions and measurements of prey–density gradients in the microhabitats selected have been undertaken, in addition to mapping thermal habitat using electronic data loggers (Sims, 2003). Laboratory experiments are used to determine metabolic costs at various levels of activity and to test specific hypotheses arising from field observations (Sims *et al.*, 1993; Sims & Davies, 1994; Sims, 1996; Sims, 2003).

Long-term acoustic tracking in *S. canicula* has been used to study individuals inhabiting Lough Hyne, a semi-enclosed sea lough in southwest Ireland that covers an area of only 0.6 km² (Fig. 8.2b). The lough has a maximum depth of 50 m and is connected to the Atlantic Ocean via a narrow and shallow rapids (width <20 m, depth 1–3 m) with a distinct raised sill, which acts to limit immigration and emigration of fish (Sims *et al.*, 2001a). Tagged fish have been recaptured in the same bay where they were originally caught after periods of up to 7 years. Thus, this sheltered site provides a benign and relatively 'controlled' natural environment in which to investigate dogfish behaviour and sexual segregation over appropriate spatio-temporal scales.

Male behaviour

Male *S. canicula* tracked continuously in Lough Hyne showed similar patterns of low activity during the day in deep water (12–24 m) followed by more rapid movements into shallow areas (<4 m) at dusk (Sims *et al.*, 2001a). Males returned to the core space in deep water at dawn (Fig. 8.3). In contrast to the crepuscular activity peaks, nocturnal distances moved and rates of movement were similar to those during daytime. Male dogfish generally remained in shallow-water areas nocturnally, but saltatory activity decreased to daytime levels before the return to deep water at dawn. A conventional mark–recapture study and underwater observations, showed fish tagged at night in shallow areas returned to deep water, confirming the telemetry results (Sims *et al.*, 2001a).

The question of why males were making nocturnal excursions into shallow areas from deeper water occupied during daytime was addressed by measuring relative prey abundance in each location. Dogfish are considered generalist feeders and opportunists on a wide range of benthic invertebrate and fish prey (Lyle, 1983; Ellis *et al.*, 1996). This is reflected in the diet of *S. canicula*, which usually contains the most abundant and readily available prey items in the habitat occupied (Wetherbee *et al.*, 1990). The dogfish studied by Sims *et al.* (2001a) in Lough Hyne consumed primarily decapod crustaceans (swimming crabs, *Liocarcinus* spp.; the prawn, *Palaemon serratus*) and small teleost fishes. Day and night deployments of baited traps in the deep (18 m depth) and shallow (1.3 m) areas showed that irrespective of light phase the abundance of crabs, prawns and small fish was between 17 and 72 times higher in shallow water compared to deep habitat (Sims, 2003). This indicates that the dogfish movements into shallow areas were foraging related and demonstrates that males prefer cooler thermal habitat close to highly productive feeding areas for their daytime resting phases.

The energetic costs and benefits of this behavioural strategy were investigated using electronic data loggers to record the thermal regime

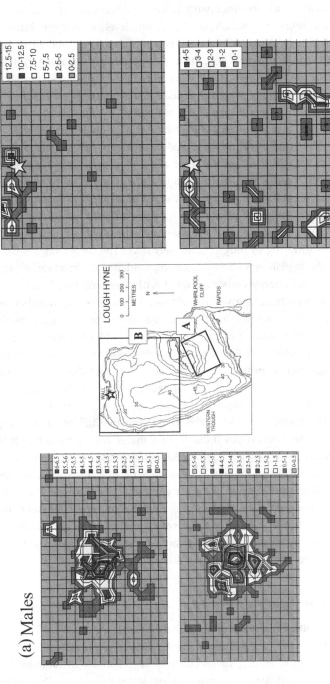

Figure 8.3 Contour plots of activity spaces of acoustic-tracked male and female dogfish over a seven-day period in Lough Hyne. Female plots show location frequency during movements away from refuges (marked with stars).

in different areas, together with laboratory respirometry studies to determine rates of metabolism at different levels of activity. Temperature data loggers recording every two minutes were moored in the shallow (1.5 m depth) and deep (18 m depth) habitats occupied by male dogfish in the sea lough at the same time of year (August–September) when tracking studies were undertaken. Daytime temperatures in the shallow, prey-rich areas ranged from 16.0 to 17.7 °C, whereas at dusk, the temperature decreased rapidly such that during night time the shallows were <15.7 °C. Temperatures recorded in the deep habitat ranged from 14.9 to 15.7 °C. Thus, it appears that male dogfish move into the shallow, prey-rich habitat when the temperature of water there converges with that found in their deeper, daytime habitat, suggesting males may be sensitive to higher temperatures. In support of this, preliminary laboratory studies indicate males presented with a choice between two chambers differing by only 1 °C, actively select the colder side.

Male dogfish may choose to occupy cold, deep habitat during the day to reduce energy costs associated with standard metabolism (R_S: defined as metabolic rate at zero swimming speed), feeding metabolism (R_F) and active metabolism (R_A). In the laboratory, a temperature increase of 10 °C (from 7 to 17 °C) more than doubles oxygen uptake in *S. canicula* (Butler & Taylor, 1975), so males moving into warm water would experience raised metabolic rates. For dogfish, the standard metabolism to body mass relationship at 15 °C (Sims, 1996) and the Q_{10} value of 2.16 (Butler & Taylor, 1975) were used to calculate the energy costs attributable to R_S for males remaining in shallow versus deep habitat. These calculations indicate that males remaining for 24 h in cooler water (15.3 °C) save 1.23 kJ, or 8.8% of R_S costs compared to energy expenditure in warm-shallow habitat (16.5 °C). In reality, the observed male strategy would result in higher energy savings because R_F and R_A costs are between two and three times higher than standard rates (Sims et al., 1993; Sims & Davies, 1994). Therefore, male dogfish reduce energy expenditure by conducting activity and digestion in the coolest water available near to preferred feeding areas (Fig. 8.3).

Female refuging behaviour

Female dogfish exhibited a different behavioural strategy to males (Sims et al., 2001a; Sims, 2003). Females preferentially spent between 62 and 73% of the time resting in female-only aggregations in labyrinthine caves located primarily in warm, shallow water (0.5–1.5 m). This refuging behaviour resulted in habitat segregation from males during the day and for some periods of the night. Female dogfish were, however,

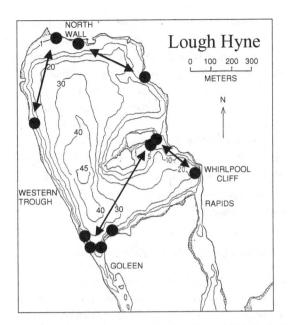

Figure 8.4 Locations of female-only refuges (solid circles) in Lough
Hyne and the movements between them by individuals (arrows).
Refuges contain between 1 and 25 females.

active for a few hours every night, or every second or third night, principally in deep water (Sims *et al.*, 2001a; Sims, 2003). Hence, there was
some degree of overlap in nocturnal habitat usage although areas of
activity were often quite distinct (Fig. 8.3). Between one and four cave
refuges were used by individual females, but generally each showed
pronounced philopatry to a preferred refuge (Fig. 8.4). Up to 25 females
have been observed to aggregate in one of the most popular refuges.
This pattern of usage indicates that the locations of refuges are known
by individuals and that these caves represent long-term 'homes' for
female dogfish. Taken together, the observations strongly suggest male
and female dogfish exhibit alternative strategies that act to segregate
them in both space and time. But what are the causes of these different
patterns of behaviour?

HYPOTHESIS TESTING

Sexual dimorphism – body size (forage selection) hypothesis

The prediction is that for species lacking sexual size dimorphism, segregation between the sexes should not occur, or should be limited,
because nutritional requirements are the same for males and females.

Indeed, a trawling study in the North Aegean Sea showed an absence of sexual segregation in *S. canicula* (D'Onghia *et al.*, 1995). In contrast, the studies by Sims *et al.* (2001a) indicate that male and female dogfish exhibit fine-scale segregation by occupying different habitat for between 62 and 73% of the time. This pattern appears stable throughout different seasons. The hypothesis that monomorphic species of shark do not segregate over various spatio-temporal scales can therefore not be supported. Dogfish, however, exhibit less pronounced sexual segregation over broad spatial scales because although male and female habitats are different, they are not necessarily well separated. Females have a similar diet to males (Lyle, 1983) presumably because there is sufficient overlap in nocturnal foraging habitat.

Reproductive strategy (predation risk) hypothesis

One prediction arising from the reproductive strategy hypothesis, states that males should exploit areas where prey are most abundant. The strategy used by male dogfish was broadly consistent with this idea. Male activity space in Lough Hyne was located in, or very close to, prey-rich areas (Kitching & Ebling, 1967; Sims, 2003). Under this hypothesis, females, in contrast, are predicted to segregate so as to reduce predation risk and increase offspring security. Refuging in caves for long periods presumably reduces predation risk for adult female dogfish. Therefore, superficially at least, the behaviour of males and females appears to correspond well with this hypothesis.

Dedicated studies are needed to determine the effects of predator activity on adult sharks to test this hypothesis fully. This is because predators of dogfish are few, especially in the tidal lough where they were studied (Sims *et al.*, 2001a). Adult dogfish (~0.7 m total length) are likely to be at risk from larger sharks, rays and seals. In Lough Hyne, the larger species of shark (nursehound, *Scyliorhinus stellaris*; angel shark, *Squatina squatina*) and ray (thornback, *Raja clavata*) are rare (Minchin, 1987). However, grey seals (*Halichoerus grypus*), sometimes in small groups, have been observed to enter the lough, and may remain there for some weeks. Although dogfish carcasses attributable to characteristic seal predation have not been encountered regularly, the potential for predation by seals cannot be discounted, although it is likely to be low. *H. grypus* have been observed hunting teleost fishes and a high proportion of calcified otoliths in seal scats in the lough indicates a diet dominated by bony fishes (D. W. Sims, unpublished data). In support of the suggestion that predation risk is low, female dogfish were observed refuging even when seals were not present. Fish are known to exhibit less caution when immediate predation risk decreases (Milinski, 1993), which taken together with the observations for dogfish, raises the possibility that refuging to avoid predators may be a minor proponent of sexual segregation in adult sharks.

Thermal niche – fecundity hypothesis

While female dogfish in the lough have an apparently low risk of predation they still refuge whereas males do not. The reproductive strategy hypothesis states that females should segregate to improve offspring security, which could account for this behaviour because females produce two eggs every two weeks that are fertilized from stored sperm. Refuging effectively protects this energetic investment. The results from studies on dogfish indicate that choice of refuge may also depend on habitat selection due to abiotic factors that may act to increase fecundity.

All the favoured, most well-populated refuges used by female dogfish in Lough Hyne are in shallow, warm water (Fig. 8.4) (Sims et al., 2001a). Despite the availability of deeper refuge habitat, this is used much less frequently. This suggests that selection of an appropriate thermal niche may be an important mechanism in refuging, and hence, sexual segregation. Like most other fish, S. canicula are ectothermic, with body temperatures determined by ambient water temperature. The rates of egg production could be increased by females selecting warmer waters for daytime resting behaviour. Furthermore, the increased metabolic costs associated with increased body temperature (see section Male behaviour) would be minimized by females remaining inactive. Preliminary thermal-choice chamber experiments indicate female S. canicula are relatively insensitive to increased thermal conditions compared with males, which always choose the coldest available habitat (D.W. Sims, unpublished data). Hence, sex differences in thermal preferences may have an as yet unappreciated role in determining sexual segregation in sharks. However, for this hypothesis (which I term the 'thermal niche – fecundity' hypothesis) to be broadly applicable, a key question is whether thermal conditions influence sexual segregation in other shark species, including dimorphic ones?

It has been proposed that social factors may be more important in determining sexual segregation than environmental factors (Conradt, 1999). Environmental factors such as temperature are critical for ectothermic species such as fish and invertebrates, because even relatively small changes in temperature can have a disproportionately large effect on many aspects of behaviour and ecological function (Wood & McDonald, 1997; Sims *et al.*, 2001b). A reappraisal of the available literature in light of the new studies with dogfish (Sims *et al.*, 2001a; Sims, 2003) shows the importance of thermal conditions for the occurrence of sexual segregation. It was suggested by Springer (1967) for pelagic sharks that sexual segregation occurs in part due to males moving towards colder waters while gravid females remain in warmer shelf waters. For sharks of the genus *Carcharhinus*, which are live-bearers displaying sexual size dimorphism, the proportion of males is greater in the southern zone off South Africa, which is colder, than the northern region where females are more abundant (Bass *et al.*, 1973). Similarly, female grey reef sharks (*Carcharhinus amblyrhynchos*) in the Pacific form female-only aggregations in shallow, warm water away from males (Economakis & Lobel, 1998). Interestingly, the lack of sexual segregation in dogfish in the Aegean Sea compared with the Atlantic appears to be linked to the homeothermic conditions of this region of the Mediterranean, which remain between 13.5 and 14.0 °C where the dogfish are found (D'Onghia *et al.*, 1995). Therefore, it seems that thermal habitat may play an important role in sexual segregation in both monomorphic and dimorphic shark species, and could occur to increase fecundity.

Social factors hypothesis

Sexual segregation resulting from aggressive behaviour of males towards females is an aspect of the social factors hypothesis stated earlier. Mating in sharks is known to be aggressive (Tricas & Le Feuvre, 1985; Carrier *et al.*, 1994) where males may bite females repeatedly during courtship and copulation, often leaving substantial wounds (Stevens, 1974; Pratt & Carrier, 2001). To combat this, females of species such as the blue shark possess skin more than twice as thick as that of males (Pratt, 1979). Courtship and copulation in *S. canicula* is also protracted and aggressive, usually consisting of many males pursuing the female, tugging violently and biting her (Dodd, 1983). It was hypothesized by Sims *et al.* (2001a) that females form female-only aggregations in refuges to reduce energetically demanding mating activity. Claspers of male dogfish during August had reddened tips indicating they had recently been used for mating. However, the caves used by females in Lough Hyne represent an effective refuge from males because they are labyrinthine, with very narrow entrances and little room inside, making copulation within them difficult. Furthermore, females store sperm, permitting egg-laying throughout much of the year (Metten, 1939a,b;

Harris, 1952) so, theoretically, constant access to males by females is not necessary.

If male aggression causes sexual segregation in dogfish the prediction is that males should concentrate their activity at female-only refuges when females are most likely to be available. Furthermore, when females are encountered, males should engage in courtship and mating. Direct observations of a female refuge in Lough Hyne have been made continuously over numerous 24-h periods. Results support the hypothesis, because male dogfish only appeared outside the refuge when the largest numbers of females were leaving or returning (Sims, 2003). Unsuccessful mating attempts were also seen, with females exhibiting a fast-swimming escape response into a refuge to avoid courtship from multiple males (V. J. Wearmouth & E. J. Southall, unpublished data). Refuging in shallow water by females to avoid copulation by dominant males occurs in nurse sharks (Pratt & Carrier, 2001), so refuging as a male-avoidance strategy may be a principal determinant of sexual segregation in sharks during the breeding season. Dogfish, however, do not have a distinct breeding season (Metten, 1939b) with females producing eggs for much of the year. This species characteristic may in part contribute to the observed inter-seasonal persistence of refuging by female dogfish. This species may therefore be a temporally stable model for examining the role of male aggression in structuring sexual segregation in shark populations.

CONCLUSIONS

Until recently, the causes of segregation by sex in sharks were not the subject of systematic, hypothesis-led investigations of behaviour. This chapter presents a new look at old results from a broad range of species, and discusses new results that use monomorphic dogfish as a model species. So what have we learnt?

Competing hypotheses have been put forward to explain sexual segregation in animals. A critical look at the available information presented here indicates that some or all of these may contribute to sexual segregation in sharks. The early idea that sex differences in body size accounts for the degree of sexual segregation in sharks, is not supported by recent work showing that pronounced sexual segregation occurs also in species without sexual body-size dimorphism. However, live-bearing shark species showing distinct dimorphism may segregate, at least in part, because of sex differences in selection of thermal and/or foraging habitat. Recent behavioural studies show that sex differences in thermal niche contribute to observed habitat segregations in both dimorphic and monomorphic species, differences which may be linked to enhancing fecundity. What emerges from all this is that the causes of sexual segregation in sharks are complex and unlikely to be explained by a single existing hypothesis. Dogfish sexes, for example, segregate due to a combination of social, reproductive and sex-specific physiological factors. Segregation in dogfish appears to be driven principally by aggressive courtship and mating behaviour of males towards females, which apparently respond by forming female-only aggregations in caves. However, another component that determines which refuges are selected appears to be sex differences in thermal-habitat choice related to maximizing surplus power for egg production. Taking another example, juvenile female scalloped hammerheads appear to segregate from males by moving offshore to exploit high-energy prey that confer high growth rates. These rates are important to maintain, because reproductive success of females that are live-bearers is influenced by body size to a greater degree than it is for males. It is suggested that females move to offshore habitats and accept higher predation risk to attain higher rates of growth at the juvenile stage. Clearly, this scenario for hammerhead sharks comprises elements of both the sexual dimorphism – body size hypothesis and the reproductive strategy hypothesis. Because of this complexity, in the future a fruitful approach would be to integrate physiological and behavioural methods within each study to help identify the relative importance of contributing factors. By determining the relative energy cost to each sex of behaviours such as male–female interactions, it may be easier to assess the causes of observed segregations. For example, determining why female dogfish segregate by refuging together in caves will require combining measures of how temperature influences egg production rates, the extent to which females in laboratory thermal-choice studies select particular thermal habitat, and how the presence of males affects either or both of these, in addition to how predation risk modifies responses. Integrating energetics and behaviour in this way should provide an insightful 'cost–benefit' component to studies of sexual segregation in sharks and probably other species as well.

ACKNOWLEDGEMENTS

I am grateful to J. Nash, D. Morritt, E. Southall, V. Wearmouth, J. Hill and P. Moore for helping to undertake the research and for many useful discussions, and to the director and staff of the Marine Biological Association (MBA) for providing a stimulating research environment. The work on dogfish was supported by the Natural Environment Research Council (NERC), The Fisheries Society of the British Isles, The British Ecological Society, The Percy Sladen Memorial Trust and the Jubilee Trust. It was conducted under licences granted by the UK Home Office, the Republic of Ireland Department of Health, and the Irish Heritage Division (Duchas). D. O'Donnell is particularly thanked for his continued support of our research at Lough Hyne. J. Stevens and J. Carrier provided valuable comments on an earlier version of this chapter. This chapter's author, D. W. Sims is supported by an NERC-funded MBA Research Fellowship.

9

Sex differences in reproductive strategies affect habitat choice in ungulates

OVERVIEW

Behaviour patterns governing habitat use, predator avoidance and foraging effort ultimately influence the reproductive success and survival of
individual animals. Seasonal energy demands, the distribution of food
and water, climatic factors, security constraints, physiological mechanisms, breeding patterns and social structure are all expected to influence behaviour patterns in different ways at different times of the year.
Furthermore, such variables are interactive, and their effects may vary
by age and sex class (Clutton-Brock, 1988).

In polygynous ungulates, reproductive roles differ between males
and females as do the behaviour patterns that promote reproductive
success. Females select habitats and compete for resources that promote offspring growth and survival, whereas males select habitats that
maximize pre-rut energy reserves and promote successful competition
with other males for mates (Clutton-Brock et al., 1982; Main & Coblentz,
1990, 1996; Main et al., 1996; Trivers, 1972). Sex differences in behaviours
and habitat use that influence reproductive success may, therefore,
be viewed as outcomes of sex differences in reproductive strategies.
Furthermore, behaviours that influence reproductive success are not
limited to periodic mating opportunities, but occur throughout the
year for both sexes in different ways via effects on offspring survival
or the accumulation of energy reserves important to future reproductive efforts. Consequently, patterns of behaviour and habitat selection
exhibited by males and females are expected to differ during different times of the year and reproductive cycle. In this chapter, we focus

Sexual Segregation in Vertebrates: Ecology of the Two Sexes, eds. K. E. Ruckstuhl and P. Neuhaus.
Published by Cambridge University Press. © Cambridge University Press 2005.

on polygynous ungulates to illustrate how differences in reproductive strategies of males and females arise in response to differing selective pressures, which ultimately influence habitat segregation.

THE REPRODUCTIVE STRATEGY HYPOTHESIS

Mating systems, resource distribution patterns and predation risks interact differently in different species and in different study areas. Consequently, the manner in which habitat segregation is expressed also varies among species and populations. It is not surprising, therefore, that multiple explanations have been put forward to explain habitat segregation in ungulates, and that explanations for this behaviour remain a topic of lively debate (Main & Coblentz, 1990; Main et al., 1996; Gross, 1998; Main, 1998; Ruckstuhl & Neuhaus, 2002). Here we focus on the reproductive strategy hypothesis, which contends that the different reproductive objectives of males and females provide the impetus for sex differences in habitat use in polygynous ungulates (Main & Coblentz, 1990, 1996; Main et al., 1996). The reproductive strategy hypothesis argues that habitat use by females is dictated by decisions that influence offspring survival, whereas habitat use by males is dictated by energetic decisions related to accumulation of energy reserves in preparation for rut and competition with other males for access to mates. The reproductive strategy hypothesis, therefore, attempts to explain habitat segregation at the ultimate level of reproductive success, which is influenced by selective pressures that operate differently on each sex. Other factors, such as sex differences in body size and digestive efficiency, may contribute to fine-scale differences in diet and activity patterns within the broader context of the reproductive strategy hypothesis, but for polygynous ungulates we contend that differences in reproductive strategies ultimately explain sexual segregation at the habitat scale.

Habitat selection by females

Much debate has focused on whether males or females utilize superior foraging habitat, and this criterion has sometimes been used to support or refute the reproductive strategy hypothesis (Bleich et al., 1997; Main et al., 1996; Post et al., 2001). The reproductive strategy hypothesis, however, is influenced by multiple factors that include, but are not limited to, forage resources. Female lifetime reproductive success is primarily dependent upon the ability to raise offspring successfully to the age of independence (Clutton-Brock *et al.*, 1982). Consequently, habitat choice by females is the outcome of a trade-off between the individual resource needs of mothers and the needs and security of their offspring. Adult females, therefore, are predicted to position themselves in the landscape to satisfy, but not necessarily optimize, the acquisition of forage resources (Bunnell & Gillingham, 1985), and to optimize the net effect of sufficient forage, water and security resources to increase the likelihood of offspring survival.

The argument that habitat choice by females with young may sometimes reflect a compromise between foraging opportunities and offspring security is well documented (but not universally supported) by field studies (Main *et al.*, 1996; Ruckstuhl & Neuhaus, 2002). Where risk of predation is important, habitat choice and/or behaviours by females with offspring should reflect decisions designed to reduce these risks. For example, females with young may use habitats that offer relatively poor foraging opportunities in exchange for increased offspring security from predators (Edwards, 1983; Bergerud *et al.*, 1984; Bleich *et al.*, 1997; Festa-Bianchet, 1988; Berger, 1991; Miquelle *et al.*, 1992; Main & Coblentz, 1996). An analogous scenario is the formation of large nursery groups that provide increased individual security, but represent a trade-off in the form of greater intraspecific competition, which can influence fecundity and offspring survival (Clutton-Brock *et al.*, 1982, 1987b; Prins, 1989; Wrona & Dixon, 1991; Komers *et al.*, 1993). Predation risk to offspring may also dictate habitat use by limiting distribution and movement patterns of females, particularly when offspring are young and in species whose offspring rely on concealment and maternal protection as defence against predators. Large groups or limited movement patterns increase foraging pressure in localized areas and may reduce forage biomass or the availability of preferred diet items (Clutton-Brock *et al.*, 1982; Main & Coblentz, 1996). Female habitat choice, therefore, must consider both predation risks to offspring and the availability of forage and water resources throughout the period when offspring are most vulnerable and least mobile.

The reproductive strategy hypothesis predicts that, where risk of predation is important, security constraints influence selection of habitat by females with offspring. It is not surprising, therefore, that the reproductive strategy hypothesis has been described as a predation risk hypothesis (Ruckstuhl & Neuhaus, 2002). However, risk of predation is only one of several factors that may influence habitat selection. Factors other than risk of predation, such as availability of water, may also dictate distribution and habitat choice by females with young (Becker &

Ginsberg, 1990; Berteaux, 1993; Bowyer, 1984; Corfield, 1973; Main & Coblentz, 1996). In situations where risk of predation is not important (Clutton-Brock *et al.*, 1987b; Post *et al.*, 2001) the reproductive strategy hypothesis predicts female groups will select the highest quality habitats with the best combination of available resources (e.g. Watson & Staines, 1978), unless behaviours and habitat choice are strongly influenced by a history of predation pressure. Females without young, or females that lose young, may modify patterns of habitat use to maximize accumulation of energy stores in preparation for winter and the raising of offspring during the following year. These options, however, may be mediated by site fidelity or the existence of social bonds with other females. Consequently, habitat selection by females may be influenced by predation pressure, but is also influenced by specific resource requirements (e.g. food and water) and, in many species, by historical patterns of range use, philopatry and the existence of stable matrilineal groups (Clutton-Brock *et al.*, 1982; Main, 1994).

Habitat selection by males

Although males and females of a species may depend on similar broad categories of food resources, the factors influencing reproductive success and habitat selection tend to differ because the reproductive goals of males and females are different. Polygynous males are continually engaged in an energetic race to maximize body condition in preparation to defeat other males in contests for mates. Body condition prior to mating has been reported as an important determinant for reproductive success (Bergerud, 1974; Clutton-Brock *et al.*, 1982, 1988; Prins, 1989). Furthermore, in the case of highly seasonal breeders, body condition prior to mating activities can be an important determinant of post-rut survival and, consequently, lifetime reproductive success (Mautz, 1978; Clutton-Brock *et al.*, 1988). This is particularly true where harsh environmental conditions (e.g. cold winters or dry seasons) reduce availability of forage during post-rut periods. In these situations, pre-rut fat stores must provide both the energy needed for mating activities and also sustain males until post-rut environmental conditions improve (Mautz, 1978). Genetics notwithstanding, the most successful males should, therefore, be those that utilize forage resources across the landscape in an optimal manner, which includes considerations of forage quantity, quality and the energetic expenditures required to find them. A prediction of the reproductive strategy hypothesis, therefore, is that the reproductive strategies of males include behaviours and habitat use patterns that build and conserve pre-rut energy reserves (Main & Coblentz, 1996; Main et al., 1996), even when doing so results in using habitats with increased predation risks (Jakimchuk et al., 1987; Prins & Iason, 1989; Miquelle et al., 1992; Berger & Gompper, 1999).

In some cases, males have been documented using lower quality habitats than nursery groups (Watson & Staines, 1978; Clutton-Brock et al., 1982). It has been argued that males may profit energetically by segregating to habitats with more widely dispersed or poorer quality forages rather than engage in scramble competition with female groups for preferred forages with limited availability (Clutton-Brock et al., 1987a; Main & Coblentz, 1996; see also Chapter 3). Analogous examples that suggest larger bodied males are more plastic in their ability to use different habitats and different diet items are provided by Post et al. (2001) who determined male bison, Bos bison, consumed lower quality diets to avoid scramble competition with cow–calf herds, and Kie & Bowyer (1999) who reported diets of male white-tailed deer changed more than diets of females at high population densities (77 deer/km^2) within a 391-hectare predator exclosure. Whether female herbivory influences habitat segregation by males has been disputed, however, and deserves additional study (Main, 1998; Conradt et al., 1999b, 2001). Nevertheless, in terms of the reproductive strategy hypothesis, habitat choice by males reflects a balance between the ultimate individual benefits of foraging in a particular habitat (i.e. enhanced ability to compete for mates), against all immediate and potential costs associated with that decision (e.g. increased foraging effort, predation risk, insect harassment, distance to water, etc.).

Habitat use by mixed-sex groups

Males and females sometimes form mixed-sex aggregations for reasons other than mating. In such cases, the conditions required for the reproductive strategy hypothesis do not apply, in that the reproductive strategies of males and females are not better served by sexual segregation at the habitat scale. This may simply be because the range of habitat options includes one habitat with outstanding foraging opportunities for both sexes, and predation risk is uniformly low. Examples of foraging opportunities that sometimes promote mixed-sex feeding aggregations include areas with seasonally emergent vegetation (Fig. 9.1), agricultural operations and recently burned areas that produce a flush of highly nutritious vegetation (Thill et al., 1987; Main, 1994; Main & Coblentz, 1996).

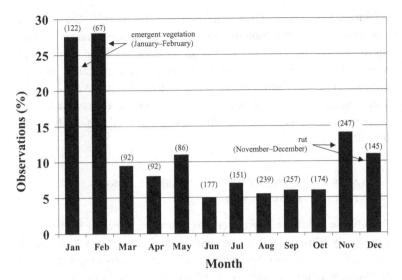

Figure 9.1   Mixed-sex groups of white-tailed deer (≥1 adult of both sexes) observed at the Welder Wildlife Foundation Refuge (Sinton, Texas, USA), and plotted as a percentage of total group observations by month (total group observations in parentheses). Mixed-sex groups were defined by proximity and behaviour of individuals (from Main, 1994). Note the peak in occurrence of mixed-sex groups in January–February, when both sexes congregate in open bunchgrass-annual forb habitat to feed on high-quality, emergent grasses and forbs.

Furthermore, habitat segregation may decline when environmental conditions limit the range of available habitats in which members of each sex are able to meet their maintenance needs. For example, in South Africa, sex differences in habitat use by greater kudu, *Tragelaphus strepciseros*, decline during the late dry season (post-rut) when palatable forage becomes scarce. During this period nursery groups increase use of wooded riverine habitat, a critical dry season food reserve for the browsing guild. Wooded riverine habitat is preferred by males throughout the year, but is avoided by nursery groups during the rainy season, presumably due to the risk of leopard, *Panthera pardus*, predation on calves (du Toit, 1995). Harsh environmental conditions may also promote the formation of mixed-sex groups when habitat segregation provides little benefit. For example, mixed-sex aggregations (yarding) of white-tailed deer may occur during winter (post-rut) in the northern United States in response to snow depth and availability of suitable thermal cover (Beier & McCullough, 1990; McCullough et al., 1989) and predation from wolves, *Canis lupus* (Nelson & Mech, 1981). Cold winters and deep snow limit mobility and increase the importance of thermoregulation for energy conservation. Furthermore, the quality and availability of forage is low during winter and white-tailed deer of both sexes reduce forage intake (Mautz, 1978; Ozoga & Verme, 1970). Consequently, males and females employ similar energy-conserving behaviours in these situations and habitat segregation provides little benefit.

A CONCEPTUAL MODEL FOR HABITAT SEGREGATION IN POLYGYNOUS UNGULATES

Despite the debate on the underlying processes, previous studies have collectively illustrated that the manner and degree to which sexual segregation is expressed in ungulates is almost certainly influenced by the interaction of multiple variables, and many of the hypotheses put forth to explain sexual segregation and sex differences in habitat use are not mutually exclusive (Main et al., 1996; Barboza & Bowyer, 2000; Ruckstuhl & Neuhaus, 2000, 2002). Consequently, it seems increasingly unlikely that a species- or population-specific explanation for why sexes segregate and utilize different habitats will hold true in all cases. Nonetheless, sexual segregation is a prevailing feature of the life-history patterns of polygynous ungulates and some type of universal, underlying basis for this behaviour seems likely. It follows, therefore, that any such universal explanation needs to be sufficiently broad to encompass and explain the different ways in which sexual segregation is expressed among different populations and species.

We suggest that asking whether and how habitat segregation benefits each sex in terms of reproductive success may provide the most robust ecological framework for understanding when, how and why sexes are likely to segregate and use different habitats. To illustrate our point, we briefly discuss components of morphological/physiological and social factors that have been proposed to explain sexual segregation in ungulates and provide examples of how these factors influence habitat segregation differently under different scenarios. Our objective is to illustrate that these factors may explain components of sexual segregation within a larger reproductive strategy hypothesis framework, but when viewed in isolation the hypotheses based on these factors fail to adequately explain habitat segregation as a universal behaviour among polygynous ungulates. We conclude by attempting to demonstrate that habitat segregation may best be viewed as consisting of a gradient or spectrum of responses, and that the reproductive strategy hypothesis provides a meaningful framework from which to understand the basis for these behaviours and the impetus for habitat segregation in polygynous ungulates.

Influence of sexual size dimorphism on habitat segregation

Sexual size dimorphism, especially as it relates to sex differences in forage intake and digestion, has been implicated as an important mechanism contributing to resource partitioning and habitat segregation by sexes in ungulates, particularly among ruminants (Clutton-Brock et al., 1987a; Gross et al., 1996; Barboza & Bowyer, 2000). Although the influence of sex differences in bite size and dental morphology on forage intake has been considered as a potential explanation for habitat segregation (Illius & Gordon, 1987; Gross, 1998), predictions of this hypothesis have not proven consistent across taxa, particularly within the browsing guild (Miquelle et al., 1992; Weckerly, 1993; du Toit, 1995; Ruckstuhl & Neuhaus, 2002).

The argument most commonly invoked in support of a morphological/physiological explanation for habitat segregation among sexually dimorphic ungulates has addressed sex differences in rumen size (Ginnett & Demment, 1997; Main, 1998; Barboza & Bowyer, 2000; Ruckstuhl & Neuhaus, 2002). The essential tenets of the hypothesis are that males possess larger rumens that confer greater digestive efficiency (Demment & van Soest, 1985; van Soest, 1994) and, because males are larger and have higher absolute energy requirements, they should segregate to habitats where they can prioritize quantity rather than quality of forage. Conversely, smaller females possess smaller rumens, are less efficient at digesting high-fibre forages, and should segregate to habitats where they can obtain high-quality forages, particularly when lactating. This hypothesis implies, therefore, that male diets should consist of greater proportions of high-fibre, low-quality forages than found in diets of females (see Chapter 3) and that forage availability in habitats used by males and females should reflect these differences.

The difficulty in extending the sexual size dimorphism hypothesis (also known as the forage selection hypothesis) upscale from the patch to the habitat is that it assumes males and females cannot select qualitatively different diets while feeding together in the same habitat. That assumption is obviously weak, and support for predictions that males use poorer quality habitats and consume lower quality diets when segregated from females has been equivocal (Main & Coblentz, 1990; Main et al., 1996; Parker et al., 1999; Ruckstuhl & Neuhaus, 2002).

The sexual size dimorphism hypothesis also fails to explain the formation of mixed-sex feeding groups attracted to areas with high-quality forage (Hirth, 1977; Main & Coblentz, 1996). In some instances, sexual size dimorphism may contribute to resource partitioning between sexes when using the same habitats, as may occur when habitat segregation is either not an option due to high population densities or provides little advantage. For example, sex differences in browsing height enable males and females to partition mutually desired resources when palatable forage is scarce (du Toit, 1990, 1995; Ginnett & Demment, 1997). In high-density populations of white-tailed deer, larger rumens enable males to subsist on poorer-quality diets than females as an energetic alternative to searching for preferred diet items in short supply (Kie & Bowyer, 1999). Consequently, although larger rumens provide a mechanism that enables males to be more efficient at digesting high-fibre forages, the consumption of poorer quality forages by males has not been consistently demonstrated and sex differences in body size alone do not appear sufficient to explain habitat segregation (Main et al., 1996). Furthermore, size dimorphism has been suggested specifically to explain habitat segregation in ruminants. Although habitat segregation is most pronounced among sexually dimorphic ruminants (Mysterud, 2000; Ruckstuhl & Neuhaus, 2002), numerous non-ruminant large herbivores also exhibit sex differences in both diet (Chapter 3) and habitat use (Sukamar & Gadgil, 1988; Becker & Ginsberg, 1990; Stokke & du Toit, 2002). It seems unlikely that an entirely different explanation for habitat segregation is needed to explain this behaviour in non-ruminant herbivores. Finally, there exists no conceivable selective mechanism that would explain male preference for low over high quality forages unless other factors are operating, such as increased energetic costs related to search time (Main et al., 1996; Main, 1998). Predictions, therefore, that larger males segregate to consume large volumes of low-quality forage are not adequately supported by the literature, nor are they particularly useful from an ecological perspective. Rather, it seems more likely that larger rumen sizes enable males to subsist on lower-quality forage when the energetic costs of obtaining high-quality forage are too demanding. This latter explanation is essentially consistent with predictions of the reproductive strategy hypothesis. Consequently, factors related to sexual size dimorphism appear to contribute to sexual segregation by habitat in some cases, but do not appear to be the sole factor explaining this behavioural pattern in ungulates.

Influence of social factors on habitat segregation

Various aspects of sex-specific social behaviours have been proposed as potential mechanisms influencing habitat segregation in ungulates. Most of these have been associated with male behaviours, such as the learning of fighting skills, establishment of dominance hierarchies or scouting of mates prior to rut. These hypotheses have been rejected as explanations for sexual segregation at the habitat scale because they either do not require sexual segregation by habitat to occur, or they are not supported by empirical data (Clutton-Brock et al., 1988; Main et al., 1996; Bleich et al., 1997).

Recent studies suggest that sex differences in activity budgets and foraging patterns lead to asynchrony between males and females and may be an important mechanistic component behind the formation of single-sex groups (Conradt, 1998a; Ruckstuhl, 1998; Ruckstuhl & Neuhaus, 2000, 2002). Sex differences in activity budgets, however, have been proposed to explain social segregation as a phenomenon independent of habitat segregation (Ruckstuhl & Neuhaus, 2002) and do not adequately explain why sexual segregation often results in the use of different habitats by males and females.

Behavioural scaling and the reproductive strategy hypothesis as a conceptual model for habitat segregation

As the previous sections demonstrate, one of the principal problems associated with resolving the debate over why habitat segregation occurs is that the various explanations based on physiological or social mechanisms work well in some but not all situations. Because sexual segregation and sex differences in habitat use are influenced by multiple variables and expressed in different ways among species and populations, it may be useful to consider sexual segregation as a behavioural pattern representing a spectrum of potential responses. This concept, described as behavioural scaling, is useful when a range of responses is associated with a larger behavioural pattern (Wilson, 1980). Viewed in this context, the reproductive strategy hypothesis provides a potential framework that helps to understand the impetus for habitat segregation in polygynous ungulates, the details of which will vary in response to the influence of environmental and behavioural factors.

Within the reproductive strategy hypothesis conceptual model, two principal factors are identified as having a major influence on reproductive strategies of polygynous ungulates and the expression of habitat segregation. These include reproductive patterns and the distribution of resources (Fig. 9.2). Reproductive patterns clearly influence the extent to which males must associate with female groups to capitalize on mating opportunities. In species where mating opportunities occur continuously throughout the year, males are compelled to associate with females more than in species with a highly seasonal rut of relatively short duration. Social structure interacts with timing of mating opportunities to influence further reproductive patterns and the expression of habitat segregation among ungulates. For example, in gregarious species that breed throughout the year and form permanent harems, as observed among equids (Berger, 1986), dominant males remain with the herd continuously and subordinate males are relegated to bachelor groups. Among species that exhibit asynchronous breeding but do not form permanent harems, such as African buffalo, *Syncerus caffer*, males defend cow–calf herds as long as body condition permits and then segregate to regain energy reserves, providing opportunities for different males to join female groups (Prins, 1989). Similar patterns are observed among some non-ruminants such as African elephants, *Loxodonta africana*, where bulls in musth prevent other adult males from joining female groups (Slotow *et al.*, 2000; Stokke & du Toit, 2002). In species exhibiting resource defence polygyny, such as pronghorn antelope, *Antilocapra americana*, males defend rutting territories against other males and associate with females attracted to resources within those territories (Kitchen, 1974). And, in rutting species that exhibit serial monogamy, harem defence or lekking behaviours, males associate with female groups during the rut and habitat segregation tends to be pronounced during non-breeding periods (Clutton-Brock *et al.*, 1982; Main *et al.*, 1996; Ruckstuhl & Neuhaus, 2002). In all cases, mixed-sex groups may nevertheless form in response to seasonal availability of preferred resources (Hirth, 1977; Main & Coblentz, 1996) or harsh environmental conditions (McCullough *et al.*, 1989; du Toit, 1995). Breeding synchrony and social structure, therefore, influence the degree to which males are compelled to associate with females to maximize reproductive success and, consequently, influence the extent to which sexes segregate and use different habitats.

The distribution and availability of resources (food, water, cover) also influence the manner in which habitat segregation is expressed

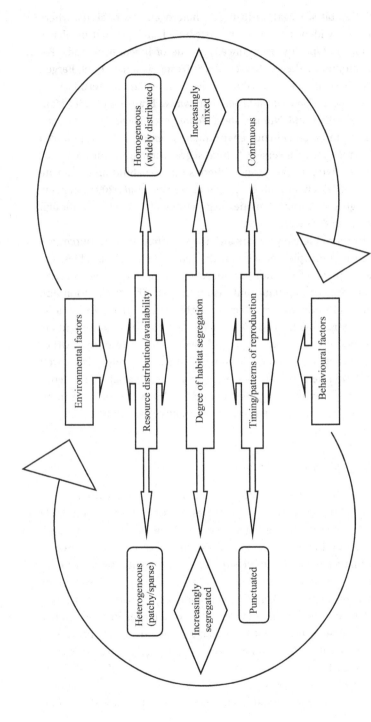

Figure 9.2 The reproductive strategy hypothesis model: environmental and behavioural factors influence the availability and distribution of resources and the timing and pattern of reproduction, the combination of which influence habitat segregation in polygynous ungulates.

(Fig. 9.2). Habitat segregation in highly heterogeneous habitats, where resources are widely and patchily distributed and population densities are low, is typically pronounced outside of breeding periods. For example, mountain sheep, *Ovis canadensis* (Festa-Bianchet, 1988; Berger, 1991; Bleich *et al.*, 1997), caribou, *Rangifer tarandus* (Bergerud *et al.*, 1984), mule deer, *Odocoileus hemionus* (Ordway & Krausman, 1986; Main & Coblentz, 1996) and Nubian ibex, *Capra ibex nubiana* (Gross *et al.*, 1995a) living in heterogeneous environments demonstrate pronounced habitat segregation. Conversely, habitat segregation may be less obvious by ungulates living in landscapes where suitable habitat and requisite resources are widely available (e.g. *Bos bison*, Post *et al.*, 2001), particularly at high population densities (e.g. white-tailed deer; McCullough *et al.*, 1989; Kie & Bowyer, 1999).

Both reproductive patterns and the distribution of resources are influenced by multiple behavioural and environmental variables. For example, as previously demonstrated, social structure influences reproductive patterns of ungulates but also influences the distribution and availability of forages via the effects of herbivory (e.g. group formation). Likewise, environmental variables, such as predation pressure and climatic factors, influence access to and availability of resources. Other important behavioural and environmental variables likely exist as well (e.g. population density). We suggest that all these variables influence habitat segregation in polygynous ungulates chiefly through the interplay between reproductive patterns and resource availability.

CONCLUSIONS

Much attention has been devoted to understanding the causative factors for habitat segregation in ungulates. It is becoming increasingly apparent, however, that habitat segregation is influenced by multiple behavioural and environmental variables. Consequently, it seems unlikely that a single, proximate explanation for habitat segregation in ungulates will be found that works universally among species and populations. It seems more likely that this complex behavioural pattern is best explained at the ultimate level of sex differences in reproductive strategies – female behaviours are adapted for raising offspring successfully and male behaviours are adapted for maximizing the energy reserves needed to compete for mates. We suggest, therefore, that the reproductive strategy hypothesis provides a robust aid to understanding habitat segregation in ungulates, and should be further investigated for other taxa.

ACKNOWLEDGMENTS

We thank the many researchers who have contributed their efforts, thoughts and interpretations to the study of sexual segregation in ungulates. We also thank the editors, Drs K. Ruckstuhl and P. Neuhaus for their efforts to organize and produce this book. This research was supported by the Florida Agricultural Experiment Station and approved for publication as Journal Series No R-10184.

Part V Sex-related activities and social factors

10

Activity asynchrony and social segregation

OVERVIEW

This chapter deals with sexual differences in activity budgets and movement rates as a proximate explanation for social segregation in ungulates. Most research on sexual segregation has been on ruminant species
where males are considerably larger than females. We will first discuss implications of body size dimorphism on the foraging ecology and
digestive abilities of the two sexes, then illustrate how sexual differences in digestive efficiencies affect activity patterns and movement
rates, and how these differences could lead to behavioural asynchrony
in mixed-sex groups. We will discuss problems of synchrony by looking
at the ontogeny of behaviour and group choice in subadult ungulate
males. This comparison will highlight the potential trade-off faced by
certain individuals between behavioural synchrony to maintain group
cohesion and following their optimal activity budgets. Finally, we will
argue that males and females of size-dimorphic ungulate species segregate socially because an asynchrony in optimal activity budgets leads
to high costs to maintain group cohesion, and that activity asynchrony
could be the ultimate explanation for social segregation in ruminants
and possibly other vertebrates as well.

SEXUAL SIZE DIMORPHISM AND FORAGING ECOLOGY

Sexually monomorphic ungulates usually live solitarily (i.e. lesser
mouse deer, *Tragulus javanicus*; Nowak, 1991), in pairs (i.e. steenbok,
Raphicerus campestris; Estes, 1991a) or in mixed-sex groups (i.e. oryx, *Oryx
gazella*; Estes, 1991a), while sexually dimorphic species in general live

Sexual Segregation in Vertebrates: Ecology of the Two Sexes, eds. K. E. Ruckstuhl and P. Neuhaus.
Published by Cambridge University Press. © Cambridge University Press 2005.

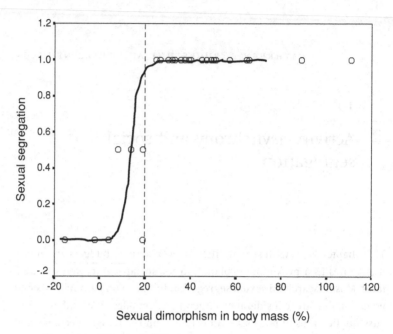

Figure 10.1 The relationship between sexual body weight dimorphism and sexual segregation in social ruminants. $x = 0$ means that no segregation occurs, $x = 0.5$ represents species that sometimes live in mixed-sex groups, temporarily segregated or sometimes solitary, $x = 1$ represents segregating species. The bold dashed line represents the 20% threshold above which segregation is expected, according to the assumptions for the activity budget hypothesis. Figure from Ruckstuhl & Neuhaus, 2002.

in sexually segregated groups outside the breeding season (Ruckstuhl & Neuhaus, 2002). Ruckstuhl and Neuhaus (2002) compared over 30 species of ruminants and hind-gut fermenters with different levels of sexual dimorphism and showed that for ruminants the threshold for social segregation is at a sexual dimorphism in body mass of about 20% (Fig. 10.1). Why is this so?

Ruminants have a very specialized digestive system, which allows them, with the help of their ruminal fauna, to digest forage more efficiently than hind-gut fermenters, such as horses, *Equus caballus* (Demment & Van Soest, 1985). Horses pass forage through their gut quickly, allowing a much higher intake rate than ruminants, but at the expense of lower extraction of nutrients from the forage. In ruminants, digestive efficiency is increased because forage is masticated and regurgitated repeatedly. Digestion, passage rate and turn-over of food depend on many variables, including the quantity and quality of food consumed, the incisor breath, rumen size, mastication effort (Pérez-Barbería & Gordon, 1998b) and microbial rumen populations (Hudson & White, 1985). These factors dictate the amount of food that can be ingested and processed per day (Hungate, 1975; Hanley, 1982; Hudson, 1985).

In many ungulates, males are considerably larger than females, affecting individual metabolic rates, digestive efficiency and foraging behaviour (Weckerly, 1993). Metabolic rate can be represented inter-specifically as an allometric function of body weight:

$$Y = aW^{0.75}$$

where Y is the metabolic rate, a is a constant and W is the body weight.

Metabolic rate per unit body weight therefore decreases with increasing body weight. Metabolic rates are allometrically related to body weight, while rumen volume and gut capacity are isometric with body weight. This means that while relative metabolic rate decreases with increasing body weight, gut capacity remains a constant fraction of body weight (Owen-Smith, 1988). Therefore larger ungulates can survive on poorer food quality than smaller ones (Jarman, 1974; Gordon & Illius, 1988), see Chapter 3 for a more detailed discussion. A large ruminant has a large rumen and a slower rumen turnover ratio, the food stays longer in its stomach and is therefore digested more efficiently than in a small ruminant, with a quicker rumen turnover and passage rate (Bunnell & Gillingham, 1985; Demment & Van Soest, 1985; Hofmann, 1989; Illius & Gordon, 1992). Body size therefore not only determines rumen size, but also food passage. Differences in rumen size should hence lead to differences in time spent foraging and ruminating/digesting food, and in the lengths of these activity bouts (Hudson, 1985; Berger & Cunningham, 1988; Schmid-Nielson, 1989; Illius & Gordon, 1992; Van Soest, 1994; Van Soest, 1996).

While studies on effects of body size dimorphism have been done looking at large scale differences between species (i.e. with respect to time budgets; Owen-Smith, 1988), it is difficult to estimate how big a sexual dimorphism in body mass/size is needed to lead eventually to marked intraspecific differences in digestive efficiency (Hudson, 1985). Van Soest (1994) estimated from physiological measures that a body size difference of at least 20% is required to significantly affect digestive capacities and to trigger a significant change in the foraging behaviour of the two sexes. Illius and Gordon (1987) also argued that, all else being equal, a sexual weight difference of over 20% could lead to competitive exclusion of the larger males by females on short swards (see Chapter 3 for the scramble competition hypothesis, but see also Chapter 2).

Accordingly, Ruckstuhl and Neuhaus (2002) found a similar level of dimorphism to affect social organization in ruminants (Fig. 10.1). Once a threshold value of 20% dimorphism is reached, social segregation, outside the breeding season, is the rule in ruminants. The degree of social segregation might actually increase gradually with an increase in size dimorphism, but data on the degree of social or habitat segregation between the sexes are lacking for most species to test this (Conradt, 1998a; see also Chapter 2 for a method on how to measure it).

THE ACTIVITY BUDGET HYPOTHESIS

Conradt (1998a) and Ruckstuhl (1998) independently proposed a new hypothesis explaining social segregation in ungulates, later termed the activity budget hypothesis (Ruckstuhl & Neuhaus, 2000). The rationale for the hypothesis is as follows: if males and females differ considerably in the time they spend foraging, moving or ruminating (due to differences in body size, energy requirements and digestive abilities), they will almost inevitably segregate over time, even if they started off in the same group. A social unit can only be cohesive if members synchronize their activities (i.e. foraging, resting and moving) with each other (Jarman, 1974; Benham, 1982; Rook & Penning, 1991; Conradt, 1998a; Ruckstuhl, 1998; Ruckstuhl, 1999; Conradt & Roper, 2003). In mixed-sex groups, where individuals differ substantially in their optimal activity budgets, individuals might have to compromise their own optimal behaviour for the sake of group cohesion. Adjusting one's behaviour to that of others may be costly due to higher energy expenditure, lower intake levels, or increased levels of stress. These potential costs could affect an individual's decision to stay in a group, thus making groups more or less stable, depending on the differences in body sizes of all group members (Ruckstuhl, 1999; Conradt & Roper, 2000). It would also lead to a tendency of similar-sized individuals to aggregate.

In a study on Rocky Mountain bighorn sheep, *Ovis canadensis*, Ruckstuhl (1998) found that females, which are about 50% smaller than males, spent much more time grazing and less time ruminating, that they travelled over longer distances, and spent more of their daytime walking than males. Females were not more selective while foraging, and did not have a higher step rate, but their bite rates were significantly higher than those of the males (Ruckstuhl, 1998). These findings led her to conclude that large differences in activity budgets and movement rates between males and females were the likely cause of social segregation in this species. That body size differences affect time spent foraging and walking in males and females was further confirmed in an interspecific comparison of activity budgets of over 30 species of ruminants (Ruckstuhl & Neuhaus, 2002). In sexually dimorphic species females spent more time foraging per day, and in several species there was a significant difference in movement rates between males and females. Ruckstuhl and Neuhaus (2002) found that in most species females had higher movement rates than males, with the exception of, for example, alpine ibex, *Capra ibex*, where females moved much less than adult males (Ruckstuhl & Neuhaus, 2001; Neuhaus & Ruckstuhl, 2002b). Sexual differences in time spent walking or distance covered per day may therefore vary between species, but these sexual differences increase with an increase in sexual size dimorphism, independent of which sex moves more.

The strongest support of the activity budget hypothesis was the observation that sexual differences in activity budgets increased with sexual size dimorphism. The bigger the males in comparison to females the lesser the proportion of daytime they spent foraging, the smaller the sexual dimorphism in body size, the more similar were male and female activity budgets (Fig. 10.2; Ruckstuhl & Neuhaus, 2002). Sexual dimorphism in body size therefore is not only an important determinant of forage selection (Chapter 3), digestive ability or predator avoidance (Chapter 9) but also strongly affects activity budgets. Time spent ruminating might be a crucial limiting factor, however, most studies only report time spent lying, not distinguishing between rumination and resting. It would be well worthwhile to investigate the relationship between time spent ruminating and body mass to find out whether sexual differences in time spent ruminating also increase with an increase in sexual size dimorphism.

ACTIVITY BUDGETS AND BEHAVIOURAL SYNCHRONY

The studies described earlier suggest that if activity budgets in ungulates differ substantially according to sex, males and females will become separated over time if they do not adjust their behaviour to each other. To test this assumption Ruckstuhl and Kokko (2002) modelled how sexual differences in the propensity to switch from an active (feeding and walking) to a passive state (lying and ruminating) alone affected group composition and social segregation. Although their model did not specify any preference for individuals of similar sex to group, it predicted that males and females were increasingly likely to segregate

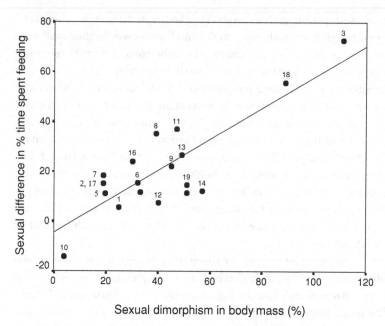

Figure 10.2 Relationship between increasing sexual body-mass dimorphism (in %) and sexual differences in % of time spent feeding in social ruminants. As sexual differences in body-size dimorphism increase, males feed increasingly less than females. Each point represents one species: (1) *Aepyceros melampus*, (2) *Antilocapra americana*, (3) *Capra ibex*, (4) *Cervus elaphus*, (5) *Connochaetes taurinus*, (6) *Damaliscus dorcas*, (7) *Gazella thomsoni*, (8) *Kobus ellipsiprymnus*, (9) *Litocranius walleri*, (10) *Oryx gazella*, (11) *Ovibos moschatus*, (12) *Ovis aries*, (13) *Ovis canadensis*, (14) *Ovis dalli dalli*, (15) *Ovis musimon*, (16) *Rupicapra pyrenaica parva*, (17) *Syncerus cafer*, (18) *Tragelaphus spekei selousi*, (19) *Tragelaphus strepsiceros*. Figure from Ruckstuhl & Neuhaus, 2002.

with increasing differences in their activity levels (Fig. 10.3; Ruckstuhl & Kokko, 2002). Animals were programmed to move randomly in space, and if there was an active animal close by (at a minimal distance to the moving animal) to move towards that animal at a specific angle. The moving animal was not more attracted to its own sex than to the opposite sex, but as males were programmed to be less active than females, moving animals were more likely to be and to encounter other females than males. Indeed, they found that the bigger the difference in the sexes' propensity to switch from being active to passive (or vice versa) was, the more they became socially segregated over time. Ruckstuhl and Kokko's model is very simple, but requires big sexual differences in activities for social segregation to occur. Improving the model to

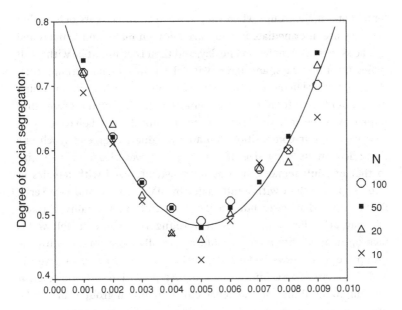

Figure 10.3 The effect of the males' (or females') propensity to switch from an inactive to an active state and vice versa at different population levels (N) on degree of social segregation observed after repeatedly running the model for 100 times. N = the number of males and females in our virtual population; N = 10 means that there were 5 females and 5 males in the population whereas 100 stands for 50 males and 50 females. The propensity (p) for females to switch from an unmotivated to a motivated state was set to be inversely related to that of the males. Figure adapted from Ruckstuhl & Kokko, 2002.

incorporate other relevant factors such as sexual differences in movement rates, or time spent moving, which can be observed in several ungulate species (see Ruckstuhl & Neuhaus, 2002). Such additional factors would likely lead to a high degree of social segregation at much smaller sexual differences in time spent active (see Yearsley & Pérez-Barbería, 2005 for an improved model and detailed discussion of Ruckstuhl & Kokko's 2002 model).

YOUNG MALES, GROUP CHOICE AND COSTS OF BEHAVIOURAL SYNCHRONY

To find out to what extent individuals synchronize their activity budgets to other group members, Ruckstuhl (1999) observed the behaviour of young bighorn rams. Young rams start switching between female and male groups when they are two or three years old. At that age they are intermediate in body size between males and females and can be seen with females on one day, and then they might be with adult males for a few days, and later switch back to a female group. Ruckstuhl (1999) predicted that to maintain contact with a group, young rams would have to adjust their behaviour to that of the group they joined. As expected, young males synchronized their behaviour irrespective of group composition, but an individual's degree of synchrony depended on its body size. If a young ram was closer in body size to that of adult females, he was more synchronized with females in female groups than with adult males in male groups, and vice versa. Being larger than ewes but smaller than fully grown rams, a trade-off may arise for young rams. If a young ram chose to follow his own optimal activity level he could potentially loose contact with the group (i.e. if he stays bedded while the group moves, or vice versa), and therefore expose himself to a higher predation risk. However, if he compromised his optimal behaviour to maintain group cohesion he may pay an energetic cost. Is there a way out for these young males? Interestingly enough there seems to be a third option for young rams. A study on population dynamics in the same bighorn sheep population revealed that young bighorn rams prefer to form groups of their own, provided enough similar-aged peers are available (Ruckstuhl & Festa-Bianchet, 2001). Young rams may hence be best off in a group of similar aged peers, where synchrony in behaviour would likely be high and least costly. Subadult male groups did only occur in high population density years when there were enough similar aged peers available. They were not found in years with only a couple of young rams (Ruckstuhl & Festa-Bianchet, 2001), which indicates that group size is a determining factor in group formation. Segregation by age has also been found in other ungulates, such as for example mouflon, *Ovis gmelini*, or ibex, *Capra ibex* (Cransac et al., 1998; Bon et al., 2001).

Body mass changes, or increased stress levels will have to be measured in young rams and other segregating social species to study costs of behavioural synchrony and to assess potential long-term impacts on an individual's performance. If an individual were to synchronize its behaviour for a longer time-period to others that display dissimilar optimal activity budgets (compared with its own), we would expect the focal individual to be more stressed, loose body mass and therefore be in poorer condition to reproduce or survive than individuals that are in body-size- and activity-budget-matched groups. A study on African buffalo, *Syncerus cafer*, reported that adult bulls lose

condition when in mixed-sex groups and gain condition when in bachelor groups (Prins, 1989). Prins (1989) could not attribute this loss in
body condition to either competition over food, or increased fighting
between bulls. Costs of behavioural synchrony could potentially explain
a loss in condition of adult bulls, as males had very similar time budgets to cows in mixed-sex groups (Prins & Iason, 1989), despite their
larger body size (19% heavier than females; see Ruckstuhl & Neuhaus,
2002). Experiments on ungulates, to test costs of behavioural synchrony,
have not yet been performed and would be difficult to carry out.

Shoaling fish could be the ideal study species to investigate the
cost of behavioural synchrony experimentally. Fish prefer to shoal with
other fish of similar size (Pitcher et al., 1985; Krause & Godin, 1994;
Krause & Godin, 1996b; Krause et al., 1996; Peuhkuri, 1999). A study
on sticklebacks, Gasterosteus aculeatus, has shown that synchronizing
behaviour can be costly (Ruckstuhl & Manica, unpublished). Fish were
put in an aquarium with several conspecifics of either the same or different body size (all shoal fish were either smaller or all were larger than
the focal fish). Dimorphism in fish length between the focal fish and
smaller shoal fish was around 40% and around 30% between the focal
fish and larger shoal fish. This range of dimorphism is well within sexual size dimorphisms found in ruminants (Fig. 10.1). Focal fish always
tightly synchronized their feeding rate and swimming speed to that
of the group, irrespective of its composition. The body mass of a focal
fish that was kept with either smaller or larger fish stayed the same or
declined over a three-day period, but it gained mass if it was with fish of
similar body size to its own. Synchrony in swimming speed is crucial for
an individual fish to maintain contact with the shoal. Failure to do so
may expose an individual to higher predation risk (see also Chapter 7
on the advantages of shoaling). This is the first study to show that
synchronizing behaviour carries a potential cost for animals that differ in body size from the majority of the other group members. Preference for size-assortativeness in fish might therefore be explained
as a means to minimize costs of behavioural synchrony in swimming speeds or feeding rates. More studies are needed to confirm this
hypothesis.

EFFECTS OF REPRODUCTIVE STATUS ON GROUPING

Given that social segregation occurs according to sex, age and body
size, it is reasonable to predict that an individual's reproductive status may also affect its group choice, as differences in reproductive

status likely lead to different energetic needs and different optimal activity budgets. Ruckstuhl and Neuhaus (2002) reported consistent differences in time spent feeding between lactating and non-lactating females of several ruminant species, with lactating females always spending more time grazing than females without young. There is some evidence for segregation according to reproductive status for ungulates, such as, for example, bighorn sheep (Festa-Bianchet, 1988), giraffe, *Giraffa camelopardalis* (Ginnett & Demment, 1999) and moose, *Alces alces* (Miquelle *et al.*, 1992). But in these cases, segregation between females of different reproductive states is usually explained as being caused by differences in forage selection or predator avoidance. Bighorn ewes, for example, show seasonal migration in spring and fall from their wintering range to lambing areas and back (Festa-Bianchet, 1988; Ruckstuhl & Festa-Bianchet, 1998). Festa-Bianchet (1988) suggested that females might go to alpine areas for lambing to avoid predation, however, by doing so they leave better feeding areas. Non-lactating ewes, or ewes who will lamb at a later date, should refrain from, or postpone, migration and exploit the abundant, better forage available at lower elevations. Indeed, observations indicate that non-parturient females of this population remain in their winter range for a few weeks longer than parturient ewes. Females whose lambs die soon after birth usually return to the low-elevation range one or two months before lactating females. In giraffe, females with young preferred more open areas to scan for predators, while non-reproducing females were found in denser habitat (Ginnett & Demment, 1999). In Alaskan moose, females with young used bushier areas than females without offspring (Miquelle *et al.*, 1992). Studies of segregation by reproductive state in species where females are using the same habitats are lacking, and the activity budget hypothesis can, therefore, not be appropriately tested.

Although social segregation by reproductive state (Miquelle *et al.*, 1992; Main *et al.*, 1996; Bleich *et al.*, 1997) has been suggested for several ungulate species, experiments are needed to test this hypothesis. In a study on Soay sheep, *Ovis aries*, on the Isle of Hirta, randomly selected male lambs were castrated to investigate the effects of reproduction on male survival and longevity (Jewell, 1997). According to osteology data analysed from captive Soay sheep and from the Hirta Island population, adult castrates were much bigger than entire males. Castrates had up to 5% longer femurs (indication of body size) than normal males and 11% longer femurs than females. Their mandibles were up to 12% longer than in either entire males or females (Clutton-Brock *et al.*, 1990). Ruckstuhl *et al.* (unpublished data) used this castration experiment to study how an alteration in reproductive state affects social segregation

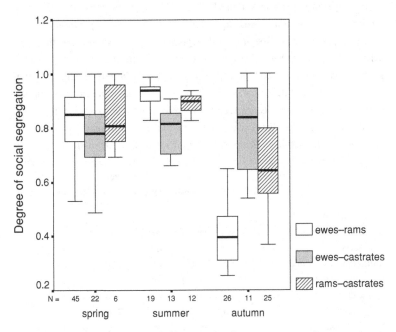

Figure 10.4 Degree of social segregation between ewes and rams, ewes and castrates, and rams and castrates according to seasons (spring, summer and autumn). Averages are displayed with quartiles and 95% confidence interval of the mean. N = numbers of censuses. Figure from Ruckstuhl, *et al.*, 2005.

and habitat use in this species. Interestingly, male castrates segregated from entire rams and formed groups of their own, although they used the same habitat types as regular males and females (Fig. 10.4). Information regarding differences in activity budgets between castrates, regular males and females is restricted to the pre-rut and rutting period (Jewell, 1997), at which time the different classes of sheep differed considerably in time spent grazing. We, however, do not know whether there were any differences in activity levels between castrates and other sheep during the rest of the year. It might be worthwhile manipulating reproductive state in males and females to investigate effects of reproductive status on activity budgets, forage selection, predator avoidance and social or habitat segregation.

ACTIVITY ASYNCHRONY AS KEY FACTOR IN THE EVOLUTION OF SOCIAL SEGREGATION?

Can incompatibilities in activity budgets between the sexes explain social segregation in ungulates? To address this question, Ruckstuhl

Table 10.1 *List of different studies with data on sexual differences in forage selection, predation risk or activity budgets. FSH = whether yes or no the forage selection hypothesis was supported by the study.* pr = *predation risk, PRH = support for the predation risk hypothesis, ABH = support for the activity budget hypothesis, graze (%) = the sexual difference in % time spent grazing (see methods). Food quality = if* m > f, *food in male habitat is better than in female habitat,* = *same quality,* m < f = *females are in better quality habitat. The same symbols are used for the predation risk.* > *means higher,* < *means lower,* = *no difference.* <> *means that both outcomes have been observed. See Ruckstuhl & Neuhaus (2002) for references to each study.*

	pr	PRH	Food quality	FSH	Graze (%)	ABH	Sexual segregation
Ruminant species							
Aepyceros melampus	m > f	yes	m < f	yes	5.49	no	1
Alces alces	m > f	yes	m = f	no			1
Antilocapra americana	m = f	no	m = f	no	18.30	yes	0
Bison bison athabascae	m = f	no	m = f	no			1
Capra ibex	m > f	yes	m < f	yes	72.73	yes	1
Capreolus capreolus	m = f	no	m < f	yes	−7.46	yes	0.5
Cervus elaphus			m <> f	no	11.63	yes	1
Cervus nippon nippon			m > f	no			1
Connochaetes gnou					−6.70	yes	0.5
Damaliscus dorcas	m = f	no			15.39	yes	1
Gazella thomsoni	m = f	no	m ≤ f	no	15.28	yes	0.5
Giraffa camelopardalis	m > f	yes	m < f	yes	23.29	yes	1
Kobus ellipsiprymnus	m < f	yes	m = f	no	35.23	yes	1
Litocranius walleri					22.06	yes	1
Madoqua kirkii	m = f	no	m = f	no	20.00	yes	0
Odocoileus hemionus	m <> f	no	m = f	no			1
Odocoileus virginianus			m < f	yes			1
Oreotragus oreotragus	m = f	no	m = f	no	25.30	yes	0
Oryx gazella	m = f	no	m = f	no	−14.00	yes	0
Ovibos moschatus					37.14	yes	1
Ovis aries (Soay sheep)			m = f	no	7.50	yes	1
Ovis canadensis	m > f	yes	m = f	no	26.66	yes	1
Ovis dalli dalli	m > f	no	m = f	no	12.32	yes	1
Ovis musimon			m = f	no	11.39	yes	1
Rangifer tarandus	m > f	yes	m > f	no			1
Rupicapra pyrenaica parva			m < f	yes	24.10	yes	1
Syncerus cafer	m > f	no	m = f	no	18.32	yes	0.5
Tragelaphus spekei selousi	m = f	no			55.85	yes	1
Tragelaphus strepsiceros					14.74	yes	1

and Neuhaus (2002) examined data on activity budgets, predator avoidance and forage selection in over 30 species of ruminants. They found that sexual differences in activity budgets were a more likely explanation for why the sexes segregate than differences in forage selection or predation risk. They found strong evidence to support the activity budget hypothesis, but conflicting evidence for either the forage selection or predation risk hypotheses (Table 10.1).

In their model Ruckstuhl and Neuhaus (2002) showed that, while asynchrony in activity budgets likely is the main cause of social segregation in size-dimorphic ruminants, sexual differences in forage selection and predation risk could indirectly increase social segregation and may, if sufficiently large, lead to segregation by habitat on top of the initial social segregation (Fig. 10.5). This model potentially explains how sexual segregation evolved in ungulates and reveals how social segregation could occur in the absence of habitat segregation or predator avoidance. Most researchers studying sexual segregation in ungulates had assumed that social segregation was a consequence of habitat segregation. However, a study on red deer and Soay sheep by Conradt (1999) supports Ruckstuhl and Neuhaus' (2002) model that social segregation occurs independently of habitat segregation (see also Chapter 2).

OUTLOOK

In this chapter we have presented a possible explanation for social segregation in ruminants, but can asynchrony in activity budgets also explain social segregation in other vertebrates? We know very little about the effects of reproductive status, sex and age on activity budgets and social segregation, and a lot needs to be done before we can be certain that there is a direct link between them. The idea that differences in activity levels can lead to social segregation is an appealing one, but it will need to be tested experimentally (in the field or in the laboratory), and empirically by comparing different animal classes within and across populations. Unfortunately, it is difficult to work experimentally with large mammals in the field, however, shoaling fish (as shown in this chapter), and to some extent domestic ungulates (see Michelena et al., 2004) might serve as model species to test the activity budget hypothesis and the costs of synchrony hypothesis more rigorously. Questions to be addressed with regard to the activity budget hypothesis are as follows:

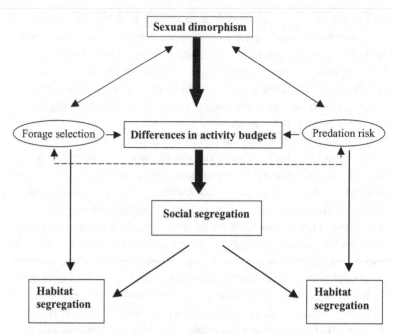

Figure 10.5 Model of proposed effects of sexual body weight dimorphism on sexual differences in activity budgets and ultimately social and habitat segregation. Two-way arrows indicate a correlation between different variables; one-way arrows indicate a hypothetical causal relationship. Figure adapted from Ruckstuhl & Neuhaus, 2002.

(1) How costly is synchrony in behaviour for an individual and does it have any short-term or long-term consequences such as mass loss or reduced reproductive performance?

Ruckstuhl (1998) proposed that synchrony in activities is potentially costly if an individual differs in its optimal activity levels from other group members. Only one study on sticklebacks (Ruckstuhl & Manica, unpublished) has so far investigated behavioural synchrony and associated short-term costs (see also in this chapter). Reduction in body condition might lead to lower survival or reproductive success. Sticklebacks would be an ideal study organism to investigate short- and long-term costs associated with behavioural synchrony, because males become territorial during the breeding season, with males in good condition likely being better at attracting females, holding and defending a territory, and caring for their offspring, than males in poor body condition.

(2)   What are the short-term and long-term costs of group choice and behavioural synchrony in young, growing individuals (i.e. young bighorn rams) that do not find an optimal group (their own body size) to associate with?

As explained in this chapter, young subadult male ungulates may face a trade-off between following their own optimal time budgets and adjusting their behaviour to that of the group. This might explain why young bighorn rams, for example, switched between male and female groups before settling down in male groups (Ruckstuhl, 1998, 1999). Their decision on whether to stay in female groups further or to switch to male groups, and the need for behavioural synchrony in both group types, might be a crucial one, and affect their body condition and development (body growth). If they make a sub-optimal choice they could be in poorer condition, grow slower and be inferior competitors in the future, than individuals that made a switch at the optimal time. Domestic sheep could be used to investigate these potential costs by rearing young rams in different group types.

(3)   How are decisions reached within groups of animals?

We only know little on how decisions are reached in groups and who decides when and where to forage (Conradt & Roper, 2003; Rands et al., 2003). If it is a subset of individuals within a group or a single individual who reaches such a decision, then animals that are under-represented might pay a higher cost of staying in such a group, than animals that have similar needs and activity levels to the decision-makers. If it is a democratic process, as proposed by Conradt and Roper (2003) then costs will be moderated for all animals in the group. Shoaling fish species and domestic ungulates would be good study organisms for an experimental approach to these questions.

ACKNOWLEDGEMENTS

This work was supported by an NSERC post-doctoral fellowship and NSERC Discovery Grant, an SNF Marie Curie PDF and a Swiss Academy of Science travel grant to K. E. Ruckstuhl. We would like to thank J. Peréz-Barbería, Larissa Conradt and Marco Festa-Bianchet for comments and input on this chapter.

RICHARD BON, JEAN-LOUIS DENEUBOURG, JEAN-FRANÇOIS
GERARD AND PABLO MICHELENA

11

Sexual segregation in ungulates: from individual mechanisms to collective patterns

OVERVIEW

Sexual segregation is an integral aspect of the socio-spatial organization of ungulate populations. Very often, the social, spatial and ecological components have been confounded (Bon, 1992) and we have argued that it is necessary to define and distinguish between each of them (Bon & Campan, 1996; see also Chapter 2 by Larissa Conradt). In the present chapter, we point out that sexual segregation is a complex phenomenon that can be produced by distinct mechanisms. One of the main issues is to know whether segregation by habitats necessarily derives from sexual difference in habitat choice, or can derive from alternative causes, i.e. spatial and social mechanisms (see also Chapter 2). Habitat segregation implies heterogeneous habitat (Miquelle et al., 1992), which we assume not to be obligatory for social and spatial segregation to occur. We distinguish hypothetical mechanisms relevant only in a heterogeneous environment from those relevant in both heterogeneous and homogeneous environments. We focus on behavioural mechanisms that may generate social and spatial segregation/aggregation, and the problem of the scale at which segregation may occur. Finally, we suggest that segregation cannot only be considered as a result of individuals behaving independently of each another, but also as a result of interactions between individuals on a larger (population) scale.

Habitat versus social segregation

Miquelle et al. (1992) noted that differences in habitat selection often lead to sexual segregation and resource partitioning between the sexes.

Sexual Segregation in Vertebrates: Ecology of the Two Sexes, eds. K. E. Ruckstuhl and P. Neuhaus.
Published by Cambridge University Press. © Cambridge University Press 2005.

For many researchers, sexual segregation is the differential use of space by the sexes (Kie & Bowyer, 1999; Weckerly et al., 2001). This statement is motivated by the conviction that the sexes are segregated because of differences in habitat choice (see Habitat segregation, Chapter 2). However, social and habitat segregation are not necessarily linked: segregation can occur by using the same resources but at different times (Francisci et al., 1985; Jakimchuk et al., 1987) or by using different resources within the same areas (Staines et al., 1982; Bowyer, 1984) (see Chapter 2 and 3). In addition, it is difficult to decide whether segregation derives from habitat choices or is a consequence of differences in social or spatial behaviour in the wild (Shank, 1982; LaGory et al., 1991). The idea that sexual segregation is strictly determined by differences in habitat choice can be challenged by a point of semantic and by empirical evidence.

Understanding sexual segregation implies defining its meaning (Bon & Campan, 1996; Main et al., 1996). Segregation comes from the Latin segregare 'separate from the flock, isolate, divide' or the Greek σε 'apart from' and γρεξ 'herd, flock' which means separating, isolating an individual or a group from conspecifics (Chambers Encyclopedic English Dictionary, 1994; Grand Usuel Larousse, 1997). Thus, etymologically speaking, segregation refers to a socially motivated action, although this social component has been considered secondary by most authors. The phrase 'males and females live apart' can mean living in distinct groups, living in distinct areas, or living in distinct habitat types (Bon, 1998; Bon et al., 2001). While sexual segregation is most often considered as ecologically determined (Polis, 1984), we argue that it is relevant to recognize the social, spatial and habitat components/dimensions that sexual segregation may involve as well (Main & Coblentz, 1990; Bon, 1991; Weckerly, 1993; Miquelle et al., 1992; Bon & Campan, 1996; Conradt, 1999).

Sexual segregation is an outcome at a population level, resulting from several possible mechanisms (Fig. 11.1). It is therefore necessary to define the components of sexual segregation as objectively as possible, and without inferring from the supposed individual mechanisms (see Chapter 2). In this chapter we will mainly be concerned with social segregation and the mechanisms supposed to be involved in it.

We propose to define social segregation as the trend for individual animals to aggregate with animals or subjects belonging to the same social category, e.g. sex and age (Bon & Campan, 1996; see also Conradt, 1998b). Before developing hypotheses involving mechanisms that may generate social and spatial segregation, we state some basic conditions to consider gregariousness.

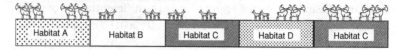

Figure 11.1 Theoretical population of dimorphic and social ungulates in which sexual segregation involves high levels of habitat, spatial and social segregation. (1) Habitat segregation: habitat A is only used by males, while habitat B is only used by females. (2) Spatial segregation: the females are located near the centre and the males at the periphery of the area occupied by the whole population; furthermore, habitat C is used by females or males according to the location of the corresponding patches. (3) Social segregation: single-sex groups are more frequent than expected by chance and this remains true within a portion of space supporting a single habitat and used by both sexes (habitat D patch).

It is important to recognize that animals could aggregate simply as a result of individuals of solitary species being attracted by the same environmental stimulus (feeding patches, refuge or migration corridors, for instance) without any social attraction. However, when the attractive environmental stimulus disappears, the groups will dissolve. Accordingly, all hypotheses discussed in this chapter implicitly assume that the species concerned are social, i.e. individual animals aggregate in more or less stable groups.

Some assumptions must be met for animals to aggregate in groups (see also Krause & Ruxton, 2002). Aggregation occurs by interattraction between mobile individuals, via visual contact for most of the wild ungulate species (see Gerard *et al.*, 2002), even though olfactory or auditory stimuli may also be involved (Barrette, 1991). In addition, interactions between individuals are necessary to keep group cohesiveness, which implies co-ordinated activities and thus allelo-mimetism (Deneubourg & Goss, 1989). See Box 11.1.

Animals are classically considered to associate at random (Grubb & Jewell, 1966; Geist, 1971; Langman, 1977; Hillman, 1987; Hinch *et al.*, 1990), with grouping depending on food distribution (Lott & Minta, 1983; Lawrence, 1990). However, ecological factors alone cannot account for phenotype assortment according to body size, sex, age or social status (Estes, 1991b; Bon *et al.*, 1993; Villaret & Bon, 1995, 1998; Cransac *et al.*, 1998). Social segregation between sexes is a particular case of social aggregation, as two categories of individuals are found together less often than expected if they were associated at random (Conradt, 1998b). Concerning the mechanism involved, this means that the sexes

Box 11.1 Social mechanisms conditioning group formation and cohesiveness

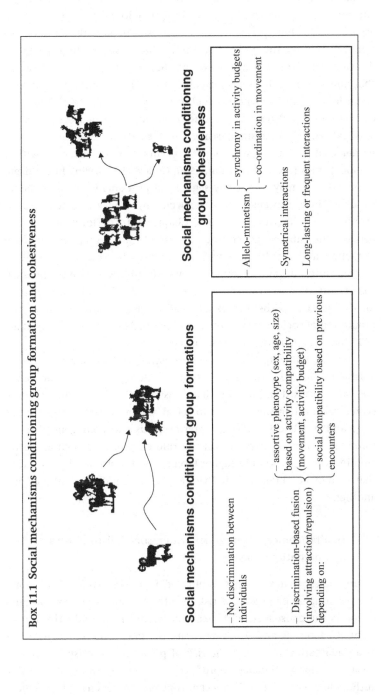

Social mechanisms conditioning group formations

- No discrimination between individuals

- Discrimination-based fusion (involving attraction/repulsion) depending on:
  - assortive phenotype (sex, age, size)
  - based on activity compatibility (movement, activity budget)
  - social compatibility based on previous encounters

Social mechanisms conditioning group cohesiveness

- Allelo-mimetism
  - synchrony in activity budgets
  - co-ordination in movement

- Symetrical interactions

- Long-lasting or frequent interactions

(social classes) can differ in the degree of social attraction to the opposite sex, move at a velocity or/and have activity rhythms that impair association for long periods of time (see also Chapter 10). Social attraction can rely on the activity of animals but also on the capacity to discriminate the sex or age of conspecifics. Experiments have revealed that domestic ungulates were capable of social discrimination between juveniles and adults (Kendrick *et al.*, 1995; Porter *et al.*, 2001). The sexes can be segregated on a large scale, involving high degrees of spatial segregation, or on a small scale. On a small scale, subgroups can be detected within larger groups as occurs between juveniles within female groups (Richard & Pépin, 1990; Gerard *et al.*, 1995). In children, girls and boys may be socially segregated at a smaller scale by being closer to their own gender than expected by chance in a school playground (see Pellegrini *et al.*, 2003; Chapter 12). The local interactions between neighbours generating this small-scale segregation include attraction and repulsion as well as mutual adjustment of activities. Furthermore, the sorting process may be facilitated by simple physical constraints such as crowding of one sex that reduces available space for the other sex. These rules can lead to group splitting and so the promotion of social segregation at a larger scale (see Deneubourg *et al.*, 1991 for an explanation in social insects).

In the following sections, we present some habitat-related mechanisms leading to sexual segregation in heterogeneous environments. We then forward arguments that illustrate the importance of non-ecological mechanisms leading to social and/or spatial segregation at large and small scales in both heterogeneous and homogeneous environment. Lastly, we suggest that unexplored processes such as 'social amplification' can produce higher level of habitat or spatial segregation that cannot be obtained if individuals behaved independently of one another.

Habitat choice, parental behaviour, social disturbances and predation risks

In ungulates, males are not implicated in raising young. The reproductive strategy or predation risk hypothesis states that sexual segregation is the consequence of sexual difference in reproductive investment (Chapters 3 and 9). Females' use of safe habitat is considered as an adaptation to reduce the risk of predation on offspring and as a way to improve females' reproductive success. However, only a few authors have considered the mechanisms promoting changes of female behaviour around parturition. The following examples will illustrate the proximate explanation of social and habitat segregation.

In mountain ungulates, females restrict themselves to steep slopes just before parturition, meanwhile non-parturient females and males use better feeding habitats (Shank, 1982; Bergerud et al., 1984). Spatial segregation is less marked when involving females without offspring in mountain and non-mountain areas (Festa-Bianchet, 1988; Miquelle et al., 1992; Bon et al., 1995; Ginnett & Demment, 1999). But the choice of habitat type depends on the local context, including prey and predation types (Ruckstuhl & Neuhaus, 2002). For instance, females with calves in Masai giraffes, Giraffa camelopardalis tippelskirchi, and kudu, Tragelaphus strepsiceros, avoid woodland habitats and use open habitats probably because higher visibility provides better risk detection (Du Toit, 1995; Ginnett & Demment, 1999). The choice of secure areas around parturition suggests that females are more sensitive to disturbance and predation risks at this period (Frid, 1999; Weckerly et al., 2001). In areas where large predators are scarce or absent, sexual segregation in habitat use vanishes (du Toit, 1995). Berger et al. (2001) showed that naïve female moose, Alces alces, experiencing predation on their calves were able very quickly to exhibit anti-predator behaviour. Kohlmann et al. (1996) reported that female Nubian ibex, Capra ibex nubiana, with young kids temporarily confined in a predator safe canyon, differ in habitat use from females followed by kids. The former move farther from escape terrain, use better feeding habitat and spent more time feeding than the latter.

However, there is also a much more proximal explanation why females search out particular areas. Increased habitat and spatial segregation is also the consequence of marked modification of parturient females' social behaviour (Poindron et al., 1988). A few days before parturition or when followed by neonates, females become asocial (Alexander et al., 1979; du Toit, 1995) and aggressive (Gosling, 1969; Cederlund, 1987; Estep et al., 1993) in several ungulate species. By withdrawing from groups, female taruca, Hippocamelus antisensis, near parturition segregated by habitat from male and mixed-sex groups commonly found year round (Merkt in Frid, 1999). Miquelle et al. (1992) reported that female moose with offspring seemed to avoid areas already used by other moose. Cliffs and forested areas provide physical obstacles disrupting visual contact and so can be chosen by parturient females as they allow seclusion from conspecifics of both sexes (Cransac et al., 1998) and animals of other species, including men and predators. Seclusion would facilitate the mother–young bonding (Poindron et al., 1988) and allow avoiding social perturbations or the possibility of adoption by other parturient females (Arnold *et al.*, 1975). In addition, other physiographic characteristics can be key factors involved in the selection of areas to give birth (see Bon *et al.*, 1995).

Social and habitat segregation is supposed to be determined by gestation and the presence of young. Thus, segregation might peak during the birth season. Behavioural changes around parturition, associability and aggressiveness, are not caused by habitat heterogeneity, although females can use it at that period to satisfy isolation, as discussed earlier. Thus, we can expect social and spatial segregation between parturient, lactating and non-lactating females or males in homogeneous environments. Because the maternal behaviour is hormonally induced, social and habitat segregation should vanish with the end of maternal care and the physiological weaning. If it persists outside the period of maternal care, it is necessary to consider other mechanisms than those invoked by the predation risks hypothesis. It remains also to be explained what causes females to venture farther from safe areas when offspring become older (Bergerud *et al.*, 1984; Bon *et al.*, 1995).

IS HABITAT SEGREGATION EQUAL TO SEXUAL DIFFERENCES IN HABITAT CHOICE?

Up to now, with a few exceptions, sexual segregation was mostly attributed to different habitat selection by the two sexes (see Bon & Campan, 1996). However, is habitat segregation necessarily caused by different habitat choices between the sexes? Some authors argued that habitat segregation might result from other mechanisms, such as, for example, social mechanisms (Shank, 1982; LaGory *et al.*, 1991; Bon, 1991).

Social mechanisms involved in segregation

Social segregation versus habitat segregation

Conradt (1998b, Chapter 2) proposed an index that allows measuring and comparing the degree of social, spatial and habitat segregation between the sexes. Social segregation here refers to the presence of males and females in single-sex groups, while spatial segregation refers to the use of exclusive quadrates by one sex. Using long-term studies on red deer, *Cervus elaphus*, and Soay sheep, *Ovis aries*, Conradt (1999) showed that the degree of social and spatial segregation was always higher than habitat segregation. This allowed the author to conclude that at least one part of social segregation cannot result from habitat segregation, and that each component was probably the result of different causes (Bon & Campan, 1996). Social segregation seems to be a rule in social dimorphic ungulates, and independent of the size of populations (Bon et al., 2001), density of males or females (Conradt et al., 1999b) and spatial segregation (Kie & Bowyer, 1999).

After a control of predators in an enclosed population of white-tailed deer, *Odocoileus virginianus*, Kie and Bowyer (1999) reported that spatial segregation decreased at high density of deer, whereas the level of social segregation was unchanged. As a consequence of higher spatial overlap between the sexes, dietary differences were lower than at moderate density and diet impoverished more in females than males. This is inconsistent with a prediction of the scramble competition hypothesis (see Chapter 2) that females actively or passively exclude adult males from preferred areas (Bleich et al., 1997; Romeo et al., 1997). In their study, Kie and Bowyer (1999) also rejected the social factors hypothesis, namely, that social segregation was driven from males avoiding costly social interactions linked to female proximity in mixed-sex groups. This hypothesis has been criticized because sexual interactions are dependent on sexual hormones that are produced seasonally, and so it is unlikely to apply outside the mating period (Main et al., 1996). However, this does not exclude the relevance of other social mechanisms.

Social segregation based on age or social status

More rarely considered, age is a factor that is implied in the degree of sexual segregation (Bon et al., 1993; Bon & Campan, 1996). Yearling males are most often observed in female groups while the oldest ones are rarely associated with females outside the rut (Nievergelt, 1967; Geist, 1971; Bon & Campan, 1989; Festa-Bianchet, 1991; Miquelle et al., 1992; Ruckstuhl, 1998; Ruckstuhl & Festa-Bianchet, 2001). Bon et al. (2001) found a gradient of social segregation linked to male age in Alpine ibex, *Capra ibex*, even when spatial segregation was low in winter, rendering ecological mechanisms a very unlikely cause of social segregation. Age difference was also found to be an important factor of social and spatial segregation among males splitting up into groups of similar-aged individuals (Bon et al., 1993; Villaret & Bon, 1995, 1998).

Social segregation within the sexes is not only observed as a function of age, but may also be dependent on events occurring early in ontogeny. Jewell (1986) reported that castrated Soay sheep males not only formed groups of their own but also used distinct home ranges. They socially and spatially segregated from entire males and females (see also Ruckstuhl *et al.*, submitted). It is likely that the lack of male hormones that is implicated in the male-like behaviour influenced the nature of interactions and levels of behaviour of early castrated males. As a consequence, these males set up a social network among themselves, which made them socially segregated from non-castrated males and females.

The assumption that aggregation is based on a general interattraction between conspecifics must be modulated because the force of attraction may vary during ontogeny. When ageing, individuals seem to be less sociable in some populations of European mouflons (*Ovis aries*), bighorn sheep, chamois (*Rupicapra rupicapra*) and isard (*R. pyrenaica*) (Pfeffer, 1967; Geist, 1971; Shank, 1985; Richard-Hansen, 1992; Hass & Jenni, 1993). According to Shank (1985: 122), social and spatial segregation of old male chamois is a trade-off between dependence on feeding resources and a 'need for solitude', reflecting social intolerance. However, old animals might more often be alone because they are less sociable or because they lack similar-aged peers (Villaret & Bon, 1998). When available, old non-reproductive buffaloes were reported to group together, apart from younger males (Sinclair, 1977). Population density and hence the probability of meeting conspecifics is also a factor contributing to the chance of both sexes to be found alone (Fig. 11.2; see also Gerard & Loisel, 1995 and Gerard *et al.*, 2002 for a discussion on mechanisms underlying aggregation).

BEHAVIOURAL MECHANISMS INDEPENDENT OF HABITAT HETEROGENEITY

Recently, new hypotheses were proposed, suggesting that sexual segregation may be explained by different mechanisms, including movement or spatial behaviour, activity budgets, and social behaviour (Bon & Campan, 1996; Conradt, 1998b; Ruckstuhl, 1998; Bon *et al.*, 2001). These hypotheses differ notably from previous ones in the sense that they do not depend upon habitat heterogeneity, and that they propose mechanisms that can produce social and eventually spatial segregation in heterogeneous but also in homogeneous environment. If social segregation is observed in controlled and homogeneous habitats, observers have either not detected an ecological heterogeneity that animals do detect, or non-ecological mechanisms are at work.

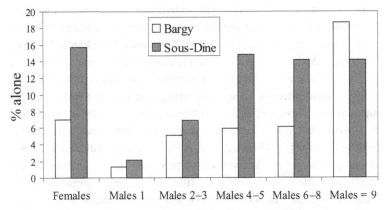

Figure 11.2 Proportion of observations corresponding to alone individuals during census of groups in two populations of Alpine ibex, Bargy and Sous-Dine, according to the sex and male age. Bargy population contained c. 120 adults while Sous-Dine population only 27 adults. Data from the mating and the birth periods were omitted because males are often alone during the rut and females less social at parturition. Males were found more often alone when getting older in both populations. Note, however, that the probability of being found alone in both sexes is higher in the smaller than the larger population. Numbers in brackets correspond to the total number of observations of groups (lone individuals included).

The activity budget hypothesis

Recently, it has been proposed that sexual dimorphism in body size could lead to sexual differences in activity budgets in ungulates (Conradt, 1998a; Ruckstuhl, 1998; Chapters 2 and 10). One basic prediction of the activity budget hypothesis is that, because of their smaller size, higher energy requirements and lower efficiency in processing forage, females would spend more time feeding than males (Ruckstuhl, 1998; Ruckstuhl & Neuhaus, 2000; see also Ruckstuhl & Neuhaus Chapter 10). The resulting asynchrony in activity is considered a major constraint for mixed-sex groups to be maintained, resulting in social segregation.

The activity budget hypothesis predicts segregation without implying differences in forage selection. It is theoretically possible to find social segregation without habitat/spatial segregation: both sexes can form separate groups, use overlapping ranges in which they can exploit the same habitat patches at different times (Francisci et al., 1985) or at the same time without mixing. The activity budget hypothesis assumes that to stay together, the individuals belonging to the same groups must share similar activity rhythms allowing activity synchrony. Allelo-mimetism is implicit to activity synchrony, i.e. when individuals in a group exhibit patterns of individual activity that would not occur if individuals were independent (Deneubourg & Goss, 1989). The hypothetical possibilities of animals meeting and having similar activity rhythms are illustrated in Fig. 11.3(a) and (b). Consider a population of individuals with two states, active and inactive, and independent from each other in their activity. A stable group will depend upon the probability of two individuals to be synchronized in order to stay together. If the activity budgets differ too much between both animals (as suggested to occur between females and males in dimorphic species), the probability of staying for a time longer than a bout of activity or inactivity is unlikely (Fig. 11.3(a)). If two animals having the same activity rhythm meet, the probability of associating for a lasting period will depend on both individuals being in the same phase (Fig. 11.3(b)). This probability would be higher for same-sex than for opposite-sex animals. However, synchrony in activity between the sexes is possible if males and females do not vary too much in activity budgets, and if at least one sex adjusts its activity rhythm to that of the other sex. Although we do not know how overall activity synchrony in a group is achieved, it can be assumed that having the same activity as surrounding animals can result in a high degree of overall synchrony. It is not necessary for individuals to adjust their behaviour to the entire group. In most ungulate species, groups are unstable in size and composition (Marchal *et al.*, 1998). However, data obtained from wild populations indicate that activity synchrony in single-sex groups is higher than expected by chance (Côté *et al.*, 1997; Ruckstuhl, 1999). This suggests that animals belonging to groups of the same sex are either synchronized by the same external releaser or possess the same internal clock. It is more parsimonious thinking that individuals with activity budgets not too different can tune their activity to each other through interactions (Fig. 11.3(c)), such as allelo-mimetism allowing to be in phase. Ramírez Ávila *et al.* (2003) have shown that interactions between oscillators that differ in their intrinsic period enable individuals to adopt the same period. This individual ability could lead to clustering of individuals having similar activity periods, and social segregation between individuals having dissimilar periods. Note that the possibility for individuals to aggregate based on similar activity period and interaction does not mean that all groups are synchronized.

(a) Individual animals with different rhythms

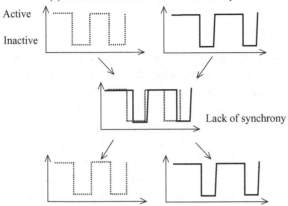

Lack of synchrony

(b) Individuals with the same but unsynchronized rhythm

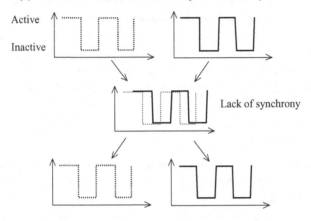

Lack of synchrony

(c) Individuals with similar rhythm, and which can synchronize activity to each other through interaction

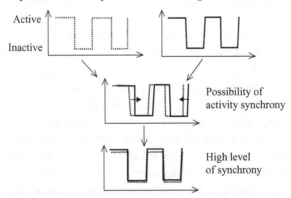

Possibility of activity synchrony

High level of synchrony

Figure 11.3 Hypothetical cases of meeting between two individuals and possibility of activity synchrony: (a) both individuals exhibit distinct or (b) similar activity rhythms without possibility of synchronization and (c) both individuals can synchronize their activity.

Dispersal behaviour

The impact of sexual differences in dispersal behaviour on social and spatial segregation has rarely been examined (Bon & Campan, 1996). In dimorphic ungulate species, as in many mammals, juvenile females usually settle in or near their maternal or natal range while males often disperse from their natal group or area (cervids: Clutton-Brock *et al.*, 1982; Bunnell & Harestad, 1983; Nelson & Mech, 1984; Cederlund *et al.*, 1987; Cederlund & Sand, 1992; Hölzenbein & Marchinton, 1992; wild sheep: Festa-Bianchet, 1986; Dubois *et al.*, 1994). Based upon long-term radio-tracking of mouflons, Dubois *et al.* (1993) found two categories of two- to three-year-old males regarding their dispersion outside the rut. Some males still used their natal range, while most of them gradually dispersed until using non-maternal and stable ranges. Even when males and females spatially overlapped, they were still socially segregated (Dubois *et al.*, 1993). The difference in spatial dispersion between males and females often results in males using a higher diversity of habitats (Ordway & Krausman, 1986; Villaret *et al.*, 1997).

Motion behaviour

Dimorphism in body size can also be at the origin of other differences in behaviour, such as motion. Larger or powerful individuals probably walk or move more rapidly than smaller individuals, or travel a larger distance per unit time. If so, group splitting and spatial segregation, or structuring at a small scale, can arise even if the individuals in the population, whatever their size, do not differ in their habitat choices (Gueron *et al.*, 1996; Couzin & Krause, 2003). Ruckstuhl (1998) found that bighorn ewes and rams had the same step rate per unit time, but ewes dedicated more time to walking, had longer walking bouts and travelled larger distances than rams. Miquelle *et al.* (1992) also reported that in moose females and subadult males moved more during the feeding periods than large males. On the other hand, Michelena *et al.* (2004) have shown experimentally that merino rams, when placed within the same pastures were walking twice as rapidly than ewes within the same pastures. However, no social segregation at a large scale was found since both sexes were together in a single group for several weeks. During the eight-week experiment, males were more often found in the front of the group than females. Same-sex pairs of nearest neighbours were significantly more frequent than mixed-sex ones, which might result from higher step rates in males than in females. However, when sheep of both sexes were distributed at random within the group, pairs of nearest neighbours of same sex still outnumbered opposite-sex pairs. These results suggest a high level of inter-sex attraction, explaining the lack of social segregation at a large scale, compensating for the difference in activity budget and moving velocity between the sexes, and higher intra-sex than inter-sex affinity accounting for the social segregation at a small scale.

The social affinity hypothesis

Mechanisms proposed to explain social segregation in children, and data collected on behavioural development and social interactions in other mammal species inspired the social affinity hypothesis (Bon & Campan, 1996; Bon et al., 2001). Jacklin and Maccoby (1978) suggested that the differences in the level of activity (Eaton & Enns, 1986) and in how both sexes get socially involved could lead to problems of social matching between girls and boys (see Pellegrini & Long, 2003 for a recent discussion). They proposed the notion of behavioural incompatibility that Legault and Strayer (1991) extended by defining it as 'a set of differences in the overall composition of behavioural repertoire' to account for social segregation between the sexes in children. Bon and Campan (1996) argued that sexual differences in behaviour and social motivation lead to behavioural and social incompatibility, and thus social segregation between different sex–age classes in social ungulates. Behavioural compatibility would be necessary for social cohesion to occur.

In mammals, sexual differences are found in levels of motor activity (Holekamp & Sherman, 1989) and type of behaviour (Cheney, 1978; Sachs & Harris, 1978; Moore, 1985; Meaney, 1988) from an early stage of life. In dimorphic ungulates, social behaviour and morphology mature more gradually in males than in females, long after reaching sexual maturity (Geist, 1968, 1971; Grubb, 1974; Jarman, 1983; Rothstein & Griswold, 1991; Shackleton, 1991). Juvenile males are more often engaged in rough-and-tumble or pseudo-sexual plays, while females are more often engaged in locomotor play and also spend more time in feeding activities than males (Bon & Campan, 1996). The difference between the sexes in the amount of interactions still persists into adulthood (Le Pendu et al., 2000). Owing to the differences of social motivation, behavioural style and morphology, females could avoid or be indifferent to male social interactions. Although mixed-sex groups of mouflons were infrequent, Le Pendu et al. (2000) found a high rate of inter-sex interactions, initiated by males over two years old when males and females co-occurred at attractive feeding sites. Males interact much more frequently than females in sheep, even when the latter are involved in the interactions (Michelena *et al.*, 2004). Several authors have argued that females avoid interacting with dominant males (see Bon & Campan, 1996). Weckerly *et al.* (2001) found that Roosevelt elk (*Cervus elaphus roosevelti*) females displayed slightly higher aggression rates in mixed-sex groups when males were more prevalent, possibly as a consequence of females approaching one another when avoiding males. Female and mixed-sex groups also walked away when approached by male groups exceeding six individuals. Because females avoided only large male groups, Weckerly *et al.* (2001) concluded that this social mechanism is unlikely to account for high degrees of social segregation.

From a physiological point of view, behavioural dimorphism in social behaviour and dispersal between the sexes is induced by perinatal androgens (Hinde, 1974; Goldfoot *et al.*, 1984; Moore, 1985; Meaney *et al.*, 1985; Holekamp & Sherman, 1989). For example, Jewell (1986, 1997) showed how castrated Soay lambs formed self-contained groups, avoided interacting with other sheep and used ranges distinct from ewes and rams as adults (see also Clutton-Brock *et al.*, 1982). These results indicate how physiological mechanisms and the type of behavioural style can affect social and spatial segregation. The social affinity hypothesis thus predicts that grouping will probably occur between animals of the same sex and age (Bon *et al.*, 2001). Yet, even if groups persist when individuals share the same motivation to associate, they may contain individuals with very dissimilar behaviour such as females and offspring. This is made possible because of the shared motivation to stay together and because individuals can carry out their maintenance activities within such groups.

SYNERGY BETWEEN DIFFERENT MECHANISMS

All populations occurring in the wild face a certain degree of heterogeneity in their habitat. How an individual animal chooses its home range will depend on some basic or vital requirements, but also on phenotypic constraints or cognitive abilities.

Habitat segregation is most often considered as the result of an active choice or compromising between conflicting factors. However, it is worth noting that an experimental design is necessary to ascertain that females and males differ in habitat choice. For instance, Morton (1990) demonstrated experimentally that habitat segregation observed in the hooded warbler (*Wilsonia citrina*) was founded on sexually distinct preferences of physical characteristics of habitat. However, Desrochers (1989) showed that male and female black-capped chickadees were segregated by habitat because males excluded females from preferred microhabitats. Such demonstrations remain scarce for ungulates. Pérez-Barbería and Gordon (1999) carried out an experiment with Soay sheep and showed that both sexes preferred high quality to low quality grazing patches. However, contrary to the predictions of the forage selection hypothesis, females spent more time foraging on the low quality swards than males.

The activity budget and the social affinity hypotheses do not exclude the contribution of mechanisms linked to reproduction or ecological factors to sexual segregation, but they state that differences in behaviour and social motivation are basic mechanisms of social segregation. Food quality and distribution, predation risks (Jarman, 1974) but also population density and habitat openness are causal factors of animal grouping (Barrette, 1991; Gerard et al., 1995, 2002). In the wild, it is difficult to set apart the impact of ecological factors from that of social factors and we argue that sexual segregation probably involves several mechanisms. The question of synergy or antagonism between different mechanisms, in particular social and ecological ones, is poorly documented (Bon & Campan, 1996). To illustrate the importance of this topic, we present a model where slight differences in habitat use between the sexes can be amplified by social attraction (see Appendix 11.1 for details of the model).

Finally, we would like to point out that experimental studies are needed to test non-ecological mechanisms in controlled habitat, with the underlying idea that a better knowledge of behavioural/cognitive mechanisms and interactions between individuals will provide new insight into aggregation and segregation dynamics. We also recommend considering the quantitative aspect and interplay of mechanisms, and the dynamics of social and spatial structures that are difficult to tackle if one only considers the individual's perspective.

APPENDIX 11.1

Let us consider a population of solitary animals (there are no interactions between individuals whatever their sex), composed of two

Table 11.1 *Probability of individual males or females moving between habitats A and B, depending on whether they are solitary or social.*

	Solitary		Social	
	Probability of moving from A to B	Probability of moving from B to A	Probability of moving from A to B	Probability of moving from B to A
Males	α_M	β_M	$\alpha_M/(1 + M_A)^*$	$\beta_M/(1 + M_B)^*$
Females	α_F	β_F	$\alpha_F/(1 + F_A)^*$	$\beta_F/(1 + F_B)^*$

* M_A and M_B (F_A and F_B) are respectively the total numbers of males (females) in habitats A and B. $M = M_A + M_B$ and $F = F_A + F_B$ are the total subpopulations of males and females.

subpopulations: males and females. The individuals can travel freely between two contiguous habitats A and B. F_A and F_B (M_A and M_B) are respectively the total numbers of females (males) in habitats A and B. $F = F_A + F_B$ ($M = M_A + M_B$) is the total subpopulation of females (males).

Solitary individuals

At each time-step, any female on habitat A or B has the probabilities α_f and β_f to move from A to B and from B to A, depending on the characteristics of habitats A and B (Table 11.1). Accordingly, the number of females moving from A to B is $\alpha_F F_A$ and from B to A is $\beta_F F_B$. At the equilibrium, the number of individual females moving from A to B equals that moving from B to A. This can be written:

$$\alpha_F F_A = \beta_F F_B \quad or \quad \alpha_F F_A = \beta_F (F - F_A)$$

It is then easy to find that:

$$F_A = \frac{\beta_F F}{(\alpha_F + \beta_F)}$$

If f_A is the proportion of females in A (F_A/F) and f_B their proportion in B (F_B/F) and we define $r_F = \alpha_F/\beta_F$, then:

$$f_A = \frac{1}{(r_F + 1)} \quad and \quad f_B = 1 - f_A$$

Similarly the fraction of males in habitat A is:

$$m_A = \frac{1}{(r_M + 1)} \quad and \quad m_B = 1 - m_A$$

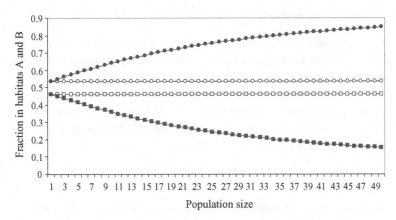

Figure 11.4 Theoretical fractions of animals in two distinct habitats, A (circles) and B (squares) as a function of population size F. White symbols correspond to solitary individuals, and black symbols to social individuals. For example, if solitary females have a small individual preference for habitat A, the proportion of females f_A in habitat A will remain to 0.54 and in habitat B f_B to 0.46. If males' preference for habitat B equals females' preference for habitat A, then the proportion of males will be such that $m_B = f_A$ and $m_A = f_B$. In the case of social species, with the same individual preference for habitat A than solitary animals, the proportion of females (males) will however increase with the female (male) population size. This figure shows how social interaction may amplify the initial individual habitat choice and, in the case of different individual habitat choice between the sexes, how the degree of habitat segregation increases with the population size of females and males.

Whatever the size of the female subpopulation, the proportion of females in habitats A and B at the equilibrium will depend on the relative preference of females for habitats A (α_F) and B (β_F) and so on the value taken by r_F. If there is no preference for any habitat in both sexes, then $r_F = r_M = 1$, and there are proportionately as many females and males in both habitats ($f_A = f_B = 0.5$ and $m_A = m_B = 0.5$) leading to a lack of habitat segregation between the sexes. If females prefer the habitat A and males prefer the habitat B, then $\alpha_F < \beta_F, \alpha_M > \beta_M, f_A > m_A$ and $f_B < m_B$. For example, consider the initial population is composed of individual females with a ratio $r_F = 0.86$, leading to a slight preference for A, then $f_A = 0.54$ and $f_B = 0.46$ (Fig. 11.4). The greater the difference between the ratio r_F and r_M, the greater will be the habitat segregation between the sexes.

Social individuals

Let us consider now the case of a social species composed of subpopulations of females and males of size F and M with an intra-sex attraction and no inter-sex attraction or repulsion. The probability of individual animals moving from habitat A to B (B to A) is the same as earlier but, in this case, as the number of same-sex individuals in the same habitat increases (e.g. A), the probability of leaving this habitat decreases for any individual (see Table 11.1; $n = 1$) as:

$$\frac{\alpha_F}{(1 + F_A)} \quad \text{and} \quad \frac{\beta_F}{(1 + F_B)}$$

The proportions of females in habitats A and B are also independent of what occurs for males, if there are no limits in space available (no indirect competition).

At the equilibrium, the number of individual females f_A moving from A to B equals the number f_B moving from B to A, so that:

$$\frac{\alpha_F F_A}{(1 + F_A)} = \frac{\beta_F F_B}{(1 + F_B)}$$

or:

$$\alpha_F F_A(1 + F_B) = \beta_F F_B(1 + F_A)$$
$$\alpha_F F_A - \beta_F F_B + F_A F_B(\alpha_F - \beta_F) = 0$$

As $F_B = F - F_A$, we obtain:

$$\alpha_F F_A - \beta_F(F - F_A) + F_A(F - F_A)(\alpha_F - \beta_F) = 0$$
$$(\alpha_F - \beta_F)F_A^2 - (\alpha_F + \beta_F + F(\alpha_F - \beta_F))F_A + \beta_F F = 0$$

Dividing by F^2, we obtain:

$$(\alpha_F - \beta_F)f_A^2 - \left(\frac{\alpha_F + \beta_F}{F} + (\alpha_F - \beta_F)\right) f_A + \frac{\beta_F}{F} = 0$$

If $\alpha_F = \beta_F$, or $r_f = 1$ then $f_A = f_B = 0.5$ whatever the total population F. If $\alpha_F < \beta_F$ or $r_f < 1$:

$$f_A = 0.5(D + \sqrt{D^2 - 4E}) \quad \text{and} \quad f_B = 1 - f_A$$

and if $\alpha_F > \beta_F$ or $r_f > 1$:

$$f_A = 0.5(D - \sqrt{D^2 - 4E}) \quad \text{and} \quad f_B = 1 - f_A$$

with:

$$D = 1 + \frac{r_F + 1}{(r_F - 1)F} \qquad E = \frac{1}{(r_F - 1)F}$$

Taking into account the intra-sex attraction and $r_F \neq 1$, the proportion of females f_A and f_B will evolve as a function of the total subpopulation F. For instance, if we set $r_F = 0.86$ (as in the case of solitary animals), that is to say a small preference of females for habitat A, we can see that $f_A > 0.54$ and grows with F (Fig. 11.4), which can be assimilated as an effect of amplification. For large value of F, all the females will be in habitat A ($f_A \approx 1$) and environment B is neglected. In contrast, if $r_F > 1$, we will observe an amplification of the preference for the environment B.

If the males have the same preference for habitats A and B, and the same degree of attraction among each other as females, the habitat segregation is nil. Besides, if males have a small and steady individual preference for habitat B ($\beta_M = \alpha_F$), then the proportion of males in habitat B increases with the number of males found in this habitat and the degree of habitat segregation between the sexes will increase with the increasing size of female and male subpopulations respectively in habitats A and B. The curves of f_A and f_B are symmetrical if $F = M$. This model shows how habitat segregation may be amplified by social interactions and the population size of females and males, despite no modification of individual habitat choice (r_F, and r_M constant).

It is also possible to show that the disequilibrium of proportions of females in habitats A and B can be theoretically obtained without initial differences of habitat preference, i.e when $r_F = 1$. This may occur if the probability for individual animals to move from A to B (B to A) are very sensitive to the number of individuals in the habitat A (B) and for example decreases with the square of the population (see Table 11.1, $n = 2$):

$$\frac{\alpha_F}{\left(1 + F_A^2\right)} \quad \text{and} \quad \frac{\beta_F}{\left(1 + F_B^2\right)}$$

In this case, most of the individuals are in habitat A ($f_A \approx 1$, $f_B \approx 0$) or in environment B ($f_A \approx 0$, $f_B \approx 1$). The selection of the habitat A or B is a random process and each environment has an equal probability to be selected.

ANTHONY D. PELLEGRINI, JEFFREY D. LONG
AND ELIZABETH A. MIZEREK

12

Sexual segregation in humans

OVERVIEW

As the chapters in this volume suggest, sexual segregation appears to be an important component in the social organization of many species. Indeed, some of the arguments made for sexual segregation among human children are similar to those made for sexual segregation in adults of other mammalian species (Ruckstuhl, 1998; Ruckstuhl & Neuhaus, 2002). For example, some authors suggest that male and female humans (Maccoby, 1998) and adult ungulates (Bon & Campan, 1996) segregate due to differences in behavioural 'styles'. From this view, males interact with each other because their behaviour is physically vigorous and rough. Females, being more sedentary, avoid this sort of behaviour.

In this chapter we examine two hypotheses ('energetics' and social roles) proffered to explain the existence of social sexual segregation in human juveniles. This involves documenting what has been labelled 'energetics' (e.g. Ruckstuhl, 1998) or physical activity (e.g. Maccoby, 1998; Pellegrini et al., 1998), of the behaviours and social roles in male and female juveniles in humans. We will argue that segregation is based, ultimately, on male and female reproductive roles and can elegantly be explained by sexual selection theory (Darwin, 1871). As in most animals, human males experience more reproduction variation, relative to females (Trivers, 1971). This leads to males being the larger, more active and more competitive sex because they must compete with each other for access to mates. The social roles enacted correspond to reproductive roles and there are corresponding

Sexual Segregation in Vertebrates: Ecology of the Two Sexes, eds. K. E. Ruckstuhl and P. Neuhaus.
Published by Cambridge University Press. © Cambridge University Press 2005.

differences in levels of physical activity. We also examine segregation during adulthood of interactions in the workplace and in other social groupings.

Lastly, this chapter aims to complement the current exhaustive human developmental psychological analysis of sexual segregation proffered by Maccoby (1998) with principles of sexual selection (Darwin, 1871; Clutton-Brock, 1983) and parental investment (Trivers, 1972) theories. Specifically, we will show how explanations proffered in the human developmental psychological literature can be integrated with these theories. To that end, we will present our model, based on these theories of sexual segregation in humans.

ONTOGENETIC DEVELOPMENT OF SEXUAL SEGREGATION

In this section we outline the degree to which human groups are segregated from the juvenile through the adult periods. As we will demonstrate, segregation is most persistent during the juvenile period, though it does continue into the adult period, with obvious exceptions being related to heterosexual relationships.

Segregation, especially during the juvenile period, for humans may be explained by sexual selection theory (Darwin, 1871). That is, during the juvenile period males and females segregate into same-sex groups to learn and practise skills consistent with their respective adult reproductive roles. As we will demonstrate later, males learn competitive and aggressive skills useful in securing and maintaining status in their juvenile and adult groups. Females, on the other hand, learn nurturing skills. It may be the case, however, that adult integration is a result of the wide variety of ecological niches that humans inhabit. For example, in severe ecological niches both males and females are needed to rear their offspring successfully (Alexander *et al.*, 1979) thus, both parents stay together and do not segregate into same sex groups. In less severe conditions there may be more segregation.

Human juveniles (i.e. children before they reach puberty) have been observed to spend most of their time with same sex peers, beginning at around three years of age (Pitcher & Schultz, 1983; Fagot, 1994). For example, Fagot (1994) observed preschool-age children in their classrooms and found that toddlers (18 months of age) began to interact with other-sex peers and by three years they spent most of their time in same-sex groups. These patterns continue through the remaining preschool years (three to five years; Fabes, 1994) and there are few differences in the degree to which boys or girls segregate themselves (Fabes, 1994).

Figure 12.1   Juvenile male segregated group.

Observations of children during the primary school years (c. five to twelve years) typically take place on the school playgrounds at break time, or recess as it is called in the USA, as illustrated in Figs. 12.1 and 12.2. The reason for this practice is that children of this age, unlike preschool-age children, are not given opportunity for peer interaction during 'teaching times'. Only during breaks is peer interaction sanctioned in schools. For example, in a study of English primary school children's playground behaviour, segregation was measured as presence in 'mostly' segregated (where 60% or more were of the same sex) or 'mixed' (where 40% or less were of the same sex) groups (Blatchford *et al.*, 2003); both boys and girls were observed in the 'mostly' segregated groups in 80% of the observations.

With the onset of puberty and as youngsters begin to approach adolescence they become more interested in heterosexual relationships, and the degree of sexual segregation begins to wane, but this seems to be sensitive to the norms associated with different social contexts. For example, when young adolescents (c. 12 to 14 years of age) move from primary to middle school we observed sexual segregation in the school halls (as they changed from one class to another) and in the lunch room. We found that the degree of segregation ranges from 0.92–0.77 (where

Figure 12.2   Juvenile female segregated group.

a male segregated group = #males/#males + #females) (Pellegrini & Long, 2003) and this proportion of segregation did not decrease across two years of observation.

However, when these same youngsters were observed at monthly school dances across the same two-year period there was a significant decrease in segregation in time (Pellegrini & Long, unpublished data). It is probably the case that the social norms of a dance, where heterosexual interaction is expected, motivated that change away from segregation. Interestingly, the degree to which there are statistically significant decreases in segregation across time varied, depending on our metric of segregation. When we used a proportional index of segregation (e.g. #males/#males + #females) there was a decrease but it was not statistical significant. When we used a dichotomous measure (that is, the group was either segregated or not) we found that the predicted odds ratio of total segregation decreased significantly, $\chi^2 = 12.57$, $p < 0.02$.

We also found a sex difference in the odds ratio where males were more likely to remain segregated than females $Z = -2.81$, $p < 0.004$.

These analyses suggest that groups do not integrate incrementally, as suggested by the lack of significant change in the proportional index. Instead, there seems to be rather abrupt integration, where groups of girls approach groups (not individuals) of boys to initiate cross-sex interaction. Further, that boys are more likely to remain segregated than girls is probably related to the observation that males, more than females, sanction each other for cross-sex interaction, perhaps because they tend to be the higher status sex (Leaper, 1994). Group identity theory would suggest that these sanctions are a way to maintain in-group solidarity.

In short, sexual segregation is a robust phenomenon in human juveniles. The patterns of sexual segregation are robust among juveniles across many human societies. For example, in Whiting and Edwards' (1973) study of ten cultures, young children (four to five years of age) spent most of the time in segregated groups, and children of six to ten years of age, 75% of their time.

SEXUAL SEGREGATION IN ADULTS

Though sexual convergence in adulthood can take place through heterosexual relationships, social outings of heterosexual couples and some integration in the workplace, segregation continues into adulthood. Many teenagers begin to become attracted to the opposite sex in adolescence, but continue to segregate into same-sex groups for the majority of their free time. When they do form cross-sex groups, they are guarded and exploratory (Maccoby, 1998). For example, Pellegrini and Long (2003, 2005) found that adolescent males and females remained segregated during much of their school day, but integrated at monthly school dances. As noted above, the social norms of a dance, which have been described as a lek for adolescent males (Low, 2000), favour sexual integration (Sroufe *et al.*, 1993). Maccoby (1998) notes that segregation may never really disappear, and perhaps adults continue to segregate for many of the same reasons that they segregate during childhood. In the remainder of this section we review segregation in the workplace and in other social groupings.

Segregation in the workplace

The evidence for continued sexual segregation in the workplace is abundant. This segregation occurs both horizontally and vertically

(Maccoby, 1998). Many jobs are dominated by one sex or the other, and even when men and women work at the same office, they often do not interact with each other. Maccoby (1998) refers to the gender hierarchy, in which men tend to work at the higher-level jobs with women assisting them. When women are in a managerial position, they tend to supervise other women. Though the sexes may work in the same office, they are often distanced by this placement at different levels in the hierarchical structure.

There are many theories as to why men and women still segregate in terms of occupation. Maccoby (1998) gives a summary of the factors leading to this segregation. Firstly, the sexes deviate in terms of interest in certain occupations and expectations of success in childhood and adolescence. These expectations and interests are encouraged by adults, and lead to training and education in certain areas. Institutional barriers, household duties and childrearing responsibilities, and access to high-paying jobs can also play a role in the sex division.

There are conflicting theories on the nature of resource networks that males and females use to enter the workforce as well. However, the evidence seems to point to the sex of the social contact as having a major impact on obtained jobs. Straits (1998) showed that males are more likely than females to use social contacts to find a job, and these contacts are primarily male. Regardless of the sex of the job seeker, female contacts tend to suggest female-dominated occupations and vice versa. Further, Menchen and Winfield (2000) analysed the Metropolitan Employer–Worker Survey, which revealed that women are significantly less likely to hold jobs in female-dominated occupations when their social contact is a man.

Segregation may continue into adulthood because many had the belief that men and women had different talents, and therefore worked in those areas that were appropriate for their strengths (Maccoby, 1998). However, these beliefs have changed and most would agree that men could perform in traditionally female occupations and vice versa. In addition, it remains impossible for both men and women to leave their attitudes towards the other sex out of the workplace. Correspondingly, male and female adults have different interactional styles, and form same-sex groups that function much the same way as childhood groups discussed earlier in this chapter. For example, female groups are more communal and less quarrelsome than male groups (Moskowitz et al., 1994).

Moskowitz et al. (1994) tested the social role hypothesis in a sample of men and women at work. They found, consistent with this theory, that social roles rather than sex influenced assertive behaviour. Both men and women were more dominant when in a supervisory role, and displayed when they were being supervised. However, social role theory did not account for gender differences that were found in communal behaviour. Women were found to be more communal than men regardless of status, and were especially communal with other women.

Segregation in less formal social groups

Men and women often form same-sex groupings in social settings as well. Sexual segregation may be thought of as simply due to divergent interests between the sexes, but the implications for segregation in the workplace suggest otherwise. Though men and women in industrialized societies often spend social time in mixed-sex couples, there is a strong pull to remain in same-sex networks. For example, women usually take the primary responsibility for raising young children, and rely heavily on their female friendships for support (Maccoby, 1998) (Fig. 12.3).

Men and women can view the costs and benefits of opposite-sex relationships in different ways. Bleske and Buss (2000) found that women report protection, while men report sex, as being a key benefit from opposite-sex friendships. However, both men and women felt that receiving information about the opposite sex regarding how to attract a potential mate was a key advantage to their mixed-sex relationships. Bleske-Rechek and Buss (2001) later found similar results in another study of opposite-sex friendships. Men reported that sexual attraction and desire were important for initiating mixed-sex friendships, and said that lack of sex was an important reason for ending these relationships. Women, on the other hand, reported that physical protection was important for initiating an opposite-sex friendship, while the lack of it could dissolve the relationship.

Also, and similar to the data on segregation in childhood, men and women display vast differences in interactional style, which makes it more comfortable for them to interact with the same sex (Maccoby, 1998). For example, the topics that men and women discuss tend to be different, and often would not be discussed if someone from the opposite sex joined the group. Men and women have different ways of using humour, as men tend to tease each other and women self-deprecate. This pattern is confusing for the other sex, and results in the feeling that men are critical or women have no confidence. Additionally, men and women have different ways of exchanging compliments.

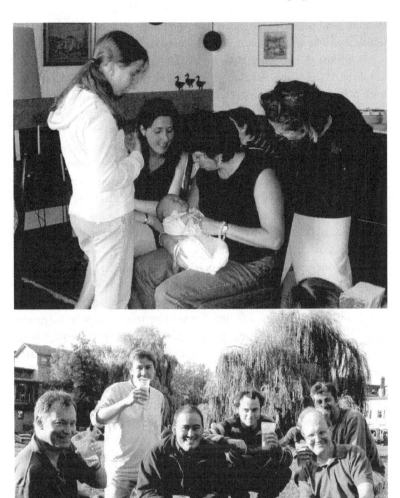

Figure 12.3 Women in their nurturing roles and men socializing outside a pub.

Women tend to apologize more than men and are more collaborative, while men like to have their efforts noticed.

It should be noted that sex differences in friendship are not absolute. Wright and Scanlon (1991) found that women's friendships with other women were reported to be especially strong and rewarding compared to their friendships with men. However, they also showed that women's same-sex friendships were found to be stronger and more rewarding than men's friendships with men or women. In addition, these authors reported an effect of gender role orientation, in that women's same-sex friendships with androgynous females were strongest and were the most rewarding. Further, Wright (1988) warns that researchers have an inclination towards exaggerating differences with a focus on statistical significance and overlooking within-group differences in same-sex friendships.

To conclude this section, social and occupational segregation in adulthood mirrors patterns seen in juvenile humans (Maccoby, 1998). Firstly, the simple fact that widespread segregation exists follows patterns developed early in life. Borderwork, or the attempt to integrate sexes, involves sexual teasing in adulthood much the same way as in childhood. The attitudes that men and women have developed regarding the other sex tend to persist, and friendly overtures in the office and social situations are often mistaken for romantic interest, similar to patterns in adolescence. Finally, the vast divergence of interest in childhood between boys and girls persists into adulthood, as seen in same-sex adult peer groups.

Nevertheless, men and women must learn to interact together in social and occupational situations. Heterosexual couples often spend time together socially, and therefore must learn to communicate effectively. Maccoby (1998, p. 144) reminds us 'childhood culture is far more gender-segregated than is adult culture'. In the workplace, men and women deal with the complicated task of fitting into a joint hierarchy, something that was rarely dealt with in childhood. Maccoby (1998) asserts that adults simply lack the scripts to work with people to whom they are not romantically interested. This leads to implications regarding the encouragement of cross-sex interaction in childhood. As shown earlier in this chapter, boys and girls learn valuable skills in same-sex groups that are utilized in their roles as adults. Perhaps if young children are expected to interact more frequently in cross-sex groups as well (e.g. in school or in their neighbourhood) they will learn an additional set of valuable skills that will carry over into adulthood.

A MODEL OF SEXUAL SEGREGATION IN HUMANS

In this section we present a model for the origins of sexual segregation in humans, and how this model can reconcile the energetic and social roles hypotheses put forth to explain segregation. The proximal environments in which an individual develops, starting with conception, most probably influences morphological (sexual dimorphism) and behavioural developmental traits (physical activity and social roles). Bateson and Martin (1999) used a jukebox metaphor to describe this process. Individuals within each species have a genetic endowment that can be realized through a wide variety of options (similar to the collection of records in a jukebox) but the specific developmental pathway taken (similar to the specific record selected) by an individual is influenced by the perinatal environment (i.e. from conception through infancy) of the developing organism.

Perhaps the most relevant aspect of the proximal environment to consider is nutrition, as it affects sexual dimorphism and the behaviours associated with the energetics and social roles hypotheses under review. Specifically, the nutritional environment of the fetus bearing mother impacts subsequent mating patterns and sexual dimorphism (Alexander et al., 1979; Bateson & Martin, 1999). Nutritional stress reduces the difference in body size between males and females, primarily by acting on male fetuses (Alexander et al., 1979).

Ecological abundance or stresses affect mating patterns such that in severe ecological niches where resources are scarce (e.g. high Arctic) monogamy among humans is ecologically imposed, as it requires both parents to provision the offspring. This mating arrangement reduces male–male competition and, in turn, reduces sexual dimorphism (Alexander et al., 1979). Reduced sexual dimorphism, in turn, may lead to attenuated sex differences in competitive and physically vigorous behaviour.

This model is presented in Fig. 12.4.

TWO HYPOTHESES FOR THE ORIGINS OF SEXUAL SEGREGATION

The energetic/physical activity hypothesis

This hypothesis, like the activity budget hypothesis presented in Chapter 10, states that males and female segregate because of differences in size, where larger males are more physically active and competitive than females and these behaviours lead to segregation.

Nutrition (*Abundant*/Thrifty)

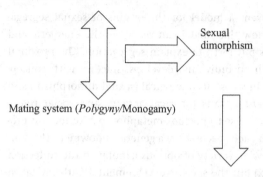

Sexual
dimorphism

Mating system (*Polygyny*/Monogamy)

Social/Sex-segregation

Male groups	*Female groups*
Active	Sedentary
R&T/Competitive	Nurturing
Physical aggression	Indirect/Social aggression

Figure 12.4 A model for differences in sexual dimorphism and sexual segregation. R&T, rough and tumble.

A number of meta-analyses of sex difference in physical activity across a wide variety of settings showed that males are indeed more active than females, with differences being observed prenatally as well as during childhood (Campbell & Eaton, 1999; Eaton & Enns, 1986; Eaton & Yu, 1989). That differences appear prenatally, suggests that differences in activity were not exclusively due to socialization. Instead, exposure to androgens prenatally is probably the mechanism by which selection pressure is exerted, resulting in the initial sex differences in physical activity (Archer & Lloyd, 2002). These differences, in turn, may be responsible for male infants' preference for high activity behaviours (Campbell *et al.*, 2000).

If differences in activity motivate sexual segregation, we would also expect sex differences in activity during those periods when sexual segregation is peaking, beginning at three years and peaking at eight to eleven years (Maccoby, 1998). We present naturalistic data on physical activity during childhood, drawn from a study comparing the activity of primary school boys and girls on the playground during school recess. Activity was assessed using a behavioural checklist indicative of caloric expenditure, developed by Eaton and colleagues (Eaton *et al.*, 1987). Two groups of children (mean ages of 7.3 and 9.5) were assessed on the rating scale measure, but children's resting caloric expenditure was subtracted

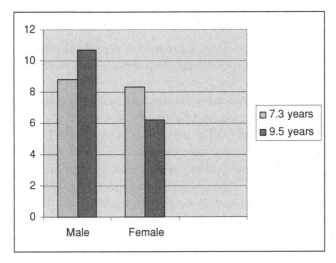

Figure 12.5   Caloric expenditure for males and females – resting rate
by sex and age. $N = 23$ males/23 females and 15 males/16 females for
7.3- and 9.5-year-olds respectively.

from the caloric expenditure during play. As displayed in Fig. 12.5,
a sex by age statistical interaction was observed; $F(19, 81) = 9.81$,
$p < 0.001$. Boys were more active than girls; boys became more active
with age and girls declined with ages.

The sex by age interactions suggest that the sex differences in
activity may have been due to differences in maturation, as males
mature slower than females. As we will discuss later, these sex differences in activity and in maturation are consistent with the argument
that sexual selection theory can explain differences in physical activity
and sexual segregation.

That sexual dimorphism is minimal during humans' juvenile
period, when segregation and differences in activity are observed,
requires explanations. It may be that energetic behaviour may be especially important for juvenile boys because it may be a sensitive period
for the development of brain and muscle systems implicated in actions
associate with later hunting and fighting (Byers & Walker, 1995). Specifically, the energetic behaviour characteristic of juvenile males may help
develop cerebellar synapse distributions and muscle fibre differentiation, important in economical and skilled movements associated with
fighting and hunting (Byers & Walker, 1995). These findings may also
help explain why humans segregate in the juvenile period even when
they may integrate at maturity.

Different social roles of males and females

The social roles hypothesis posits that adult reproductive sex roles are responsible for segregation during the juvenile period. Specifically, and following sexual selection theory, the competitive bias of males, with development, takes the form of learning and practising roles associated with dominance, hunting and fighting for males, and maternal roles for females. These sex-specific roles are learned in segregated groups by observing adults and older peers and practised by playing with conspecifics. Males, for example, segregate around dominance-related activities, such as rough-and-tumble play (R&T), competitive games (Pellegrini *et al.*, 2003) and aggression (Pellegrini, 2003). These behaviours are related to affiliation with a variety of peers and physical conditioning during childhood (Pellegrini & Smith, 1998).

Groups of human male juveniles, as noted earlier, are very active and competitive. These specific behaviours, besides having energetic antecedents and consequences, are related to reproductive roles. Specifically, the quasi-aggressive behaviours, such as R&T, observed in male groups may have their origins in their ancestral roots in intra-sexual competition for securing and defending mates.

Further, males are often socialized by their fathers to express their physical activity in the form of rough play and aggression. Fathers of young males spend significantly more time with sons, relative to their daughters (Parke & Suomi, 1981). During this time, fathers often engage in R&T and other forms of vigorous play with their sons (Carson *et al.*, 1993).

These rough and competitive behaviours exhibited by males in their groups have implications for their peer group status. Male, not female, groups are hierarchically organized in terms of dominance status. Dominance in male groups is determined by a combination of agonistic and affiliative strategies used in the service of resource acquisition. Thus, in male groups, juveniles are using R&T, aggression, reconciliation and facility in competitive games as means of achieving status (Pellegrini, 2003). Males tend to use aggressive strategies when their social groups are in the formative stages. After status is achieved, instances of aggression decrease and dominant members use more affiliative strategies to reconcile former foes (Pellegrini & Bartini, 2001).

Following a social dominance argument, we would expect more sexual segregation at transition points between social groupings, such as when youngsters change schools. Specifically, male groups should be segregated as dominance issues are being sorted as males enter a

Table 12.1 *Males' and females' segregation, dominance and aggression,*
predicting dating. Note: The criterion variable is dating popularity;
*p < 0.05, **p < 0.01. The rows labelled Segregation, Dominance and
Relational aggression are predictors (correlation coefficients, separately for
males and females) of dating popularity at four time points.

	Dating popularity			
	6th Fall	6th Spring	7th Fall	7th Spring
Males				
Segregation	−0.15 (54)	−0.30* (43)	−0.11 (69)	−0.24 (37)
Dominance	0.07 (46)	0.34** (59)	0.42** (64)	0.46** (64)
Relational aggression	−0.04 (42)	−0.05 (27)	0.11 (61)	0.05 (63)
Females				
Segregation	0.10 (44)	−0.11 (43)	0.06 (51)	0.17 (28)
Dominance	−0.21 (27)	0.16 (53)	0.06 (50)	−0.03 (51)
Relational aggression	0.15 (40)	0.37* (30)	0.07 (49)	0.15 (54)

new school. Our observational data of youngsters interacting with each
other during non-structured time (e.g. in the lunch room and in hall-
ways) across the first two years of middle school (13–15 years of age)
showed that both boys' *and* girls' groups remained segregated (Pellegrini & Long, 2003). However, groups became integrated as youngsters
became more active in heterosexual dating. Specifically, there was a
dynamic relation between sexual segregation and opposite-sex dating
popularity across time. Increases in dating popularity lead to decreases
in sexual segregation and vice versa. Boys' dominance status also predicted opposite-sex dating popularity (Pellegrini & Long, 2003). The
intercorrelations between segregation and dating popularity for males
and females separately are presented in Table 12.1.

The data in Table 12.1 display the bivariate correlations, by sex,
between an index of group segregation (expressed as a proportion) and
the number of nominations to a hypothetical party (i.e. individuals
were ask to name opposite sex peers they would invite to a hypothetical party). The data indicate that for males only, the level of group
segregation decreased as they became more popular with opposite sex
peers; thus, for males at least, interest in heterosexual relations may
motivate group integration.

The results of the dynamic covariate relations are presented in
Table 12.2. Briefly, dynamic covariate analyses illustrate how sex,
segregation, dominance, relational aggression and dating interact

Table 12.2 Results of dynamic covariate analysis. Note: The criterion variable is dating popularity; ***p < 0.001; Model 2 is identical to Model 1 with the exception that γ_4 is set to zero; omnibus χ^2 d.f. = 3; fixed effects t-tests d.f. = 111. These separate models test the extent to which segregation, dominance and relational aggression each (noted by a main effect) and in interaction with sex, predict dating popularity. The significant interaction terms indicate that dominance and relational aggression predict opposite sex dating popularity differently for males and females.

Model	χ^2	Segregation		Dominance		Relational aggression	
		Main effect	Sex interaction	Main effect	Sex interaction	Main effect	Sex interaction
		$\hat{\gamma}_0$(SE)	$\hat{\gamma}_1$(SE)	$\hat{\gamma}_2$(SE)	$\hat{\gamma}_3$(SE)	$\hat{\gamma}_4$(SE)	$\hat{\gamma}_5$(SE)
1	18.33***	−1.34*** (.22)	1.63*** (.41)	0.83*** (.11)	−0.84*** (.21)	0.04 (.07)	0.51*** (.13)
2	20.52***	−0.33*** (.22)	1.64*** (.41)	0.85*** (.11)	−0.87*** (.21)	0	0.57*** (.12)

across time (Long & Pellegrini, 2003). These analyses indicate the ways in which three variables (segregation, dominance and relational aggression) change across time and interact with sex to predict dating popularity. The analyses show that, for males, their cross-sex popularity was predicted by their dominance status and the degree to which they integrated female groups. Relational, or indirect aggression, was not a significant predictor for males. Indirect aggression, not dominance, was, however, important for females' opposite-sex popularity, along with their integrating male groups. These findings are consistent with Campbell's (1999) notions that male groups are organized around dominance displays, while female groups use less direct agonistic strategies. Campbell suggests that females' less direct confrontation style may have its origin in their higher level of parental investment. That is, females' higher level of parental investment, relative to males, means that they minimize risks associated with direct aggression and dominance displays. Instead, they take a safer, less direct route to gain influence over their peers, by damaging their reputations and by gossiping and spreading rumours about them.

Girls' interactions with their peers, after the second year of life (Power, 2000), reflect their roles in adult society. Specifically, the literature on children's pretend play is unequivocal in documenting sex differences in themes enacted in fantasy (see Power, 2000 for an exhaustive review). When fantasy play begins to emerge, at around 18 months of age (Fein, 1981), girls' play reflects domestic and nurturing themes (McLoyd, 1980). For example, in a classic study of preschool children's fantasy, Saltz and colleagues (Saltz et al., 1977) labelled the sorts of fantasy play that girls enacted as 'social dramatic' and boys' play as 'thematic fantasy'. In social dramatic play domestic and familial themes are enacted, with girls 'mothering' or teaching younger children. Thematic fantasy, on the other hand, is rooted in the world of superheroes and has themes associated with dominance, fighting and competition.

These biases are reinforced in numerous ways by parents and teachers (e.g. in the toys provided for each and the style of interactions), though fathers seem to be most concerned with shaping sons' actions and choices (Power, 2000). Additionally, children themselves, but especially boys, are reluctant to cross sex-role stereotypic boundaries (Gottman, 1983). For example, in an experimental study of preschoolers' play with sex-typed toys, older boys' play with female-preferred toys was *less* sophisticated than younger boys' play with the same toys (Pellegrini & Perlmutter, 1989).

Females may avoid males because they represent threats in the form of general aggression. Consistent with this idea is the finding that the presence of a familiar adult is likely to increase the likelihood of juvenile girls interacting with juvenile boys (Maccoby, 1998). Young girls may see familiar adults as protectors from aggressive and boisterous boys. While male children seem to be more concerned than females with maintaining segregated groups, at least during childhood, females actively withdraw from interactions with males, often to seek protection of their parents (Maccoby, 1998). For example, in experimental studies of unacquainted toddlers where children were observed in mixed-sex dyads, girls often withdrew from the interaction to stand next to their mothers (Maccoby, 1998). This sort of withdrawal was not observed in female dyads. Similar findings are reported for older children (five- and six-year-olds). When quartets were observed in an experimental playroom, girls tended to stay near an adult when in mixed-sex quartets but not when in all-girl groups (Maccoby, 1998).

Throughout the juvenile period, females' groups tend to be smaller and more intimate than males' groups (Maccoby, 1998). Where males use dominance-related strategies to access resources, females tend to be more indirect. Specifically, female strategies involve manipulating social relationships with other females, through gossip and rumour mongering (Campbell, 1999). During adolescence, these strategies are predictive of females' opposite-sex dating popularity during early adolescence (Pellegrini & Long, 2003). As displayed in Table 12.2, females', not males', indirect, relational/social aggression predict heterosexual dating.

In short, boys' and girls' peer groups are segregated in preschool, and males' and females' interaction is organized around different activities and toys by both adults and peers. The themes of interactions in these groups, in turn, reflect males' bias toward competitive and dominance-oriented roles, and females' toward nurturing and domestic roles. It may be the case that the activity dimensions of behaviours initially bias males and females to each other. Then, with subsequent social cognitive development, children learn the social roles associated with males' and females' roles in society. The roles associated with these behaviours, in turn, predict heterosexual relationships.

CONCLUSION

In conclusion, there is clear evidence for sexual segregation of juveniles in most human societies, as Maccoby (1998) has documented in her integrative review. As Maccoby acknowledges, however, socialization theories alone cannot account for the origins of the behaviours associated with segregation.

In the human developmental psychology literature differences in activity, competitiveness and maturation have been pitted as alternative hypotheses for proximate causes of sexual segregation. We suggest that these differences can be explained by sexual selection theory. Specifically sex differences in activity, competitiveness and maturation are ultimately caused by different reproductive roles where males are the more competitive of the two sexes. That males mature slower than females too is consistent with the fact that their reproductive success is more varied than females. Slower maturation results in the slower introduction of males in the reproductive breeding pool and, consequently maintaining an optimal sex ratio (Alexander *et al.*, 1979).

Part VI Sexual differences in ecology: comparisons within different taxa

‹

13

Ecological divergence between the sexes in reptiles

OVERVIEW

Living reptiles display an immense diversity in morphological, phys-iological, behavioural and ecological traits. Indeed, 'reptiles' actually constitute a highly polyphyletic assemblage, encompassing four major lineages that diverged >200 million years ago. The extant taxa com-prise 7390 species of squamates (lizards and snakes), 295 of turtles, 23 of crocodilians and 2 species of sphenodontians (the New Zealand tuatara) (Uetz, 2000). Correspondingly, the ways in which ecological divergence is manifested between the sexes, and the selective pressures and prox-imate mechanisms responsible for such divergence, take many forms. This chapter will outline the kinds of ecological traits known to differ between males and females in reptiles, and then explore the degree to which such differences are consistent with alternative hypotheses. His-torically, conceptual models to predict and explain segregation between the sexes have been developed by workers with a primary focus on endothermic vertebrates, especially ungulates (see Chapters 2, 3, 9, 10, 11 and 19). Inevitably, the kinds of arguments that have been devel-oped rely upon specific features of these organisms, so that attempts to interpret analogous sexual segregation phenomena in reptiles cannot be neatly subsumed within the same framework. However, the analo-gies in many cases are clear enough, and the framing of the models broad enough, that strong parallels can be drawn. Thus, we use the conceptual framework developed by ungulate biologists to categorize and investigate causal mechanisms for segregation between the sexes

Sexual Segregation in Vertebrates: Ecology of the Two Sexes, eds. K. E. Ruckstuhl and P. Neuhaus. Published by Cambridge University Press. © Cambridge University Press 2005.

within reptile populations. We then propose a framework with which to analyse studies of sexual segregation.

BACKGROUND

Firstly, we briefly review aspects of reptile biology that bear directly upon ecological divergence between the sexes.

Sex-determining systems and sex ratios

Although parthenogenesis (female-only reproduction) has evolved in some lineages of lizards and snakes (Moritz, 1993), most reptile populations contain both males and females. However, the mechanisms that determine an individual's sex are diverse, with substantial phylogenetic lability. Heteromorphic sex chromosomes occur in many reptile species, sometimes with males as the heterogametic sex (XX-XY, as in many lizards) and sometimes with females as the heterogametic sex (ZZ-ZW, as in many snakes: Bull, 1983). However, many reptile species lack heteromorphic sex chromosomes. In some of the latter group (and even one of the former: Shine et al., 2002c), an individual's sex is determined by the incubation temperatures that it experiences prior to hatching (temperature-dependent sex determination: Bull, 1983; Shine, 1999). Temperature-dependent sex determination is not restricted to egg-laying reptiles; in two lineages of viviparous (live-bearing) scincid lizards, a pregnant female's selection of thermal regimes determines the sex of her offspring (Robert & Thompson, 2001; Wapstra et al., 2004). One consequence of this linkage between sex and incubation conditions is that spatial or temporal variation in nest (or maternal) temperatures can generate correspondingly wide fluctuations in sex ratios of hatchling cohorts (Bull & Charnov, 1989). In turn, such variation can influence the apparent (or real) degree of sexual segregation in natural populations (see Chapter 2).

Sexual dimorphism

Ectothermic organisms are not subject to the same strong constraints on surface-area to volume as those experienced by endotherms; thus, ectotherms can exploit the ecological opportunities offered by small as well as large body sizes, and a population of ectotherms typically contains a wider size range of free-living individuals than is the case for endotherms (Pough, 1980). Hence, the degree of sexual divergence in mean adult body sizes (as well as the range of adult body sizes in each sex) often is greater in ectotherms than in endotherms. Females are larger than males in most reptile species, probably due to fecundity selection for large body size in females, but males attain larger sizes in at least some species within most major lineages. In crocodilians and sphenodontians (both of which have temperature-dependent sex determination) males grow much larger than conspecific females (Dawbin, 1982; Webb & Smith, 1984). The same is true for many terrestrial, but not aquatic, turtles (Berry & Shine, 1980) and for lizard and snake taxa in which males engage in physical combat with each other for mating opportunities (see below; Shine, 1978, 1994b; Stamps, 1983).

Reproductive modes

Squamate reptiles display more than 100 separate evolutionary transitions from oviparity (egg-laying) to viviparity (live-bearing), mostly in association with invasion of cold climates (Blackburn, 1985; Shine, 1985). Sometimes, oviparity and viviparity co-occur within a single species (but not within a single population: Qualls et al., 1996; Heulin et al., 1999). Because viviparity strongly affects a female's patterns of feeding, habitat selection and movements (Shine, 1980a; Madsen & Shine, 1993a; Bonnet et al., 1999a), these multiple phylogenetic shifts in reproductive mode introduce significant complexity into any ecological comparison between conspecific males and females.

Mating systems

In some reptile species, adult males engage in combat bouts to determine social dominance and hence, mating opportunities. In others, rival males virtually ignore each other and instead focus their efforts on 'scramble' competition to locate and inseminate receptive females (Berry & Shine, 1980; Stamps, 1983). Male tactics may change ontogenetically within a single individual's lifetime (Madsen et al., 1993; Shine et al., 2001) and can vary dramatically even among populations within a single wide-ranging species. For example, males grow larger than females and fight vigorously with other males in some populations of Australian carpet pythons (Morelia spilota) whereas in other populations of the same taxon, males are smaller than females and avoid physical struggles with other males (Shine & Fitzgerald, 1995; Pearson et al., 2002a,b). Because the mating system influences male-specific behaviours such as mate-searching, this diversity in male tactics necessarily influences the ways in which ecological features differ between males and females.

Behavioural thermoregulation

In ectotherms, higher body temperatures typically facilitate rates of activities such as locomotion, digestion and embryogenesis (Huey & Slatkin, 1976; Hertz et al., 1993). Ectotherms typically inhabit environments with substantial ambient thermal heterogeneity (which is greater in terrestrial compared to aquatic systems) and thus organisms often are able to select appropriate body temperatures for specific activities (Peterson et al., 1993). Hence, sex differences in thermal optima generate sex differences in habitat use.

ECOLOGICAL DIFFERENCES BETWEEN THE SEXES, UNRELATED TO REPRODUCTIVE ACTIVITIES

The following discussion divides cases of sexual segregation into two categories: those that persist throughout 'non-reproductive' phases of the organism's life, as opposed to those that are expressed only, or most strongly, during the reproductive season. This division is arbitrary: from an evolutionary perspective, all activities link ultimately to reproductive success. From that view, it is incorrect to label as 'non-reproductive' a foraging bout for resources that ultimately will be expended in reproducing. However, the link with reproduction is much more direct for some traits than for others, especially where the specific sex differences appear only during the time when an individual is actively engaged in reproductive activities (mate-searching, pregnancy, etc.).

Ecological differences between the sexes that are not strongly linked to reproductive activities include the following.

Habitat use

Many ecological studies report significant differences between conspecific males and females. Some of the best examples come from aquatic reptiles, perhaps because water depth presents an easily measured abiotic variable that correlates strongly with important biotic traits (e.g. the types, sizes and abundance of potential prey and predators, and the locomotory, ventilatory and thermal challenges imposed by diving to various depths). For example, in most aquatic snake species females grow to substantially larger body sizes than do conspecific males, and

larger animals often consume larger prey, are less vulnerable to predators and are physiologically better suited to exploiting deeper water
(Shine, 1986; Houston & Shine, 1993). Perhaps reflecting such size-
related factors, females of several species have been reported to forage
in deeper water than do males within the same populations. Importantly, the species involved come from a wide range of phylogenetic
lineages, so that any sex-based habitat partitioning must have evolved
independently in natricine colubrids, acrochordid filesnakes and laticaudid sea kraits (see Table 13.1). The evidence for sex differences in
foraging depths is indirect in most of these cases, usually based on
sex differences in prey sizes and types (see below), but has been confirmed by direct measurements in filesnakes (*Acrochordus arafurae*; Shine,
1986). Indeed, the association between sex and water depth within this
species is evident not only within a single water body, but in a correlation between a water body's depth and the sex ratio of filesnakes that
it contains: males (specialists in shallow-water foraging) are disproportionately common in shallower billabongs (Houston & Shine, 1994).

Information on other types of aquatic reptiles is less detailed,
but clear sex differences in diet, especially in highly dimorphic species
such as map turtles (*Graptemys* spp.), support a strong inference of sex-
specific partitioning by water depth in these taxa also (Table 13.1). Galapagos marine iguanas (*Amblyrhynchus cristatus*) are often regarded as the
only truly marine lizards, but in fact it is only one sex (males) that
utilizes marine resources. Male iguanas attain much larger body sizes
than females and dive into the cold ocean to feed on marine algae. Perhaps because of their smaller body size (and consequently, lower thermal inertia) female iguanas forage on the fringes of the ocean rather
than by diving (Buttemer & Dawson, 1993; Wikelski & Trillmich, 1994).
The available evidence thus suggests that males and females of many
aquatic reptile species may differ in the water depths at which they
forage, in turn reflecting both prey availability and sexual size dimorphism. We note, however, a cautionary tale from a recent study: the seasnake *Emydocephalus annulatus* displayed an association between snake
body size and water depth despite a lack of size-related dietary shifts
(Shine *et al.*, 2003a).

Similar but less clear-cut cases of sex divergence in foraging habitats occur in many other types of habitats. For example, males may
be more or less arboreal in their habits than are females, perhaps
depending upon the relative body sizes of the two sexes (large body size
precludes utilization of slender branches: Henderson & Binder, 1980;
Lillywhite & Henderson, 1993). Although there are anecdotal reports

Table 13.1 Reptile taxa exhibiting ecological sexual segregation in a variety of characteristics. H = habitat, PS = prey size, PT = prey type, FF = feeding frequency, FB = foraging behaviour, A = activity, AP = anti-predator behaviour and T = thermoregulation. All differences noted occur in adults unless otherwise specified. To save space, sexual differences in home-range size are not included.

Family	Species	Larger sex	Difference in	Nature of difference	Authority
Crocodilians					
Alligatoridae	Alligator mississippiensis	M	PT	Males more cannibalistic	Rootes & Chabreck, 1993
Crocodylidae	Crocodylus johnstoni	M	A	Males move and disperse farther	Tucker et al., 1997, 1998
Turtles					
Bataguridae	Cuora flavomarginata	F	H, A	Gravid females prefer forest edge, males deep forest; males more active April–July, females August–October	Lue & Chen, 1999
Carettochelyidae	Carettochelys insculpta	F	H, A	Females prefer deeper water and sand flats, males shallower water and isolated logs; females more active and move farther	Doody et al., 2002
Cheloniidae, Dermochelyidae	All species	F	H, A	Females come ashore to lay eggs; males remain at sea	Ernst et al., 1994; Cogger, 2000
Chelydridae	Chelydra serpentina	Same	A	Males more active in breeding season (May), females after nesting (July); females move farther	Brown & Brooks, 1993; Pettit et al., 1995
	Macroclemys temminckii	M	H, A	Subadult males move more often and farther; subadult females more associated with logs	Harrel et al., 1996

226

	Species				Reference
Emydidae					
	Clemmys insculpta	M	H, A	Males more aquatic and move more	Ross *et al.*, 1991; Kaufman, 1992
	C. marmorata	M	PS, PT, A	Males more carnivorous, take larger prey, and move farther; females more herbivorous	Bury, 1972, 1986
	C. muhlenbergii	F	A	Males move more	Chase *et al.*, 1989
	Deirochelys reticularia	F	A	Males move farther	Gibbons, 1986
	Emydoidea blandingii (Illinois)	M	A	Males move farther	Rowe & Moll, 1991
	E. blandingii (Wisconsin)	M	A	Females move farther	Ross, 1989
	Graptemys barbouri	F	PS, PT	Males eat more insects, females more molluscs	Sanderson, 1974; Lee *et al.*, 1975; Ernst *et al.*, 1994
	G. caglei	F	H, PS, PT	Females prefer deeper, slower pools; females eat molluscs, males insect larvae	Porter, 1990; Craig, 1992
	G. ernsti	F	PS, PT	Males eat more insects, females more molluscs	Shealy, 1976
	G. flavimaculata	F	H	Females prefer deeper, faster water farther from shore	Jones, 1996
	G. geographica	F	H, PS, A	Females prefer deeper, faster water farther from shore and take larger prey; males move more and more active in winter; females move more during nesting season	Gordon & MacCulloch, 1980; Pluto & Bellis, 1986, 1988; Graham & Graham, 1992

(*cont.*)

227

Table 13.1 (cont.)

Family	Species	Larger sex	Difference in	Nature of difference	Authority
	G. nigronoda	F	H, T	Females select larger basking branches and bask for longer periods	Waters, 1974; Lahanas, 1982
	G. ouachitensis	F	PT	Females more herbivorous	Webb, 1961; Shively, 1982; Shively & Jackson, 1985
	G. pseudogeographica	F	PT	Females more herbivorous	Vogt, 1981
	G. pulchra	F	H	Females prefer deeper water	Ernst et al., 1994
	G. versa	F	H, PS, PT	Females prefer deeper, faster water farther from shore; females eat molluscs, males insect larvae	Lindeman, 2003
	Malaclemys terrapin	F	H, PS, A	Females prefer deeper water farther from shore; females take larger prey and move more often	Tucker et al., 1995; Roosenburg et al., 1999
	Pseudemys floridana	F	A	Males move more in spring and fall, females more in summer	Gibbons, 1986
	P. scripta	F	H, PT, A	Females prefer deeper water and are more herbivorous; males move farther; more in spring and fall, and more overland; females move more in nesting season	Moll & Legler, 1971; Hart, 1983; Morreale et al., 1984; Gibbons, 1986
	Trachemys scripta	F	PT, T	Males more insectivorous; females bask more in spring and summer	Hammond et al., 1988; Moll, 1990
Kinosternidae	*Kinosternum subrubrum*	Same	A	Males move farther and more in spring and fall; females move more in summer	Gibbons, 1986

Testudinidae				
Sternotherus carinatus	Same	A	Males move farther	Mahmoud, 1969
S. depressus	Same	A	Males move farther	Dodd et al., 1988
S. odoratus	Same	PT	Females more carnivorous	Ford, 1999
Gopherus berlandieri	M	A	Males move more	Rose & Judd, 1975
Testudo horsfieldi	F	A	Males eat less, rest less and move farther in mating season, patrolling more limited territory; males aestivate earlier in summer	Lagarde et al., 2002, 2003
Trionychidae				
Apalone ferox	F	PS, PT	Females eat more snails, fish and turtles, and fewer insects	Dalrymple, 1977
A. muticus	F	H, PT, A	Females prefer deeper, more stable water and take mostly aquatic prey; males eat mainly terrestrial prey; females move farther, and males active earlier in spring	Plummer, 1977; Plummer & Farrar, 1981
A. spinifera	F	H, PT, A	Females prefer more open water; females eat more fish and fewer insects, and move more	Williams & Christensen, 1981; Cochran & McConville, 1983; Galois et al., 2002
Lizards				
Agamidae				
Acanthocercus atricollis	Same	PT	Females eat more orthopterans	Reaney & Whiting, 2002
Agama agama	M	H	Males use elevated perches more	Anibaldi et al., 1998
Uromastyx philbyi	?	FF, T	Gravid females feed less and have higher body temperatures	Zari, 1998

(cont.)

Table 13.1 (*cont.*)

Family	Species	Larger sex	Difference in	Nature of difference	Authority
Anguidae	*Elgaria coerulea*	M	H, A	Males shelter under larger, thicker rocks; females emerge from retreats more often in summer	Rutherford & Gregory, 2003
Chamaeleonidae	*Chamaeleo dilepis*	F	H, A	In dry season, females perch higher and in greener shrubs; males more sedentary and prefer brown shrubs	Hebrard & Madsen, 1984
Cordylidae	*Platysaurus capensis*	M	FB	Males process large food items more efficiently	Whiting & Greeff, 1997
	P. intermedius	M	A, AP, T	Males have higher body temperatures and flee in response to predators sooner	Lailvaux et al., 2003
Gekkonidae	*Gonatodes humeralis*	F	H, PT	In mating season, males perch higher; females eat snails, males beetles	Miranda & Andrade, 2003
	Hoplodactylus maculatus	M	T	Gravid females have higher body temperatures	Rock et al., 2000
	Ptyodactylus hasselquistii	M	PS, PT, FF	Males take larger and more diverse prey items, and feed less often	Perry & Brandeis, 1992
Iguanidae	*Amblyrhynchus cristatus*	M	H	Males forage in subtidal as well as intertidal, females only in intertidal	Buttemer & Dawson, 1993; Wikelski & Trillmich, 1994
	Iguana iguana	M	H, A	In mating season, males move more, spend more time in vegetation, and use more energy	Van Marken Lichtenbelt et al., 1993

Lacertidae	*Acanthodactylus erythrurus*	M	PS	Males eat larger prey	Busack & Jaksic, 1982
	Lacerta lepida	Same	PS	Males eat larger prey	Brana, 1996
	Podarcis bocagei	M	A	Males move more	Galan, 1999
	P. hispanica	M	H, PS	Females shift microhabitats according to season; males don't and eat larger prey	Brana, 1996; Diego-Rasilla & Perez-Mellado, 2003
	P. milensis	M	PS	Males eat larger prey and have broader diet	Adamopoulou & Legakis, 2002
	P. muralis	Same	PS, T	Males eat larger prey; gravid females have lower body temperatures	Brana, 1993, 1996
	P. sicula	M	PT, FF	Females take fewer types of prey and eat more often; males eat more hard-bodied prey	Burke & Mercurio, 2002
	Psammodromus algirus	?	H, A, AP	Males prefer leaf litter, move more, and flee in response to predators sooner; females prefer grass	Martin & Lopez, 1999
Phrynosomatidae	*Crotaphytus collaris*	M	PT	Males eat more plant matter, more tenebrionid beetles and more curculionid beetles	Best & Pfaffenberger, 1987
	Holbrookia propinqua	Same	H	Females stay closer to cover	Cooper, 2003
	Phrynosoma douglassi	F	PT	Females take broader range of prey sizes, more orthopterans and fewer ants	Powell & Russell, 1984

(cont.)

Table 13.1 (cont.)

Family	Species	Larger sex	Difference in	Nature of difference	Authority
	Sceloporus clarkii	M	PS	Males take larger prey	Brooks & Mitchell, 1989
	S. cyanogenys	F	T	Males have higher body temperatures	Garrick, 1974
	S. gadoviae	M	T	Females thermoconform more than males do	Lemos-Espinal et al., 1997
	S. grammicus	M	T	Gravid females have lower body temperatures	Andrews et al., 1997
	S. jarrovi	M	PS, FF	Females take larger prey and feed more often; gravid females have lower body temperatures	Simon, 1975; Beuchat, 1986; Ruby & Baird, 1994
	S. occidentalis	F	H, T	Males more often choose exposed, less rocky retreats	Sabo, 2003
	S. undulataus	F	H, T	In mating season, males perch higher; females have higher body temperatures	Pounds & Jackson, 1983; Gillis, 1991
	S. scalaris	F	T	Males have higher body temperatures	Smith et al., 1993
	S. virgatus	F	H, FF, A, T	Males prefer ground and logs, females rocks; in mating season, males move more, eat less, and use up more fat; gravid females have lower body temperatures	Rose, 1981; Smith & Ballinger, 1994; Smith, 1996
	Uma inornata	M	PT, A	Males move farther and more frequently; in May, males eat more flowers and fewer arthropods	Durtsche, 1992
	U. paraphygas	M	PT	In winter, females eat more orthopteran nymphs, males more plants	Gadsen & Palacios-Orona, 1997
	Urosaurus ornatus	Same	H	Males perch higher	Zucker, 1986

Uta stansburiana	M	PT, A, AP	Females take more lepidopterans, males more insect eggs; males move more; females lose tails more easily	Irwin, 1965; Best & Gennaro, 1984; Fox et al., 1998
Polychrotidae				
Anolis acutus	M	H	Males prefer higher, larger perches	Ruibal & Philibosian, 1974
A. aeneus	M	PS, PT, FF	Males take larger prey, eat more ants and plants, and feed less often	Schoener & Gorman, 1968; Stamps, 1977
A. angusticeps	M	PS	Males take larger prey	Schoener, 1968
A. carolinensis	M	H, PS, FF, FB	Males prefer thicker and higher perches and take larger prey; subadult males forage more actively and feed more often than same-size females	Sexton, 1964; Schoener, 1968; Lister, 1976; Preest, 1994; Jenssen & Nunez, 1998; Lovern, 2000
A. conspersus	M	H, PS	Males prefer higher, larger perches and take bigger prey	Schoener, 1967; Lister, 1976
A. cristellatus	M	H	Males prefer grey tree trunks; females distributed more randomly with respect to trunk colour	Heatwole, 1968
A. cupreus	M	H, FF, A	Females forage more; in mating season, males perch higher	Fleming & Hooker, 1975
A. distichus	M	H, FF	Males perch higher; subadult males eat more often than same-size females	Schoener, 1968; Lister, 1976
A. evermanni	M	H	Males perch higher	Moll, 1990
A. frenatus	M	H, PS, FF	Males prefer higher and wider perches, take more large and small insects, and feed more often	Scott et al., 1976

(cont.)

233

Table 13.1 (cont.)

Family	Species	Larger sex	Difference in	Nature of difference	Authority
	A. grahami	M	H	Males perch higher	Lister, 1976
	A. gundlachi	M	H	Males perch higher	Moll, 1990
	A. humilis	F	H, FF, A	Males perch higher and are more arboreal; females forage more	Talbot, 1979; Parmelee & Guyer, 1995
	A. limifrons	Same	H, FF	Males perch higher and feed less often	Sexton et al., 1972; Talbot, 1979
	A. lineatopus	M	H	Males perch higher	Lister, 1976
	A. monensis	Same	H	Males perch higher	Lister, 1976
	A. nebulosus	M	H, FF, A	Males perch higher and are more arboreal; females forage more, males hide and rest more; in wet season, males switch from ambush foraging to active searching (females ambush year-round)	Jenssen, 1970; Lister & Aguayo, 1992; Ramirez-Bautista & Benabib, 2001
	A. polylepis	M	H, PS, PT, FB, FF	Males prefer higher and wider perches and take smaller and different types of prey; females forage more actively and feed more often	Andrews, 1971; Perry, 1996
	A. richardi	M	H, PS, PT, FF	Males perch higher, take larger prey, eat more ants and plants, and feed less often	Schoener & Gorman, 1968
	A. roquet	M	PS, PT	Males eat more large insects and more ants	Schoener & Gorman, 1968
	A. sagrei	M	H, PS	Males prefer higher and wider perches and eat larger prey	Schoener, 1968; Lister, 1976
	A. scriptus	M	H	Males perch higher	Laska, 1970
	A. stratulus	Same	H	Males perch higher	Lister, 1976; Moll, 1990

Family	Species				Reference
Scincidae	*Chalcides ocellatus*	?	T	Gravid females have higher body temperatures	Daut & Andrews, 1993
	Eulamprus tympanum	Same	AP, T	Gravid females rely more on crypsis to avoid predation and bask more	Shine, 1980a; Schwarzkopf & Shine, 1991, 1992
	Eumeces elegans	M	T	Males select higher body temperatures	Du et al., 2000
	E. laticeps	M	PS	Males eat larger prey	Vitt & Cooper, 1986
	Mabuya brevicollis	?	FF	Gravid females eat less	Zari, 1987
	Oligosoma grande	F	PS, PT, FF, A	Males eat larger, stronger-flying insects and less fruit; females feed more frequently and make more frequent but shorter movements	Eifler & Eifler, 1999a,b
	Pseudomoia coventryi	F	AP	Gravid females more vulnerable to predation	Shine, 1980a
	P. entrecasteauxii	F	A, T	Gravid females bask more	Shine, 1980a
	Tiliqua rugosa	F	A	Males home more faithfully	Freake, 1998
Teiidae	*Ameiva plei*	M	FF, A	In mating season, males forage less often	Censky, 1995
	Cnemidophorus costatus	M	PS	Males take larger prey	Brooks & Mitchell, 1989
	C. lemniscatus	M	T	Males have lower body temperatures in mating season	Magnusson, 1993
Varanidae	*Varanus albigularis*	M	A	Males move more during mating season	Phillips, 1995
	V. niloticus	Same	PT	Males eat more beetles and vertebrates, and fewer molluscs	Luiselli et al., 1999
Tropiduridae	*Leiocephalus viriscens*	M	PT	Males more herbivorous	Schoener et al., 1982
	Microlophus albemarlensis	M	AP	Females flee sooner in response to predators	Snell et al., 1988; Miles et al., 2001

(cont.)

235

Table 13.1 (cont.)

Family	Species	Larger sex	Difference in	Nature of difference	Authority
	Tropidurus flaviceps	M	A	Males become active earlier in morning	Vitt & Zani, 1996
	T. melanopleurus	M	PS, PT, T	Males eat more diverse and larger prey; females eat more ants; males thermoregulate, females thermoconform	Perez-Mellado & de la Riva, 1993
Snakes					
Acrochordidae	*Acrochordus arafurae*	F	H, PS, PT, FF	Females forage in deeper water, eat larger fish species and feed less often	Shine, 1986; Houston & Shine, 1993
Boidae	*Charina bottae*	F	A/AP	Females' tails more often scarred and shortened	Hoyer & Stewart, 2000
	Eunectes murinus	F	PS, PT	Females eat mammals, reptiles and birds, males only reptiles and birds	Rivas & Burghardt, 2001
Colubridae	*Boiga irregularis*	M	PS, PT	Males eat more birds and mammals, females more lizards	Savidge, 1988
	Coluber constrictor	F	H, PS, PT	Males more arboreal; males eat more insects, females more mammals	Fitch & Shirer, 1971; Fitch, 1982, 1999
	C. viridiflavus	M	A	Males move more	Ciofi & Chelazzi, 1994
	Coronella austriaca	Same	PS, PT, FF, A	Females eat mammals and snakes, males lizards; females feed less frequently; males emerge earlier in spring	Phelps, 1978; Luiselli et al., 1996
	Diadophis punctatus	F	A	Females move farther	Fitch, 1999
	Elaphe obsoleta	M	H, A	Females more arboreal and use shelter more often; males move more	Durner & Gates, 1993

Species				
Geophis nasalis	F	PS	Females eat larger worms	Seib, 1981
Masticophis lateralis	F	A	Males more active during breeding season	Hammerson, 1978
M. taeniatus	M	A	Males emerge earlier in spring; gravid females move more frequently	Parker & Brown, 1980
Natrix maura	F	PS, PT, FF, A	Males eat small fishes, females frogs and large fishes; males forage for less of the year	Santos & Llorente, 1998; Santos *et al.*, 2000
N. natrix	F	PS, PT, FF, A	Females eat mice and adult toads, males frogs and juvenile toads; gravid females feed less often; males move more during mating season and emerge earlier in spring	Phelps, 1978; Madsen, 1983, 1984; Luiselli *et al.*, 1997; Gregory & Isaac, 2004
Nerodia cyclopion	F	H, PS, PT	Females eat larger fishes and forage in deeper water	Mushinsky *et al.*, 1982
N. fasciata	F	H, AP	In lab, females prefer conspecific-scented substrates, males unscented ones; males flatten and strike more often in response to predators	Scudder & Burghardt, 1983; Allen *et al.*, 1984
N. rhombifer	F	H, PS, PT	Females forage in deeper water and eat larger fishes	Mushinsky *et al.*, 1982
N. sipedon	F	PS, PT, FF, A, T	Females eat fishes and salamanders, males only fishes; males forage for less of the year and thermoregulate less	King, 1986, 1993; Brown & Weatherhead, 2000
Opheodrys aestivus	F	PT	Females eat more odonates, males more caterpillars	Plummer, 1981b

(*cont.*)

Table 13.1 (cont.)

Family	Species	Larger sex	Difference in	Nature of difference	Authority
	Pituophis melanoleucus	M	A	Males emerge earlier in spring; gravid females move more frequently	Parker & Brown, 1980
	Telescopus dhara	F	H, PS, PT, FB, AP	Females ambush birds and are more arboreal; males actively search for lizards, are more terrestrial, and bite more frequently in response to predators	Zinner, 1985
	Thamnophis elegans	F	FF, A	Gravid females cease feeding	Gregory et al., 1999
	T. sirtalis	F	PS, PT, A, AP, T	Females eat toads, suckers and mice, males frogs and dace; males emerge earlier in spring, stay near den longer, and flee in response to predators more often (females more often display and bite); females thermoregulate more precisely and have higher body temperatures	Carpenter, 1956; Gregory, 1974; White & Kolb, 1974; Gibson & Falls, 1979; Fitch, 1982; Gartska et al. 1982; Shine et al., 2000
	Tropidoclonium lineatum	F	FF, A	Heavily gravid females cease feeding	Krohmer & Aldridge, 1985
	Tropidonophis mairii	F	AP	Female hatchlings flee more continuously from predators (males stop more); reproduction decreases approach distances in males but increases it in females	Webb et al., 2001; Brown & Shine, 2004
Elapidae	Aspidelaps scutatus	F	PS, PT	Females eat snakes and mammals, males more frogs	Shine et al., 1996
	Hoplocephalus bungaroides	F	H, A	Gravid females remain near cliffs in summer; males and non-gravid females move into forest	Webb & Shine, 1997

Family	Species	Sex	Codes	Description	Reference
	Pseudonaja textilis	M	H, A	Males inhabit wider burrows on steeper, less north-facing slopes; males move earlier in spring and move farther and more often	Whitaker & Shine, 2003
	Pseudechis porphyriacus	M	H, PT	Females found closer to water; males eat more frogs, females more lizards	Shine, 1979, 1991a
Hydrophiidae	*Emydocephalus annulatus*	F	H, FF, A	In mating season, females spend more time amid rubble, males sand; females forage much more often	Shine et al., 2003a
Laticaudidae	*Laticauda colubrina*	F	H, PS, PT, FF	Females forage in deeper water and eat conger eels, males smaller morays; females feed more frequently, but males take more prey items at a time	Pernetta, 1977; Shetty & Shine, 2002b
	L. frontalis	F	H, PS	Females forage in deeper water and eat larger morays	Pernetta, 1977; Shine et al., 2002d
Pythonidae	*Liasis fuscus*	F	FF, A	In breeding season, males cease feeding for 6 weeks, females for 3 months	Madsen & Shine, 2000
	Morelia spilota imbricata	F	PS, PT, A, T	Females eat large mammals, males small mammals, lizards, and birds; males spend more time in ambush posture, bask less and have more variable body temperatures	Pearson et al., 2002a, 2003
	M. s. spilota	F	H, FF, A	Females prefer rocky areas, males woodland and scrub; females forage for more of the year	Slip & Shine, 1988a,b
	M. s. variegata	M	A	Males move farther	Shine & Fitzgerald, 1996

(cont.)

Table 13.1 (cont.)

Family	Species	Larger sex	Difference in	Nature of difference	Authority
	Python regius	F	H, PS, PT	Females more terrestrial, eat more mammals; males more arboreal, eat more birds	Luiselli & Angelici, 1998
	P. reticulatus	F	H, PS, PT, FF	Females prefer thick forest, males disturbed habitats; females eat larger mammals and feed more frequently	Shine *et al.*, 1998b
Viperidae	*Bitis caudalis*	F	PS, PT	Females eat more mammals, males more lizards	Shine *et al.*, 1998a
	B. gabonica	F	H, A	Males found in wider variety of microhabitats and move more	Angelici *et al.*, 2000
	Vipera aspis	M	FF, A	Males cease feeding during mating season (April), move farther, and are more active; females cease feeding when gravid (July–August)	Naulleau, 1966; Luiselli & Agrimi, 1991
	V. berus	F	H, A	Males emerge earlier in spring and move more during mating season; gravid females stay at den longer	Viitanen, 1967; Prestt, 1971; Phelps, 1978
	V. ursinii	F	PS, PT	Females eat mammals, males ectotherms	Agrimi & Luiselli, 1992
	Agkistrodon contortix	M	H, PS, PT	Gravid females prefer more open areas; females eat more invertebrates, amphibians and reptiles, males more mammals	Fitch, 1960, 1982; Garton & Dimmick, 1969; Reinert, 1984

Species				Reference
A. piscivorus	M	PT	Males eat mostly fish, females mostly reptiles	Vincent et al., 2004
Bothrops moojeni	F	PS, PT	Females eat more endothermic prey	Nogueira et al., 2003
Calloselasma rhodostoma	F	PT, FF, A	Males cease feeding during mating season, females during gravidity; in Java (but not other parts of range), females take more endotherms	Daltry et al., 1998
Crotalus cerastes	Same	A	Males move farther, especially during mating seasons	Secor, 1994
C. horridus	M	H, FF, A	Gravid females prefer more open habitats and cease feeding	Keenlyne, 1972; Reinert, 1984
C. viridis	M	A, AP	Males move more during mating season, but females more active overall; gravid females allow closer approach by predators before rattling	Duvall et al., 1985; Gannon & Secoy, 1985 Kissner et al., 1997
Gloydius shedaoensis	Same	H, FF	Males prefer vegetation as cover, females rocks; females feed more often; during mating season, females more arboreal	Li et al., 1990; Shine et al., 2003c
Sistrurus catenatus	M	A	Gravid females more sedentary	Reinert & Kodrich, 1982
Trimeresurus stejnegeri	Same	FF	Males feed more often	Creer et al., 2002

241

of major habitat dichotomies between males and females (e.g. arbor-eality in one sex but terrestrial habits in the other, as reported for the snake *Telescopus dhara*: Zinner, 1985), detailed data are rare. Most cases are more subtle; for instance, males of a wide variety of lizard species perch higher in vegetation than do females, presumably for territorial surveillance (Table 13.1). Many cases have been inferred from dietary habits: for example, female reticulated pythons (*Python reticulatus*) in Sumatran jungles grow much larger than males (maximum body masses of 75 kg vs. 19 kg) and take much larger prey. These large prey types (e.g. monkeys, porcupines, pangolins) are restricted to forest patches, whereas the rats taken by male pythons are most abundant in disturbed habitats near villages. Thus, the large adult female pythons presumably survive only in relatively thick forest (which also supports a diverse prey base) whereas the smaller males utilize the abundant rodent resources of agricultural land (Shine *et al.*, 1998b).

Prey type and size

Most snakes span a wide range of body sizes between hatching (or birth) and maximum adult size, and eat relatively large prey items which they must swallow entire (Pough & Groves, 1983; Greene, 1997). Thus, gape-limitation is common; the maximum ingestible prey size is limited by the snake's own body size, and as a snake grows larger it consumes larger and larger prey items (Shine, 1991b; Arnold, 1993). This close functional relationship between predator size and maximal ingestible prey size means that average prey sizes are likely to differ between the sexes in any snake population in which males and females attain different maximum body sizes. Because sexual size dimorphism is seen in most snake taxa, some degree of sex divergence in mean prey size is probably the rule rather than the exception.

In the simplest case, snakes continue to eat the same prey type as they grow larger, with the only change being in mean prey size. However, because any given prey taxon displays only a limited size range, it will often be true that a shift in prey size translates into a shift in prey type. This phenomenon of ontogenetic shifts in prey types is most obvious in relatively large species of snakes, because these animals pass through a wider range in absolute body sizes from hatching through to adulthood (Shine & Wall, 2005). For example, many snake species depend upon ectothermic prey (such as lizards) during juvenile life, but switch across to larger endotherms (generally mammals) as they grow larger. Such shifts occur in many different lineages of snakes, including elapids (Shine, 1980b, 1989b), pythons (Luiselli & Angelici, 1998; Shine *et al.*, 1998b), boids (Henderson, 1993; Rivas & Burghardt, 2001), vipers (Fitch, 1960), and colubrids (Plummer, 1981b; Mushinsky *et al.*, 1982). The actual shift in prey composition depends upon the snake's size and the prey resources available; for instance, some giant constrictors feed almost exclusively on mammals throughout their life, but inevitably the massive prey items taken by large adult snakes (such as wild pigs by *Python reticulatus* (Shine *et al.*, 1998b) or capybaras by anacondas, *Eunectes murinus* (Rivas & Burghardt, 2001)) are far too large to feature in the diet of juvenile conspecifics.

Spatial heterogeneity in prey resources, allied to the extreme flexibility in growth trajectories of snakes, generates massive variation in the degree of sex divergence in dietary habits. For example, female European grass snakes (*Natrix natrix*) grow much larger than males throughout most of their extensive geographic range, and take larger prey. In particular, adult female grass snakes frequently feed upon toads (*Bufo* spp.) too large to be ingested by conspecific males (Madsen, 1983, 1984). On the small island of Hallands Vadero near Sweden where toads are not found, however, both sexes of *N. natrix* feed upon smaller prey (newts) and attain very similar (and small) adult body sizes (Madsen & Shine, 1993b). Island snakes raised on abundant food in captivity grow rapidly, and exhibit the large adult size and marked sexual size dimorphism evident in mainland populations; thus, the restricted size spectrum of available prey on the island not only eliminates sex-based dietary divergence, but also directly modifies the usual pattern of sexual dimorphism in body size (Madsen & Shine, 1993b).

An analogous but more extreme case occurs in populations of carpet pythons (*Morelia spilota imbricata*) in southwestern Australia. These large constricting snakes are found both in remnant forest patches on the mainland and on small offshore islands. Females attain larger maximum body sizes than do males in all of these populations. Juveniles feed primarily on lizards, but adults of both sexes rely largely upon mammalian prey. Mammalian diversity differs dramatically among areas, with strong consequences for the snakes. In sites with a relatively continuous size spectrum of available prey (i.e. small, medium-sized, and large mammals) adult females are on average about twice as large as adult males (200 cm vs. 148 cm snout–vent length, 3 kg vs. 1.1 kg: Pearson *et al.*, 2002a). However, some islands have a depauperate mammal fauna; most notably, Garden Island near Perth has only two mammal species, and these differ dramatically in mean body mass: housemice (*Mus domesticus*, 12 g) and tammar wallabies (*Macropus eugenii*, 3 kg).

Accordingly, male carpet pythons on Garden Island feed almost exclusively on mice, and average 104 cm snout–vent length (305 g mass), whereas adult female pythons feed on wallabies and average 214 cm and 3.9 kg (Pearson *et al.*, 2002a). We do not know the extent to which this spectacular spatial divergence in the degree of sexual size dimorphism and sex-based dietary differences is due to genetic factors or to developmental plasticity. Nonetheless, the clear message from such examples is that sexual size dimorphism and dietary divergence are causally linked, and that local ecological features (especially the availability of prey of various body sizes) strongly constrain the expression of sex divergence in size as well as in dietary habits.

Sexual dimorphism in trophic morphology provides clear evidence of dietary differences between the sexes, although its functional significance is sometimes obscure. For example, neonatal pit-vipers of several species have brightly coloured tail-tips, used for luring prey from ambush positions. In one species (*Bothrops atrox*), neonatal males have yellow tail-tips but their sisters do not (Hoge & Federsoni, 1977). Perhaps the longer tails of males (a common feature of snakes: King, 1989) make luring a more successful strategy than would be the case for their less well-endowed siblings. A more widespread form of foraging-related sexual dimorphism involves the size of the snake's head relative to its body. Because head size limits maximal ingestible prey size, a sex divergence in relative head size influences prey-swallowing abilities of the two sexes. Such divergence is widespread among most major phylogenetic lineages within snakes (Shine, 1991a). The female is usually but not always the sex possessing a larger head relative to body length (Shine, 1991a). Field studies confirm that at the same body length, the sex with the larger head size takes larger prey (for filesnakes, Houston & Shine, 1993; for sea kraits, Shetty & Shine, 2002b; Shine *et al.*, 2002d).

Detailed anatomical studies confirm that the disparity in relative head size is due to a disproportionate enlargement of feeding structures (relative to other components of the head) in the sex with the larger head in some species, but simply to a general enlargement of all cranial structures in others (Camilleri & Shine, 1990). Such sex differences in relative head size are evident at birth in many taxa, suggesting a genetic causation; endocrine manipulations of juvenile gartersnakes (*Thamnophis sirtalis*) indicate that sex hormones are the likely proximate mechanism for such effects (Shine & Crews, 1988). However, an animal's feeding history can also modify relative head size: snakes given large prey items can develop larger heads (relative to body size) than do their siblings fed equal amounts of smaller prey items (Bonnet *et al.*, 2001).

Similar phenomena occur in other types of reptiles. For example, many lizard species display facultative herbivory, but generally only at larger body sizes (Pough, 1973). Thus, sex differences in mean adult body size can generate sex differences in the proportion of the diet composed of plant material (Schoener & Gorman, 1968; Schoener et al., 1982). The same pattern is seen in many turtles, especially map turtles (*Graptemys* spp.; see Table 13.1). Also, the greater physical strength contingent upon larger body size allows large turtles to crush mollusc shells. In several turtle lineages, the largest individuals within a population (always females) develop massively enlarged heads and shift their diet to this difficult-to-process prey type (Legler, 1985). Again, studies on map turtles (*Graptemys*) provide the most detailed evidence of this phenomenon (Table 13.1). Sexual size dimorphism can also influence the way plant material is eaten. For example, male Cape flat lizards (*Platysaurus capensis*) have larger heads than females, probably as a result of sexual selection for contest competition. Males can eat entire figs in a single swallow; females must break them up and therefore require a much longer processing time (Whiting & Greeff, 1997).

ECOLOGICAL DIFFERENCES BETWEEN THE SEXES, DIRECTLY RELATED TO REPRODUCTIVE ACTIVITIES

Reproduction profoundly affects the ecological attributes of reptiles, and often generates significant differences between males and females in traits such as the times, places and extent of activity, the relative priority accorded to feeding, thermoregulation and anti-predator behaviours, and so forth. The effects of reproduction are most easily seen by comparing reproductive vs. non-reproductive individuals, a task facilitated by two facts: reproduction generally occurs only at specific times of year, and females of many taxa skip years between successive reproductive episodes (and thus the population contains both reproductive and non-reproductive animals at the same time). Such comparisons reveal many sex-specific modifications of behaviour and ecology associated with reproduction.

The effects are most obvious in females. Many gravid reptiles carry a clutch or litter comprising >40% of their own pre-reproductive body mass; the developing eggs strongly constrain a female's locomotor ability, and hence increase her vulnerability to predation (Shine, 1980a; Schwarzkopf, 1994; Bonnet et al., 1999b). Perhaps reflecting this handicap, females of many reptile species reduce or cease feeding activities while they are burdened with the clutch (Shine, 1980a) and may remain close to cover (Bauwens & Thoen, 1981; Brodie, 1989). Especially in viviparous species inhabiting cool climates, pregnant females select higher and more stable thermal regimes than do non-reproductive conspecifics (Charland & Gregory, 1990; Andrews *et al.*, 1997; Rock & Cree, 2003); this behavioural thermoregulation accelerates development and may enhance the viability of their offspring (Shine & Harlow, 1993; Arnold & Peterson, 2002). Thus, pregnant females spend much of their time basking, even in relatively inclement conditions early in the morning or late in the evening while males are safely ensconced in their overnight retreat sites (Schwarzkopf, 1994).

Reproduction may modify not only activity and microhabitat selection in female reptiles, but also broader-scale patterns of habitat use. Most obviously, females of some marine reptile species must return to land to lay their eggs (sea turtles, sea kraits). In the case of sea turtles, the males remain in the ocean and it is only the females that face the very different challenges posed by a temporary sojourn on land. For laticaudid sea kraits, however, terrestrial environments are used not only for oviposition, but also for courtship, mating, sloughing and digesting food; and, thus, males and females probably spend fairly equal times on land (Shetty & Shine, 2002a). Indeed, males are more adept at terrestrial movement than are females (Shine *et al.*, 2002b).

Even in taxa that do not move between the ocean and the land, specific requirements for embryogenesis often force females to travel long distances through unfamiliar habitats to find optimal sites for oviposition or parturition. For instance, oviparous species living in cool, densely forested habitats often utilize communal egg-laying sites in relatively open, sun-exposed microhabitats that offer higher incubation temperatures than would be available elsewhere (e.g. Shine *et al.*, 2002a). Females may spend considerable time periods thermoregulating in such sites prior to oviposition, generating a strong sexual segregation in habitat types for weeks or months each year (Fitch, 1999). Long migrations from a female's usual home range to a preferred (often, communal) nesting site are widespread in tropical as well as temperate-zone reptiles, and in snakes as well as lizards (Whitaker & Shine, 2001; Madsen & Shine, 1996). For example, female Galapagos land iguanas travel long distances to volcanically produced 'hot-spots' that provide high incubation temperatures, producing strong habitat segregation between the sexes during the oviposition period (Pianka & Vitt, 2003). Such migrations may expose females to substantial danger, can result in high mortality rates and should be considered in conservation planning (Bonnet *et al.*, 1999a).

Although viviparous females do not need specific sites for oviposition, their careful thermoregulation frequently results in migration to relatively sun-exposed but predator-free microhabitats. For example, gravid snakes of several species aggregate in relatively open rocky areas throughout pregnancy (Graves & Duvall, 1987, 1995). Even when they do not aggregate, pregnant females may select sites that facilitate precise thermoregulation (e.g. Shine, 1979; Reinert, 1984; Blouin-Demers & Weatherhead, 2001).

Reproduction also affects the movements and habitat use of male reptiles, at a variety of spatial scales. For example, the basic social unit in many lizard populations comprises a territorial male plus one or more females that live within his territory. Especially in arboreal taxa, males tend to select perch sites in relatively exposed locations that provide good visibility to detect rival males. In contrast, females are more likely to be found in less exposed (and thus, safer) sites deeper within the vegetation (see references in Table 13.1). The consequent microhabitat difference may well result in males encountering different types of prey, experiencing different operative temperatures and being more exposed to visually hunting predators.

Primarily as a result of mate-searching movements, males in many reptile taxa have larger home ranges or activity areas than do females. This trend is especially prevalent in lizards (Perry & Garland, 2002) and crocodilians (e.g. Tucker et al., 1997), two groups in which male territoriality is also common. Snakes show the same pattern, though with more variability (Gregory et al., 1987; Reinert & Zappalorti, 1988; Whitaker & Shine, 2003), while in turtles females may be just as likely to cover more ground (e.g. Carter et al., 1999; Doody et al., 2002). As males move over wide areas, they inevitably spend their time in habitats different from those occupied by females. Within a habitat mosaic, such movements will take males through habitat types interspersed between those used by females. For example, dominant male freshwater crocodiles (Crocodylus johnstoni) travel considerable distances overland to move between billabongs containing receptive females (Webb et al., 1983). Male–female differences in habitat occupancy also may arise if males concentrate their mate-searching activities in areas where females are easier to find, rather than where they are most common. This may be true for the seasnake, Emydocephalus annulatus; males use visual cues to locate females, and spend much of their time in areas with light-coloured (sandy) substrates where the dark-coloured females are less cryptic (Shine et al., 2003a). Although there are no robust empirical data on the topic, we suspect that reproductive males often may spend disproportionate time in habitats where females are most easily located, or where females obtain access to some resource required for breeding. There is an obvious analogy with many species of anuran amphibians, in which males tend to wait beside water bodies (often calling to attract mates) whereas females visit such sites only briefly (for egg deposition) but spend most of their time in the surrounding matrix habitat obtaining food (Duellman & Trueb, 1986). Such reproduction-based habitat divergence can generate significant differences in dietary composition and feeding seasonality between the sexes (Lamb, 1984; Katsikaros & Shine, 1997).

Reproductive activities also cause differences in feeding habits between the sexes. The low maintenance requirements of ectothermy allow reptiles to forgo feeding for periods of months at a time, relying upon stored energy to fuel their activities. Thus, both males and females of many reptile taxa either reduce or cease feeding during reproduction. Typically, anorexia persists longer in females than in males, reflecting the longer duration of reproduction-associated activities (e.g. pregnancy, parental care) in females than in males (mate-searching). For instance, male water pythons, *Liasis fuscus*, in tropical Australia stop feeding for about 6 weeks during the mating period, losing about 17% of their body mass over this period. In contrast, females in the same population stop feeding for about three months, and lose about 44% of their body mass (Madsen & Shine, 2000). Because the reproductive activities that preclude feeding typically occur at different times of year in the two sexes (e.g. mate-searching precedes pregnancy), the climatic conditions (and, potentially, prey availability) during foraging bouts may differ between the two sexes. Seasonal asynchrony between the sexes in the primary periods of feeding also occurs in tortoises (*Testudo horsfieldi*: Lagarde et al., 2002, 2003). For many reptiles, prey availability is highly seasonal. For example, adult pit-vipers on one Chinese island feed solely upon migrating passerine birds that visit the island for only a few weeks each year (Sun et al., 2002). Thus, a seasonal shift in feeding rates between the sexes may substantially affect overall energy balance. Similarly, gape-limitation may force snakes to rely upon juveniles rather than adults of common prey species, so that a slight shift in the timing of snake feeding relative to the prey taxon's breeding season may affect body-size distributions of the prey and, hence, feeding rates (Worrell, 1958; Schwaner & Sarre, 1988; Madsen & Shine, 2000).

In summary, reproduction substantially modifies ecological parameters (habitat use, foraging ecology, etc.) in both sexes of many reptile species. Because such modifications take different forms or operate at different times of year, the end result may be spatial, habitat and/or nutritional segregation between the sexes. If such ecological factors place the sexes under different selection pressures, they may evolve different responses. For example, reproduction reduces the vigilance of male watersnakes (*Tropidonophis mairii*) but increases that of females; thus, reproduction decreases the distance at which a male will flee from an approaching observer, but increases this threshold distance for the heavily burdened females (Brown & Shine, 2004). Such sex divergences are likely to be widespread.

THE ADAPTIVE SIGNIFICANCE OF SEXUAL SEGREGATION IN REPTILES

The preceding review reveals that sexual segregation occurs commonly in reptiles, and often takes similar forms to those reported in other groups of animals. Such analogies suggest that similar processes may be at work, and hence encourage attempts to interpret sex-specific segregation within the same conceptual framework. As well as clarifying the nature of similarities and differences among phylogenetic lineages, this approach forces us to take a broader view of sexual segregation. Existing models were derived from research on herbivorous mammals (especially ungulates: see Chapter 2) and incorporating terrestrial ectotherms into our analyses may help to frame hypotheses in a broader view. We now consider sexual segregation in reptiles within the framework outlined in Chapter 2 (refer to that chapter for the original sources of the hypotheses listed).

Forage selection hypothesis

This hypothesis attributes habitat divergence to sex differences in forage or water requirements. The idea fits well with the observations that males and females of aquatic reptile species sometimes forage at different water depths to obtain different types of prey. Indeed, the same kind of sexual segregation of dive depth has been reported in water birds (cormorants, penguins) and also appears to be driven by sexual size dimorphism: that is, larger animals can dive deeper (Chapter 18). The same general explanation also accords well with cases in terrestrial systems, where females of some highly dimorphic taxa rely upon prey types that are restricted to different habitats than are those taken by males (e.g. rats vs. larger mammals for reticulated pythons). Sexual divergence in trophic morphology (such as relative head size in snakes) bears strong functional analogies to phenomena such as differences in wing loading or bill shape that facilitate foraging-habitat separation between male and female birds (Chapter 18).

However, incorporating ectotherms into the discussion suggests a broadening of the 'forage selection' model to include any ecological resources that are utilized differentially by males and females because of their different reproductive roles. The most obvious addition is the thermal environment, with males and females selecting sites that provide different operative temperature regimes and hence, different opportunities for precise regulation of their body temperatures and/or the thermal regimes available to their developing offspring. The same issue applies to aquatic ectotherms such as sharks, where females select warmer waters than males and thus, exhibit significant spatial segregation (Chapter 8). However, thermal differentials are far greater in terrestrial environments and hence, the potential costs and benefits of alternative sites are magnified. The widespread pattern for modified thermal preferenda in reproductive female reptiles thus may be a pervasive influence generating sex divergences in microhabitat selection, especially for viviparous species in cool climates.

Scramble competition hypothesis

Under this hypothesis, one sex reduces resource levels so dramatically that the other sex avoids the area. This hypothesis may be better-suited to explaining the behaviour of ungulates than of reptiles; even in herbivorous species, it is doubtful that the low offtake rates of most individuals would substantially reduce food availability for the other sex – or at least, to no greater a degree than they would reduce food supply for other individuals of their own sex. We know of no reptile study that would support this hypothesis.

Predation risk hypothesis

If there is a trade-off between food and safety (habitats with more food are also more risky), the sexes may assort based upon their energy needs and their vulnerability to predation. Reproductive strategies may also be important; for example, a territorial male lizard may accept a high predation risk by perching in the open, because it enables him to survey his territory for intruders; a female gains no such benefit and thus remains well hidden. For many ectotherms, however, the critical resource that is traded off against predation risk is thermal opportunity. Thus, gravid females may bask for long periods in open areas, thus exposing themselves to considerable risk, whereas males are more secretive because they derive no equivalent benefit from thermophily (Schwarzkopf & Shine, 1991). It is difficult to test the idea that activities such as prolonged basking (by pregnant females) or active mate-searching (by males) are risky, but analysis of sex ratios in roadkills and museum collections supports these notions. For example, adult (especially, pregnant) females usually outnumber adult males in collections of cool-climate viviparous species, whereas males are collected in large numbers as road casualties during the mating season (Shine, 1994a; Bonnet et al., 1999a; Reed & Shine, 2002).

Weather sensitivity hypothesis

This idea is framed around the energy budgets of endotherms, but in a broader view is relevant to ectotherms also. For example, the increased thermal inertia accorded by larger body size allows the larger sex to spend longer periods foraging in thermally extreme habitats (Seebacher et al., 1999). Underwater foraging by male Galapagos marine iguanas provides a good example of this phenomenon: males, which are larger, are able to forage in subtidal areas, while females are restricted to the intertidal areas (Table 13.1).

Social segregation as a by-product of habitat segregation

Depending on the spatial distribution of microhabitats, divergence in foraging sites can generate sexual segregation on a broader scale. Perhaps the best reptilian example involves highly variable adult sex ratios of filesnakes (*Acrochordus arafaurae*) within different isolated billabongs, depending upon water depths (Houston & Shine, 1994).

Social segregation due to inter-class avoidance

One sex may forcibly exclude the other from preferred habitats, thus generating ecological segregation. Sexual conflict is intense in some reptilian systems and may include forcible copulation (Shine et al., 2003b). Under such circumstances, females may actively avoid areas that contain scent cues from males (gartersnakes *Thamnophis sirtalis*: Shine et al., 2004) or leave when a male arrives. In free-ranging radio-tracked brown-snakes (*Pseudonaja textilis*), adult females vacate their usual burrows as soon as an adult male moves in (Whitaker & Shine, 2001). Female sea-snakes flee to deeper water to escape vigorous courtship by males

(M. Guinea, personal communication). Booth and Peters (1972) reported that between nesting events, female sea turtles (*Chelonia mydas*) aggregate in small areas of the lagoon where they are not exposed to harassment by courting males. Sims' observations of female sharks avoiding amorous males (Chapter 8) suggest the same basis for sexual segregation. Lastly, sexual size dimorphism may cause sexual segregation through direct agonistic encounters, independent of sex. For example, giant tortoises on Aldabra compete for limited patches of shade, with the larger males excluding the smaller females and sometimes causing their death due to over-heating (Swingland & Lessells, 1979).

Social segregation due to intra-class attraction

Aggregations of reptiles probably occur mostly because of habitat heterogeneity, with individuals clustering in restricted areas that provide optimal conditions. However, experimental studies confirm that some aggregations are actually driven by attraction to conspecifics. This is true for armadillo lizards (*Cordylus cataphractus*) and thick-tailed geckos (*Nephrurus milii*), which aggregate in the laboratory even when resources (food and shelter) are abundant (Visagie, 2001; Shah et al., 2003). For nesting females, this behaviour may provide advantages associated with communal egg deposition, such as predator swamping (Plummer, 1981a). In other cases the aggregations may confer thermal benefits (control over rates of heating and cooling) but, because such benefits are independent of sex, the aggregative behaviour does not generate sex-based segregation (Myres & Eells, 1968; Shah et al., 2003).

Social segregation due to asynchrony of activity budgets

This hypothesis relies upon specific features of energy budgets in endotherms and is unlikely to explain sexual segregation in ectotherms.

SUMMARY

In summary, sex-based divergence in reptiles is widespread both phylogenetically and ecologically. The functional significance of sexual segregation is likely to be equally diverse. Many cases of differential resource use may reflect secondary effects of pre-existing sex differences in traits (such as mean adult body size, nutritional needs, thermal preferenda) that have evolved as a result of the differing reproductive roles of males and females. Thus, ecological divergence in reptiles often may arise as a by-product of adaptations generated by sexual selection or fecundity selection. In contrast to endothermic vertebrates, exploitation of thermal heterogeneity within the terrestrial environment may be a major axis along which the sexes segregate in many reptile species. Particularly in cold climates, embryonic requirements for high, stable incubation temperatures may pose a major selective advantage for maintenance of such thermal regimes by pregnant females, in turn favouring significant shifts in habitat use and activity schedules. Direct competition between the sexes, either in terms of exploitation of resources or direct behavioural interference, seems likely to be less important in most reptiles than is commonly believed to be the case for many endotherms. However, social systems are far from irrelevant: for instance, territorial males may often select habitats that maximize opportunities for surveillance (to detect intruding rivals) rather than habitats that offer greater safety or more feeding opportunities.

The immense behavioural, ecological and morphological diversity of reptiles provides abundant opportunities for further research on the nature and determinants of ecological divergence between the sexes. Such work could usefully focus on comparisons between closely related species, or (even better) on populations within a single widespread species. It is clear that nearby populations often differ enormously in the degree of sexual divergence in ecological traits, that such differences are driven by a combination of phenotypic plasticity and adaptive change, and that careful experimentation can tease apart the relative roles of these two causal processes (e.g. Madsen & Shine 1993b). Manipulative experiments may often be more feasible with such animals than with endotherms, and have the potential to greatly expand our understanding of the general phenomenon of sexual segregation in vertebrates.

ACKNOWLEDGEMENTS

Funding was provided to Richard Shine by the Australian Research Council and to Michael Wall by a National Science Foundation Graduate Research Fellowship. We also thank J. B. Iverson, J. B. Losos, L. J. Vitt and M. J. Whiting for advice and ideas, and two anonymous reviewers for helpful comments on the manuscript.

14
_____

Sexual segregation in Australian marsupials

OVERVIEW

Most research on sexual segregation has been focused on eutherian mammals, showing that this phenomenon is widely but unevenly distributed across eutherian taxa, and is particularly prevalent amongst ungulate species that are sexually dimorphic in body size and give birth highly synchronously. Marsupials comprise a clade of mammals that has undergone extensive radiation in parallel to that of eutherians in terms of morphology, ecology and behaviour. We would then expect sexual segregation to occur in some marsupials, as it does in some eutherians, and most likely in those marsupial species that exhibit sexual dimorphism in body size and give birth highly synchronously.

We reviewed the literature for evidence of sexual segregation in 23 species from three orders of extant Australian marsupials. These species were drawn from each family and sub-family within these orders, and from distinct life-history categories within one family. We collated the incidence and form of segregation, the degree of body size dimorphism and the degree of birth synchrony in each species. We predicted that if dimorphism and synchrony were associated with sexual segregation in marsupials, then segregation should occur predominantly in species that were dimorphic and/or highly synchronous, but not species that were monomorphic and gave birth year-round. We also reviewed, in greater detail, the occurrence of sexual segregation in the genus *Macropus*, comprising the kangaroos and larger wallabies, since they are ecologically and behaviourally comparable to many ungulates.

Sexual Segregation in Vertebrates: Ecology of the Two Sexes, eds. K. E. Ruckstuhl and P. Neuhaus.
Published by Cambridge University Press. © Cambridge University Press 2005.

Sexual segregation occurred in all three extant orders of Australian marsupials and was patchily distributed among taxa: 9 of the 23 species we examined showed at least one form of segregation. Sexual segregation appears to be associated with body size dimorphism and birth synchrony in Australian marsupials, as it is in ungulates: all species that exhibited at least one form of segregation were either dimorphic in body size, highly synchronous in birth period, or both. Species that were monomorphic and gave birth either year-round or moderately synchronously, did not exhibit any form of sexual segregation. Segregation was particularly obvious amongst the larger macropods, as it is amongst their eutherian counterparts, the ungulates. These convergences suggest that similar evolutionary pressures have shaped the development of sexual segregation in eutherian and marsupial mammals.

INTRODUCTION

Sexual segregation has been identified as occurring at three distinct levels: habitat, spatial and social (Conradt, 1998b; Conradt, 1999; Bon et al., 2001; see Chapter 2 for details). Social segregation occurs when males and females live in separate groups; habitat segregation, when males and females occupy areas that differ in abiotic and biotic resources (such as forage and shelter); and spatial segregation when males and females use different ranges or areas within the same or different habitats. The proximate and ultimate causes of this phenomenon continue to be debated, but four main hypotheses have emerged in recent reviews (Main et al., 1996; Ruckstuhl & Neuhaus, 2000): (i) differences in energy requirements and offspring protection due to disparate reproductive strategies of the sexes, (ii) differences in nutritional requirements due to sexual size dimorphism, (iii) the need for each sex to engage in sex-specific social interactions, and (iv) incompatibility of activity budgets due to sex differences in body size.

Two key life-history characteristics appear to typify ungulate species that show sexual segregation: body size dimorphism and birth synchrony. Body size dimorphism, which has been associated with a polygynous mating system in many species (Darwin, 1871; Weckerly, 1998; Loison et al., 1999; Pérez-Barbería et al., 2002), has long been recognized as a key factor involved in sexual segregation (Jarman, 1983). In large browsing eutherian herbivores, for example, Mysterud (2000) found that the occurrence of segregation was positively related to sexual size dimorphism. It has been argued that sex differences in body size lead to differences in energetic requirements and food selection between the sexes, which in turn result in habitat segregation between the sexes (e.g. Pérez-Barbería & Gordon, 1998a). Sexual segregation also tends to occur in species in which mating activity and subsequent birth, is confined to a synchronous period. In most species, sexual segregation particularly occurs during the non-reproductive season, but it can also occur during the rut, as in Pere David's deer, *Elaphurus davidianus* (Jiang *et al.*, 2000). Atypically, social and spatial segregation also occurs in some mammalian herbivores that breed year-round, such as the African buffalo, *Syncerus caffer* (Prins, 1989).

Since the Cretaceous period, marsupials have evolved largely in parallel with eutherians and are regarded as a sister group (Lillegraven *et al.*, 1979). Marsupials probably originated in North America, then reached Australia via South America and Antarctica (Novacek, 1992; Springer *et al.*, 1997). Many marsupial species underwent extinctions in the Americas following invasion by eutherians in the Pliocene (Eisenberg, 1981; Springer *et al.*, 1997). The radiation of Australian marsupials, however, continued in relative isolation, allowing them to occupy ecological niches equivalent to those of eutherians elsewhere, in some cases displaying striking convergence in morphology and behaviour (Springer *et al.*, 1997).

Like eutherians, extant marsupials comprise a rich assemblage of species showing a variety of trophic types: herbivores, omnivores, insectivores and carnivores (Lee & Cockburn, 1985). Marsupials have occupied almost all types of climatic conditions, from deserts to monsoonal tropics and high alpine areas, and are represented by arboreal, terrestrial and fossorial species that show both nocturnal and diurnal activity patterns (Lee & Cockburn, 1985). Extant marsupials show dramatic variation in body size: the smallest species, the long-tailed planigale, *Planigale ingrami*, weighs no more than 4.5 g, some 20 000 times less than a male red kangaroo, *Macropus rufus* (Macdonald, 2001). Birth synchrony is also highly variable, ranging from two weeks in the mountain pygmy-possum, *Burramys parvus*, to year-round in species such as the honey possum, *Tarsipes rostratus*. In this review we examine whether body size dimorphism and birth synchrony are associated with the occurrence of sexual segregation in marsupials. We predict that if body size dimorphism and birth synchrony are associated with sexual segregation in marsupials, then segregation should predominate in species that are dimorphic and/or give birth highly synchronously, but not in species that are monomorphic and give birth year-round.

This is the first review of sexual segregation in marsupials. It is necessarily exploratory in nature because so little is known about the occurrence of sexual segregation in marsupials. This lack of information is in part because most species are nocturnal, solitary and cryptic, which makes behavioural studies of species in their natural habitat difficult. Nevertheless, enough species have now been studied in sufficient detail, at least at the level of describing their sociality and whether or not the sexes use habitat, space or food differently, to justify this review. One taxon, the genus *Macropus* (kangaroos, wallaroos and wallabies), has been relatively well-studied. We review the occurrence of sexual segregation in this genus, in greater detail, for two additional reasons. Firstly, species within *Macropus* show variation in parameters such as body size and dimorphism, birth synchrony, sociality and habitat preferences. Secondly, the gregarious species of *Macropus* are ecologically and behaviourally comparable to cervids and bovids (Jarman, 1983) and, since it is amongst the cervids and bovids that sexual segregation is most prevalent in eutherian mammals, we might expect sexual segregation to occur with similar prevalence amongst *Macropus* species.

METHODS

Taxonomic coverage

Extant marsupials comprise about 270 species within seven orders (Kirsch *et al.*, 1997). We restricted our review to the four orders of extant Australian marsupials: Peramelina, Dasyuromorphia, Notoryctemorphia and Diprotodontia. These are represented by 15 families, comprising approximately 145 species (Kirsch *et al.*, 1997). Subsequently, we excluded the Notoryctemorphia (the marsupial moles) because virtually nothing is known about their reproduction, social organization, behaviour and ecology, in either their natural habitat or in captivity (Benshemesh & Johnson, 2003). We opted not to cover all of the species within the remaining three orders, primarily because many species are very poorly studied. Instead, we selected a species corresponding to each family or sub-family. In the case of the Dasyuridae, we selected a species within each sub-family, from each of the six recognized life-history strategies (Lee *et al.*, 1982), where applicable. The dasyurid strategies are distinguished on the basis of five characters: the frequency of oestrous, the duration and timing of male reproductive effort, the seasonality of breeding and the age of females at maturity (Lee & Cockburn, 1985). Each strategy is described in Table 14.1. The resulting 23 species we

Table 14.1 Male and female body masses, body size dimorphism, birth periods and sociality for 23 Australian marsupial species, including life-history strategies of dasyurids (after Lee et al., 1982). % sex dimorphism in body mass = percentage difference of male body mass compared to female body mass. Mass, birth period and sociality source = literature (or other) source from which data on body mass, birth period and sociality was taken.

Taxon	Common name	Life-history strategy	Mean adult body mass (g) Female	Mean adult body mass (g) Male	% sex dimorphism in body mass	Mass source	Birth period (days)	Birth period source	Gregarious	Sociality source
Dasyuromorphia										
Dasyuridae										
Dasyurinae										
Dasyurus hallucatus	Northern quoll	1 or 2	460	760	65.2	Oakwood (2002)	16	Oakwood (2000)	No	Oakwood (2002)
Dasyurus maculatus	Spotted-tailed quoll	3	1700	3200	88.2	Jones & Barmuta (2000)	45	M. Jones (pers. comm.)	No	Jones et al. (1997)
Sarcophilus harrisii	Tasmanian devil	2	5400	8400	55.6	Jones & Barmuta (2000)	45	M. Jones (pers. comm.)	No	Jones (1995)
Antechinus agilis	Agile antechinus	1	18	30	66.7	Dickman (1995)	20	Lazenby-Cohen & Cockburn (1988)	Yes	Lazenby-Cohen & Cockburn (1991)
Planigalinae										
Planigale gilesi	Giles' planigale	5	6.9	11.5	66.7	Read (1995)	190	Read (1984b)	No	Read (1982)
Sminthopsinae										
Sminthopsis crassicaudata	Fat-tailed dunnart	4	16.5	19	15.2	Lee et al. (1982)	240	Morton (1978)	Yes	Morton (1978)
Sminthopsis leucopus	White-footed dunnart	3	19	26	36.8	Lunney (1995)	40	Lunney (1995)	No	Lunney & Leary (1989)
Myrmecobiidae										
Myrmecobius fasciatus	Numbat		550	700	27.3	Friend (1996)	20	Friend & Whitford (1993)	No	Friend (1989, 1995)

Peramelina									
Peramelidae									
Peramelinae									
Isoodon obesulus	Southern brown bandicoot	476	614	29.0	Stoddart & Braithwaite (1979)	180	Stoddart & Braithwaite (1979)	No	Stoddart & Braithwaite (1979)
Thylacomyidae									
Macrotis lagotis	Greater bilby	751	1705	127.0	Moseby & O'Donnell (2003)	356	Moseby & O'Donnell (2003)	No	Moseby & O'Donnell (2003)
Diprotodontia									
Vombatidae									
Lasiorhinus krefftii	Northern hairy-nosed wombat	27100	27000	−0.37	D. Taggart (pers. comm.)	180	D. Taggart (pers. comm.)	No	Wells (1978)
Phascolarctidae									
Phascolarctos cinereus	Koala	8300	11300	36.1	Martin & Handasyde (1990)	240	Martin & Handasyde (1990)	No	Mitchell (1990)
Phalangeridae									
Phalangerinae									
Trichosurus vulpecula	Common brushtail possum (NT)	1400	1600	14.3	Kerle *et al.* (1991)	300	Kerle (1984)	Not	Day *et al.* (2000)
Trichosurus vulpecula	Common brushtail possum (Tas)	3200	3600	12.5	Kerle *et al.* (1991)	90	Kerle (1984)	at	
Trichosurus vulpecula	Common brushtail possum (NSW)	2500	2700	8.0	Kerle *et al.* (1991)	90	Kerle (1984)	any	
Trichosurus vulpecula	Common brushtail possum (NZ)	2330	2460	5.6	Crawley (1973)	120	Kerle (1984)	site	

259

Table 14.1 (cont.)

Taxon	Common name	Life-history strategy	Mean adult body mass (g) Female	Male	% sex dimorphism in body mass	Mass source	Birth period (days)	Birth period source	Gregarious	Sociality source
Burramyidae										
Burramys parvus	Mountain pygmy-possum		43	40	−7.0	Mansergh & Scotts (1990)	14	Mansergh & Scotts (1990)	Yes	Mansergh & Scotts (1989)
Petauridae										
Petaurinae										
Petaurus australis	Yellow-bellied glider		556	631	13.5	Goldingay & Kavanagh (1990)	180	Goldingay & Kavanagh (1990)	Yes	Goldingay & Kavanagh (1990)
Dactylopsilinae										
Dactylopsila trivirgata	Common striped possum		439	471	7.4	Handasyde & Martin (1996)	356	Handasyde & Martin (1996)	Yes	Handasyde & Martin (1996)
Pseudocheiridae										
Pseudocheirinae										
Pseudocheirus peregrinus	Common ringtail possum		735	729	−0.8	How et al. (1984)	255	How et al. (1984)	Yes	Ong (1994)
Pseudocheiropsinae										
Petauroides volans	Greater glider		1265	1490	17.8	Henry (1984)	60	Henry (1984)	No	Henry (1984)
Acrobatidae										
Acrobates pygmaeus	Feathertail glider		12.7	14.2	11.8	Ward (1990)	210	Ward & Renfree (1988)	Yes	Fleming & Frey (1984)

Tarsipedidae									
Tarsipes rostratus	Honey possum	12.5	8.6	−31.2	Renfree et al. (1984)	365	Wooller et al. (2000)	No	Bradshaw & Bradshaw (2002)
Macropodidae									
Macropodinae									
Macropus fuliginosus	Western grey kangaroo	25 100	35 000	41.4	Coulson & MacFarlane (unpub.)	150	Norbury et al. (1988)	Yes	Coulson (1993)
Potoroinae									
Bettongia tropica	Northern bettong	1268	1220	−3.8	Vernes & Pope (2002)	356	Vernes & Pope (2002)	No	Vernes & Pope (2001)
Hypsiprymnodontidae									
Hypsiprymnodontinae									
Hypsiprymnodon moschatus	Musky rat-kangaroo	511	529	3.5	Johnson & Strahan (1982)	90	Dennis & Marsh (1997)	No	Johnson & Strahan (1982)

Strategy 1: monoestrous females, abrupt mortality of males at the conclusion of their first short mating period, a synchronized and predictable mating period that is shorter than gestation, and females attain sexual maturity at about 11 months of age.

Strategy 2: monoestrous females, some individuals of both sexes survive to reproduce in at least a second year, a synchronized and predictable mating period, and females attain sexual maturity at about 11 months of age.

Strategy 3: females are usually monoestrous but some may undergo a second oestrous, the mating period is synchronized and highly predictable and females attain sexual maturity at about 11 months of age.

Strategy 4: polyoestrous females, an extended, seasonal breeding period, and females attain sexual maturity at about 6 months of age.

Strategy 5: polyoestrous females, an extended, seasonal breeding season, and females attain sexual maturity at 8–11 months of age.

Strategy 6: since there is no detailed information available on the life-histories of sixth strategy species, we did not include any in our analysis.

261

chose were those that had been best studied, within their group, and in which information on all the parameters of interest was available from one study population (Table 14.1). If data from more than one population existed, we selected the population with the most published information available.

We also reviewed, in greater detail, the occurrence of sexual segregation in the genus *Macropus* (Table 14.2), which is represented by 14 species (Strahan, 1995). Subsequently, we excluded 5 species because little was known about their behaviour and ecology in the wild, either because they were extinct (e.g. Toolache wallaby, *Macropus greyi*) or relatively cryptic (e.g. black wallaroo, *M. bernardus*; western brush wallaby, *M. irma*; Parma wallaby, *M. parma*; black-striped wallaby, *M. dorsalis*).

We reviewed only those species for which information was available from a single study population for most, and preferably all, of the variables considered. We did this in order to enhance our ability to interpret a clear ecological relationship between body size dimorphism and birth synchrony. Our approach contrasts with some previous comparative studies of the association between ecological variables in marsupials, which have compiled data from multiple populations of the same species (e.g. Fisher & Owens, 2000; Fisher *et al.*, 2002) to eliminate gaps in comparative data sets. However, any gain in taxonomic coverage may be at the expense of data quality. Injudicious pooling of different components of data across the range of a species is likely to obscure important patterns because ecological variables are often strongly associated within a population, for example, home range size and body weight (Harestad & Bunnell, 1979), population density and group size (Southwell, 1984), birth synchrony and diet (Di Bitetti & Janson, 2000).

In order to achieve our aim of compiling data from single populations, we avoid, where possible, drawing body mass data from Strahan's (1995) *The Mammals of Australia*. A number of reviews (e.g. Weckerly, 1998; Fisher & Owens, 2000) have used this compendium, but these values are usually obtained from multiple populations and are not consistently derived from sexually mature adults.

Ecological data

We compiled data on three key variables. The two key life-history variables were *body size dimorphism* and *birth synchrony*. The third variable was the expression of *sexual segregation* in its various forms. Where possible we used the published literature, but we also obtained information

Table 14.2 Reported occurrence of behaviour consistent with sexual segregation in Macropus species. % sex dimorphism in body mass = percentage difference of male body mass compared to female body mass.

Taxon	Common name	Maximum adult body mass (kg)[a]		% sex dimorphism in body mass	Birth synchrony[b]	Type of segregation		
		Female	Male			Habitat	Diet	Social
M. agilis	Agile wallaby	15	29	99.7	All year	✓	?	×
M. antilopinus	Antilopine wallaroo	20	49	145	Moderate	✓	?	✓
M. eugenii	Tammar wallaby	7	9	28.6	High	?	?	?
M. fuliginosus	Western grey kangaroo	39	72	84.6	Moderate	✓	✓	✓
M. giganteus	Eastern grey kangaroo	40	76	88.8	Moderate	✓	×	✓
M. parryi	Whiptail wallaby	15	26	73.3	All year	?	?	?
M. robustus robustus	Eastern wallaroo	24	49	108.5	Moderate	✓	?	✓
M. robustus erubescens	Euro	24	49	108.5	All year	✓	?	✓
M. rufogriseus banksianus	Red-necked wallaby	15	27	86.2	Moderate	?	?	✓
M. rufogriseus rufogriseus	Bennett's wallaby	16	27	68.8	High	?	?	?
M. rufus	Red kangaroo	37	93	151.4	All year	✓	✓	×

[a] From Jarman (1989), Jarman (1991).
[b] From Tyndale-Biscoe & Renfree (1987).

directly from authorities on some species. The species covered and the sources used are presented in Table 14.1.

Body size dimorphism

We used Ruckstuhl and Neuhaus' (2002) index to describe sexual size dimorphism for each species. We used mean body mass of sexually mature adult males and females for this index. We acknowledge that by using mean body mass values we may understate the degree of dimorphism achieved by old individuals in species showing continuous growth (Jarman, 1983), but conclude that the influence of this effect depends on population structure, so the operational level of dimorphism is best represented by population means. The exception to this is the data presented in Table 14.2 (from Jarman, 1989, 1991), which gives maximum adult body masses for *Macropus* species. We followed Ruckstuhl and Neuhaus (2002) and considered species showing ≥ 20% difference in body mass between the sexes as functionally dimorphic. This 20% threshold is based on physiological differences in digestive efficiency between species of ruminants (Van Soest, 1994) and is thought to apply not only to differently sized species, but also to differently sized sexes of the same species (Illius & Gordon, 1987, 1992).

Birth synchrony

We expressed birth synchrony as the length of the birth period in days. Birth period is commonly expressed in the literature as the months of the year during which births occur (e.g. July to September). In such cases we multiplied the number of months that births occurred by 30 days. For example, births have been reported to occur from July to February in fat-tailed dunnarts, *Sminthopsis crassicaudata*, from southern Victoria (Morton, 1978), which we calculated as eight months multiplied by 30 days = 240 days. We did not include months where less than 5% of annual births had been reported to occur. We acknowledge that this measure of birth synchrony is relatively crude because it does not reflect the distribution of births within the birth period. For example, western grey kangaroos, *Macropus fuliginosus*, in north-west Victoria have a five-month breeding season (September to January), yet up to 87% of births occur from October to December (Norbury *et al.*, 1988). Ideally we would have used a more precise measure, such as the percentage of births occurring in the peak months (Isaac & Johnson, 2004), but the level of data required (number of births in each month) was generally unavailable for most marsupial species.

Sexual segregation

We examined the published literature for evidence of three levels of sexual segregation: habitat, diet and social. Habitat and social segregation are recognized as distinct forms of segregation (Conradt, 1999). Segregation by diet is a distinct level of segregation, occurring when males and females consume different foods, or the same foods but in different proportions. Segregation by diet thus operates at a finer scale of resource selection than does habitat segregation, so that classes of animals occupying the same habitat stratum may segregate in terms of the food types acquired within that habitat, as is seen in muskoxen, *Ovibos moschatus* (Schaefer & Messier, 1995) and giraffes, *Giraffa camelopardalis* (Caister *et al.*, 2003).

We used a modified version of Müller & Thalmann's (2000) classification of social life to distinguish between solitary and gregarious species. We defined a species as *gregarious* if it either lived in a cohesive group, or foraged mainly solitarily but exhibited social networks (i.e. formed sleeping associations and/or showed friendly interactions with conspecifics even outside the breeding season). We defined a species as *solitary* if it foraged alone and had no relations with conspecifics except during the breeding season. We did not look for evidence of social segregation in species showing a solitary social system, because solitary animals separate from conspecifics of their own sex as much as from those of the opposite sex (Conradt, 1998b). As noted by Conradt (1998b), however, solitary animals can still be inter-sexually segregated in respect to habitat if the ranges of individuals overlap, and the same could apply to segregation by diet.

Sexual segregation in Australian marsupials has rarely been measured in any formal manner, and no published studies have applied the index of segregation advocated by Conradt (1998b). Consequently, we were able to report only the presence or absence of segregation. We considered behaviour consistent with males and females living or nesting or foraging apart, for at least part of the year, as evidence of social segregation, and sex differences in diet or habitat as evidence of diet and habitat segregation, respectively. We accepted evidence if (i) a statistically significant difference, regardless of magnitude, was reported, or (ii) if a trend or pattern was described, but no statistical analyses carried out. We used 'significantly' to denote when a study reported a difference of statistical significance, or simply described the trend when a study reported a difference but did not conduct a statistical analysis. We included only cases where sexual segregation occurred between adults, since sexual segregation is not necessarily fully developed in young animals (Conradt, 1999). We also included cases where only a proportion of the population of one sex segregated from the other sex, such as between large males and females with young, or between old males and old females.

VARIATION BETWEEN POPULATIONS

The common brushtail possum, *Trichosurus vulpecula*, posed a particular problem for our analysis. This species is one of the most well-studied Australian marsupials and a number of populations have been studied in considerable detail. According to our selection criteria, any of these populations of brushtail possum could have been chosen to represent the species, but there was considerable variation between these populations in many of the parameters of interest. For example, birth synchrony varied considerably between populations, being highly seasonal in Tasmania to year-round in the Northern Territory (Kerle, 1984). Body size dimorphism also varied between populations, but to a lesser degree than birth synchrony (Isaac & Johnson, 2004). Some variation was also evident between populations in other parameters such as reproductive success, population density, dietary preferences, home range size, mating system and social structure (Kerle, 1984). Because of this variation and because more than one population has been studied thoroughly, we present the mean length of birth period and body size dimorphism of the best-studied populations from four widely separated locations: southern Tasmania (Hocking, 1981), northern New South Wales (NSW) (How, 1972), northern Northern Territory (Kerle, 1983; Kerle et al., 1991) and an introduced population in New Zealand (Crawley, 1973).

RESULTS

Body size dimorphism

The 23 selected Australian marsupial species showed wide variation in body size dimorphism (Table 14.1; Figs. 14.1–14.3). Eleven species showed male bias (males > 20% heavier than females), the most extreme example being the greater bilby, *Macrotis lagotis*, in which males were

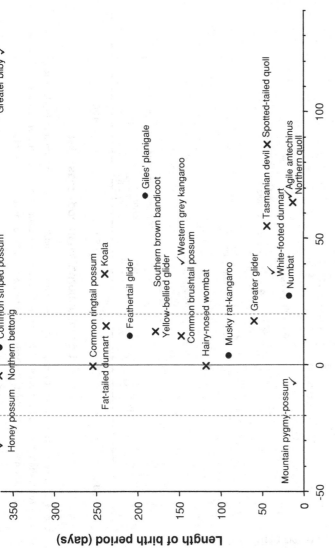

Sexual dimorphism in body mass (%)

Figure 14.1 The occurrence of habitat segregation, in relation to body size dimorphism (expressed as the percentage difference in male body mass compared to female body mass) and birth synchrony (expressed as the length of the birth period in days), in 23 selected Australian marsupial species. ✓ Evidence of habitat segregation; ✗ no evidence of habitat segregation; ● habitat segregation has not been studied. Dashed lines indicate the 20% dimorphic threshold.

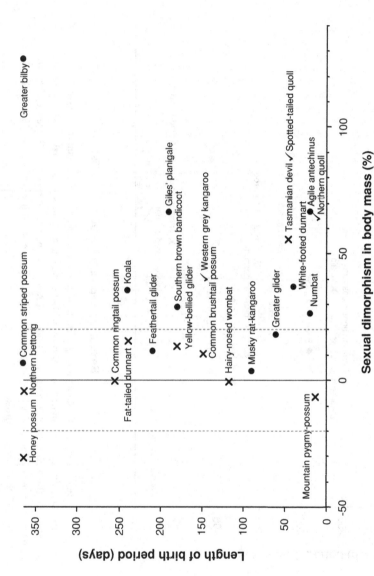

Figure 14.2 The occurrence of diet segregation, in relation to body size dimorphism and birth synchrony, in 23 selected Australian marsupial species. ✓ Evidence of diet segregation; × no evidence of diet segregation; ● diet segregation has not been studied. Dashed lines indicate the 20% dimorphic threshold.

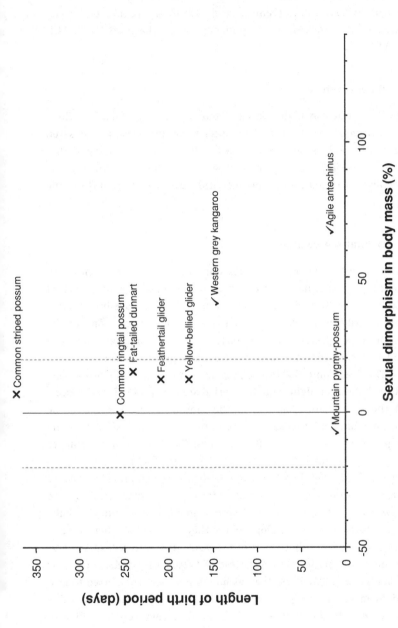

Figure 14.3 The occurrence of social segregation, in relation to body size dimorphism and birth synchrony, in eight gregarious Australian marsupial species. ✓ Evidence of social segregation; × no evidence of social segregation. Dashed lines indicate the 20% dimorphic threshold.

127% heavier than females (Moseby & O'Donnell, 2003). Female-biased dimorphism occurred only in the honey possum, in which females were 31% heavier than males (Renfree *et al.*, 1984). The remaining species showed <20% difference in body mass between the sexes (Table 14.1; Figs. 14.1–14.3).

Birth synchrony

Birth synchrony was highly variable, but the 23 species fell into three basic patterns (Figs. 14.1–14.3). Four species gave birth year-round, seven gave birth highly synchronously, with all matings occurring within a six-week period, and the remaining 12 gave birth moderately synchronously, with matings occurring within two to eight months of the year.

Habitat segregation

Data on habitat use by both sexes was available for 18 species, and habitat segregation had been detected in 7 of these (Fig. 14.1). Each of these 7 species were either dimorphic in body size, gave birth highly synchronously or showed both of these characteristics. None of the monomorphic species that gave birth year-round or moderately synchronously showed any evidence of habitat segregation (Fig. 14.1).

The direction of habitat segregation, in terms of which sex occupied the habitat of highest quality, varied among species. Females occupied the best quality habitats in the mountain pygmy-possum, the honey possum and the agile antechinus, *Antechinus agilis*. In mountain pygmy-possums on the Bogong High Plains of Victoria, females remained throughout the year at high elevations. These areas were composed of boulder-fields with low woody shrubs, which provide food and shelter of high quality and quantity (Mansergh & Scotts, 1989). During the two-week breeding season in spring, males moved into this habitat from further down the slope where they occupied a wider range of habitat types at all other times of the year (Mansergh & Scotts, 1989). In honey possums in the south-west of Western Australia, Bradshaw and Bradshaw (2002) found that, although both sexes preferred to feed at night in six-metre-high, dense stands of holly banksia, *Banksia ilicifolia*, males left this habitat during the day. They sought refuge in the surrounding low, and often quite exposed, heathland at some distance from the banksia stands, where the larger and behaviourally dominant females (Russell, 1986) remained throughout the day. In agile antechinus in south-eastern NSW, males and females had different foraging ranges (Lazenby-Cohen & Cockburn, 1991). The foraging ranges of females were stable throughout the year and, according to Lazenby-Cohen and Cockburn (1988), were of higher quality than the drifting foraging ranges of males.

In contrast, females occupied habitat of apparently lower quality in the western grey kangaroo. In semi-arid, north-west Victoria, Mac-Farlane and Coulson (in press) found that during the non-breeding season females occurred significantly more often in floodplain and less often in dune than expected, while males showed the opposite pattern. Although both habitats offered a similar amount of shelter, floodplain offered a significantly lower biomass of forage.

Although habitat segregation also occurred in the greater bilby, white-footed dunnart, *Sminthopsis leucopus*, and southern brown bandicoot, *Isoodon obesulus*, we were unable to comment on the direction of this habitat segregation because either the quality of the habitats occupied by the different sexes was not reported, or it was not clear which habitat was of higher quality. In a re-introduced population of greater bilbies in arid South Australia, Moseby and O'Donnell (2003) found that although both sexes preferred to dig and use burrows in sand dunes, the nocturnal fixes and burrows of males were found in swale habitat more often than those of females. Moseby and O'Donnell (2003) regarded swale habitat as lower quality, in terms of shelter, than dune, but its quality in terms of food was not assessed. Lunney *et al.* (1989) described habitat segregation in white-footed dunnarts on the south-coast of NSW. About 80% of individuals caught on the ridges were males, whereas half of the individuals caught on the lower mid-slopes were female. The trend described was not statistically analysed and habitat quality was not assessed. Stoddart and Braithwaite (1979) reported segregation between classes within sexes in southern brown bandicoots in southern Victoria. The authors found that young, sexually mature females and large, old males had a strong preference for newly regenerating heathland, whereas old females (and middle-aged males) occupied habitats that had more established vegetation. Again, this trend was not statistically analysed and habitat quality, in terms of food and shelter, was not discussed.

SEGREGATION BY DIET

Dietary studies that distinguished between the sexes had been conducted on 12 species. Of these, unequivocal evidence of diet segregation occurred in three: northern quoll, *Dasyurus hallucatus*, spotted-tailed quoll, *Dasyurus maculatus*, and western grey kangaroo (Fig. 14.2). Each of these species was size-dimorphic, a synchronous breeder, or both. None of the monomorphic species that gave birth year-round or moderately synchronously showed any evidence of segregation by diet (Fig. 14.2).

Jones and Barmuta (1998) found a significant difference in diet between male and female spotted-tailed quolls in central Tasmania during summer: males fed primarily on medium and large-sized mammals, while females fed mostly on small mammals and birds. A similar pattern of diet segregation has been reported, although not statistically tested, in northern quolls in the north of the Northern Territory. Both sexes consumed the same three dietary groups (invertebrates, vertebrates and plants), but males took a greater percentage of vertebrates than did females (M. Oakwood, personal communication). Dietary differences in western grey kangaroos in north-western Victoria were reported by Cutter (2002), who found a significant difference in the proportion of grass consumed by three sex–age classes. Large males consumed a significantly higher proportion of grasses than females (90% compared to 81%), while small males fell in-between (85%).

SOCIAL SEGREGATION

We sought evidence of social segregation only in the eight species that had a gregarious social system (Table 14.1; Fig. 14.3). There may have been other species that qualified as gregarious (e.g. Giles' planigale, *Planigale gilesi*) but insufficient information was available. Of these eight species, only three showed evidence of social segregation: mountain pygmy-possum, western grey kangaroo and agile antechinus (Fig. 14.3). Each of these species was either size-dimorphic, a synchronous breeder, or both. No gregarious species that were monomorphic, and either gave birth year-round or were moderately synchronous, showed any evidence of social segregation (Fig. 14.3).

Mountain pygmy-possums and agile antechinus nest communally (Cockburn & Lazenby-Cohen, 1992; Mansergh & Broome, 1994). Mansergh and Scotts (1989) found that, leading up to and during hibernation, adult female mountain pygmy-possums nested with overlapping generations of related females (aged 1–12 years), while males nested alone or with other males. Agile antechinus, in south-eastern NSW, also nested communally up to and during winter until the end of the mating period (Cockburn & Lazenby-Cohen, 1992). Males nested communally with females before the mating season, but females also nested alone or with other females. During the short mating season in late August, males aggregated inside a limited number of nest trees, which females briefly visited (Lazenby-Cohen & Cockburn, 1988). Western grey kangaroos in north-western Victoria also showed strong social segregation outside the breeding season when single-sex foraging groups were observed significantly more often than expected by chance (MacFarlane and Coulson, in press).

SEXUAL SEGREGATION IN KANGAROOS AND WALLABIES

The red kangaroo is a large, strongly size-dimorphic species, which gives birth year-round and shows two forms of segregation (Table 14.2). Newsome (1980) reported dietary segregation in this species in central Australia: grasses were eaten almost exclusively, but females consumed significantly more forbs than did males as pastures dried, and the reverse pattern occurred as drought intensified. Three studies of red kangaroos in western NSW have reported some degree of habitat segregation: population surveys by Johnson and Bayliss (1981) showed that females predominated on seasonally preferred pastures. Croft's (1991b) radio-tracking study showed that the home ranges of females encompassed more creekline habitat, where green grasses persisted in dry periods, and radio-tracking by McCullough and McCullough (2000) showed that the home ranges of females encompassed a mixed woodland habitat that males almost never used. However, studies of group composition by Russell (1979), Johnson (1983) and McCullough and McCullough (2000) did not reveal social segregation (but see MacFarlane and Coulson, in press).

The common wallaroo, *M. robustus*, is somewhat smaller in size and less dimorphic than the red kangaroo. The common wallaroo has four distinct subspecies, two with wide distributions: the eastern wallaroo, *M. r. robustus*, on the slopes of the Great Dividing Range, and the euro, *M. r. erubescens*, further west. Both show strong habitat and social segregation (Table 14.2). Adult males occupied significantly higher elevations on rocky outcrops in central and western NSW (Taylor, 1983; Croft, 1991a), and used a wider range of habitats, including more open vegetation, than did females in south-western Western Australia (Arnold *et al.*, 1994). At the social level, Taylor (1983) found that males of all size classes associated significantly with each other in larger groups (≥ 4 individuals), whereas females tended to dissociate from each other. Croft (1981) also found that medium-sized males occurred significantly more often in all-male groups than large males.

The antilopine wallaroo, *M. antilopinus*, is similar in size to the common wallaroo, but is more dimorphic (Table 14.2). In the only study

of the antilopine wallaroo, conducted in the north of the Northern
Territory, Croft (1987) found some evidence of habitat and social
segregation: the sex ratio was significantly female-biased in the high-
density stratum in both wet and dry seasons, and both small and large
males tended to associate with their own class in the high-density stratum, as did females with small pouch young.

The western grey kangaroo is a large, moderately size-dimorphic
species, which gives birth moderately synchronously and is the only
Macropus species so far to show all three levels of sexual segregation
(Table 14.2). Segregation by diet has been detected in north-western
Victoria (see earlier). Habitat segregation has been indicated by significantly skewed sex ratios during the winter (non-mating) season in western NSW (Johnson & Bayliss, 1981) and in north-western Victoria (see
earlier), and in minor differences in the vegetation types encompassed
by home ranges of radio-tracked individuals in south-western Western
Australia (Arnold et al., 1994) and western NSW (McCullough & McCullough, 2000). Social segregation has been inferred from the significant
over-representation of all-male groups in winter in western NSW (Johnson, 1983), in most habitats throughout the year in north-western Victoria (Coulson, 1993), and in groups of two emerging from cover in
south-western Western Australia (Arnold et al., 1990).

The eastern grey kangaroo, *M. giganteus*, is morphologically, ecologically and behaviourally similar to the western grey kangaroo (Table
14.2). Segregation by diet has been sought but not found in the eastern
grey kangaroo (Taylor, 1981; MacGregor & Jarman, personal communication), while habitat and social segregation have been reported. Jarman
and Southwell (1986) noted limited segregation by habitat in north-
eastern NSW: a proportion of large and small adult males decreased
their use of the high-density habitat stratum, where most females
occurred, in summer. McCullough and McCullough (2000) found that
radio-tracked females spent 95% of their time in only two habitats,
whereas males distributed their activity more evenly over those and
four additional habitats. In social terms, Jaremovic and Croft (1991a)
found that large males were significantly and positively associated with
medium-sized males and negatively associated with females during winter in southern NSW, while Jarman and Southwell (1986) found that a
significant proportion of females (40%) occurred in single-sex groups in
north-eastern NSW.

The agile wallaby, *M. agilis*, mainland red-necked wallaby, *M. rufogriseus banksianus*, Bennett's wallaby, *M. rufogriseus rufogriseus*, tammar
wallaby, *M. eugenii*, and whiptail wallaby, *M. parryi*, are smaller species,

which cover a range of dimorphism and birth synchrony (Table 14.2). The most social of this group of wallabies is the whiptail wallaby, which is at least as gregarious as the grey kangaroos and antilopine wallaroo (Kaufmann, 1974), but there is a lack of information pertinent to sexual segregation. Johnson (1989) reported some degree of social segregation in mainland red-necked wallabies from northern NSW: adult females associated significantly with their mothers, whereas males tended to associate with other males of similar body size. Birth synchrony is highly seasonal in the Tasmanian subspecies of the red-necked wallaby, the Bennett's wallaby (Curlewis, 1989). However, no systematic study of habitat use or social grouping in Bennett's wallaby has yet been conducted. The tammar wallaby is the only other species within *Macropus* that gives birth highly synchronously (Tyndale-Biscoe & Renfree, 1987), but there has been no study with sufficient precision to detect possible habitat or social segregation in this species. The agile wallaby gives birth year-round in the wet/dry tropics of northern Australia, and has been the subject of more study. Dressen (1993) reported significant differences in sex ratios among habitats in the dry (non-breeding) season in the north of the Northern Territory, but no obvious social segregation could be detected at that site by Dressen (1993), or in northern Queensland by Johnson (1980).

DISCUSSION

Sexual segregation occurs in all of the orders of Australian marsupials that we were able to review: the Dasyuromorphia, Peramelina and Diprotodontia. Of the 23 species that we reviewed, sexual segregation, at all levels, was absent in seven, but present, in at least one form, in nine. The occurrence of sexual segregation in the three orders is patchily distributed. In the Dasyuromorphia, evidence of sexual segregation occurs in the sub-families Dasyurinae and Sminthopsinae, but does not occur in all of the species examined within these sub-families (e.g. segregation occurs in the white-footed dunnart but apparently not in the fat-tailed dunnart). Sexual segregation has not been specifically studied in Planigalinae or Myrmecobiidae. In this order, all three levels of sexual segregation are expressed, although not consistently throughout. In the Peramelina, the two members we reviewed, from the families Peramelidae and Thylacomyidae, show evidence of habitat segregation. In the Diprotodontia, sexual segregation is present in the species we examined that represent the families Tarsipidae, Burramyidae and Macropodidae, but did not occur in any of the species we examined that represent the other seven families. All three levels of sexual segregation occur in this order, but they are not expressed consistently in these three families.

Our review shows that, like in ungulates, body size dimorphism and birth synchrony appear to be associated with the occurrence of sexual segregation in Australian marsupials. All examined Australian marsupials that exhibited at least one form of sexual segregation were either dimorphic in body size, highly synchronous in birth period or showed both of these life-history characteristics. Furthermore, species that were monomorphic and gave birth year-round or moderately synchronously, showed no evidence of sexual segregation. This pattern was particularly distinct in the occurrence of social segregation: only the three gregarious species that were dimorphic and/or gave birth highly synchronously exhibited social segregation.

The relationship with body size dimorphism and birth synchrony was less obvious when habitat and dietary dimensions of sexual segregation were considered. Not all species that were dimorphic in body size and/or gave birth highly synchronously also showed segregation at the habitat or dietary level. Three solitary dasyurids, the northern quoll, spotted-tailed quoll and Tasmanian devil, were intriguing: they were strongly dimorphic and highly synchronous, yet showed no evidence of habitat segregation. The solitary koala, *Phascolarctos cinereus*, also showed no evidence of habitat segregation, despite being dimorphic in body size. The mountain pygmy-possum, honey possum and Tasmanian devil showed no evidence of diet segregation, despite being synchronous, dimorphic or both. In some cases, the lack of segregation may simply reflect inadequate resolution in the study techniques used. For example, diet segregation was apparently absent in the Victorian population of mountain pygmy-possums studied by Mansergh et al. (1990), but in a second population from the adjacent alpine region of NSW, Smith and Broome (1992) found female mountain pygmy-possums to have a higher intake of arthropods than males in all seasons and locations, and attributed the lack of a sex difference in diet in Victoria to poor or incomplete seasonal representation of faecal samples. Similarly, diet segregation has not specifically been studied in our selected population of agile antechinus in south-eastern NSW, but male agile antechinus have been reported to take relatively more slow-moving prey than females (Lunney et al., 2001) in a nearby population on the south coast of NSW.

In many marsupial species, segregation by diet has not been specifically studied, and as a result, we could not be fully confident of the relationship between dimorphism, synchrony and the occurrence of diet segregation. Nevertheless, sexual differences in diet have been hinted at, or suggested as likely to occur, in a number of poorly studied dimorphic and/or synchronously breeding species. For example, the habitat segregation reported in the greater bilby (Fig. 14.1) is likely to be associated with segregation in diet as well, since shallow diggings in swale habitats suggest that males are accessing food resources that are not used by females (Moseby & O'Donnell, 2003). In planigales, Read (1984a, 1989) found evidence of sexual differences in diet between and within species on the basis of size, with the largest individuals taking the larger prey. Since males are considerably larger than females in Giles' planigales, diet segregation is likely in this species. Diet segregation may also occur in the koala, where observations suggest that adult males are constrained by their much greater weight and cannot reach outer branches where young leaves are growing (K. Handasyde, personal communication). Segregation by diet may also occur, at least seasonally, in the numbat, *Myrmecobius fasciatus*. During winter, males emerge from their nests for only four hours each day, while females feed continuously throughout the day, digging into termite galleries just below the soil surface (Friend, 1996). If different termite taxa or castes become active near the surface at different times during the day, females may have access to termite taxa not available to males.

Intraspecific variation also supports an association between the two life-history correlates and sexual segregation in Australian marsupials. Different populations of some marsupial species, living under different ecological conditions, showed variation in their degree of body size dimorphism and birth synchrony and in the occurrence and degree of sexual segregation. For example, segregation by diet was reported in our selected population of spotted-tailed quolls in Tasmania (Jones & Barmuta, 1998), where births occurred within six weeks of the year and body size dimorphism was extreme (M. Jones, personal communication), but did not occur in two populations of the same species on the mainland, where dimorphism was less pronounced (Belcher, 2003). Similarly, habitat segregation occurred in white-footed dunnarts from coastal NSW, which showed moderate male-biased dimorphism and strong birth synchrony (Lunney & Leary, 1989; Lunney et al., 1989; Lunney, 1995), but a monomorphic population of the species from coastal Victoria showed no evidence of habitat segregation (Lunney, 1995; Laidlaw et al., 1996). Cases like these are probably quite common in Australian marsupials, but their frequency of occurrence is difficult to determine when sampling of most species has been inadequate.

There is, however, a growing number of ecological studies of Australian marsupials that demonstrate intraspecific variation in a wide variety of parameters, including birth synchrony and body size dimorphism in species such as the common brushtail possum (Kerle, 1984; Fletcher & Selwood, 2000; Isaac & Johnson, 2004) and southern brown bandicoot (Heinsohn, 1966; Stoddart & Braithwaite, 1979; Thomas, 1987). Comparative studies within the same species would therefore be extremely useful to shed light on the factors that affect sexual segregation in marsupials or any other taxon.

Broome's (2001) study of the density, home range, seasonal movements and habitat use of the mountain pygmy-possum in Kosciuszko National Park, NSW, is the only study that addresses possible mechanisms causing sexual segregation in a marsupial. Broome (2001) suggested that sexual segregation in this species may be a result of female aggression during the non-breeding season, or inter-sexual competition for resources, but may also be explained by differences in energy requirements, seed availability and hibernation strategies between the sexes. Until further research is carried out on this species, we cannot resolve which of these factors, or combination of factors, is responsible for the marked sexual segregation observed in the mountain pygmy-possum.

In eutherians, sexual segregation is most prevalent in ungulates, particularly amongst cervids and bovids. Since the gregarious species of *Macropus* are ecologically and behaviourally comparable to cervids and bovids, we would expect sexual segregation to occur with similar prevalence amongst this marsupial group. Our review reveals abundant evidence of sexual segregation in *Macropus*, although often it has not been recognized or described as such. Segregation is particularly obvious amongst the larger macropods, as it is amongst their eutherian counterparts, the ungulates. These convergences provide evidence that similar evolutionary pressures have shaped the development of sexual segregation in eutherian and marsupial mammals alike.

To date, research into the behavioural ecology of kangaroos and wallabies has focused mainly on the patterns of association between individuals or sex–age classes (Croft, 1989; Jarman & Coulson, 1989; Jaremovic & Croft, 1991b) rather than on dissociation (i.e. segregation). While the larger species of *Macropus* have received most attention and show clear evidence of segregation, two smaller taxa would also be expected to show marked segregation. These taxa, the tammar wallaby and Bennett's wallaby, are quite dimorphic in body size, and are the most synchronous breeders within *Macropus*, yet are largely unstudied in terms of their behavioural ecology.

The synthesis we sought of sexual segregation in Australian marsupials has been constrained by the availability of detailed information on this phenomenon. We encourage researchers working on the behavioural ecology of Australian marsupials to be attuned to the potential of their study species to sexually segregate. Until we become more aware of sexual segregation in Australian marsupials, and measure it in a uniform manner, we cannot draw firm conclusions about its relationships with body size dimorphism and birth synchrony, or its relative strength. We require better data on sexual segregation as a continuous variable, using statistical analyses with appropriate phylogenetic control and a more comprehensive coverage of taxa.

ACKNOWLEDGEMENTS

We wish to thank the following people for providing information about their study species: Chris Belcher, Tony Friend, Kath Handasyde, Joanne Isaac, Menna Jones, Scott Laidlaw, Meri Oakwood, the late John Seebeck and Simon Ward. We also thank Peter Jarman, Chris Johnson, Peter Neuhaus, Kathreen Ruckstuhl and Simon Ward for their helpful comments on the manuscript. The Holsworth Wildlife Research Fund provided funding to Abigail MacFarlane to travel from Australia to attend the workshop at Cambridge University. A grant (S1991097) from the Australian Research Council to Graeme Coulson, and an Australian Postgraduate Award to Abigail MacFarlane, provided funding for the research that formed the basis of this review.

15

Social systems and ecology of bats

OUTLINE

Despite their small size (2–1000 g), bats have life-cycles comparable to those of much larger mammals, characterized by longevity, low fecundity and substantial parental care (Barclay & Harder, 2003). They also have very diverse life-cycles, with considerable variation in social systems and mating strategies (Altringham, 1996; McCracken & Wilkinson, 2000). There are currently believed to be almost 1100 species in 18 families, dominated by 5 very large families: Vespertilionidae, Rhinolophidae, Molossidae, Phyllostomidae and Pteropodidae (Koopman, 1993). Most species have not been studied, but enough is known about a small but varied minority to begin to understand the significance of the different social structures seen in bats.

Although a common feature in many mammals and often seen in sexually segregated species, marked sexual dimorphism in bats is rare. Female microbats are typically a little larger than males, but the differences are frequently only statistically significant with large sample sizes. One exception is *Ametrida centurio*, in which the females are markedly larger than the males (Peterson, 1965). The few instances of marked dimorphism are usually adaptations for sexual signalling, such as the enlarged nasal cavities of male hammer-headed bats, *Hypsignathus monstrosus*, and the erectile and scented hairs of the crested free-tailed bat, *Chaerephon chapini* (Altringham & Fenton, 2003).

So why do bats segregate by sex? Sexual segregation has been given relatively little attention in bats despite their diversity, their tendency to form large social units and indeed their tendency to segregate.

Sexual Segregation in Vertebrates: Ecology of the Two Sexes, eds. K. E. Ruckstuhl and P. Neuhaus.
Published by Cambridge University Press. © Cambridge University Press 2005.

Our aim in this review is to describe briefly the roosting and foraging ecology of bats and summarize what is known about social structure and social interaction. We will then explore possible explanations for the sexual segregation observed in many species before describing the few studies that have addressed the subject directly. We begin by examining the reasons why bats are so diverse.

WHY ARE BATS SO DIVERSE?

The diversification of bats into a wide range of ecological niches must be due in part to the evolution of echolocation and powered flight, which opened up a niche, the night sky, that is largely unoccupied by other vertebrates. Flight alone confers many potential benefits. Although flight is energetically expensive by the second, it is very economical by the kilometre, so bats can travel long distances cheaply (Altringham, 1996). The larger Old World fruit bats (up to 1 kg) forage 50 km or more from their roosts on a nightly basis. Most of the smaller species (2–10 g) typically forage within a few kilometres of the roost, but 10 or even 50 km foraging trips are not unknown (e.g. Williams et al., 1973; Fenton, 1990; Arlettaz, 1999). Maintaining a high body temperature is expensive, particularly in small bats, and demands a reliable and plentiful food source. Flight increases the daily energy demands still further. Bats, like other mammals, have evolved strategies to deal with seasonal changes in climate and food supply. Many temperate and subtropical bats are migratory (Fleming & Eby, 2003), making seasonal flights of hundreds or thousands of kilometres. Bats migrate either to find new food resources and so remain active throughout the year, or to find suitable hibernation sites. This ability to make long daily or seasonal flights has enabled bats to evolve varied social structures that often show interesting seasonal changes.

Flight also gives bats the ability to exploit roosting sites that are unavailable to most other mammals and that offer considerable protection from predators (Kunz, 1982; Altringham, 1996; Kunz & Lumsden, 2003). At one extreme this has allowed bats to form the largest mammalian assemblages currently found in nature. At the other extreme we see a diversity of smaller social units of varying stability and kinship, including solitary roosting.

Although the majority of bats eat arthropods (primarily insects), they exhibit a wide range of feeding habits (Altringham, 1996). Many species have diets of fruit, nectar and pollen, others are carnivores (eating fish, frogs, birds and other mammals) and three species of vampire bat feed exclusively on blood. There are also many omnivorous species, particularly amongst the Phyllostomidae of the New World. Even within the 'insectivores' there are many specialists, showing a range of feeding strategies. Optimal exploitation of different foods by different feeding strategies may require different social structures. This will be the case if group foraging, or the exchange of information about food sources, are important factors in optimizing foraging efficiency (e.g. Wilkinson, 1995). Thus, differences in feeding habits, together with differences in roosting habits, habitat and climate have all contributed to the evolution of a diverse assemblage of social systems.

Because of their mobility, bats that roost together do not necessarily feed together, since the benefits and drawbacks of sociality in the two activities may be very different. We know a good deal about roosting behaviour in bats, but far less about what they do when they leave the roost: only relatively recently has technology allowed us to track and study these elusive mammals outside their roosting places.

ROOSTING BEHAVIOUR: WHY ROOST IN COLONIES?

A characteristic common to most bats is their extreme sociability. Colony size typically varies from a few to a few hundred bats, but colonies of a few species form the largest known mammalian aggregations (Altringham, 1996). It is probable that species that form colonies do so for three main reasons: reduced predation risk, reduced thermoregulation costs and the social benefits of co-operation and information transfer.

Reduced predation risk

In most cases the bats gain some protection from predators, both in the roost and on emergence from the roost. A major factor driving coloniality in the larger aggregations is probably the dilution effect: for any individual bat, amongst hundreds or thousands, the risk of being taken by a predator is small. Bats are known to adopt a number of strategies that further reduce the risk of predation on emergence. For example, many species emerge from the roost not at random, but in clusters. This temporal clustering probably makes it difficult for predators to fix on a single bat, enhancing the dilution effect (e.g. Speakman et al., 1999). Bats frequently change roost and one reason for this may be to avoid those predators that have located their roost (Fenton et al., 1994).

The time bats emerge from the roost is variable: some species are flying before sunset, most in the first hour after sunset, but others later still. Emergence time is related to factors that are likely to determine a species vulnerability to predation, such as its flight performance, its foraging habitat and its foraging strategy (Jones & Rydell, 1994). For example, fast flying aerial hawkers emerge much earlier than slow flying species that glean prey from the ground. Emergence time varies with colony size within a species (Fenton et al., 1994). Large colonies emerge earlier, when the risk from diurnal predators is greater, than small colonies. This is probably because of the lower risk to individuals in large colonies.

Reduced thermoregulation costs

Colony formation also yields thermoregulatory benefits, because bats that cluster together reduce their exposed surface area and therefore reduce heat loss (Speakman & Thomas, 2003). This is particularly important to bats, since they have large surface-area-to-volume ratios and often use cool roost sites. Furthermore, a large colony in a small roost cavity can significantly increase local ambient temperature, so clustering gives bats the ability to alter their own environmental conditions. Clustering can also limit water loss (Thomas & Cloutier, 1992).

Social benefits of co-operation and information transfer

Colonial behaviour can facilitate information transfer about good foraging sites (e.g. Wilkinson, 1992). Several studies provide good evidence for information transfer and these are discussed later. Coloniality will also facilitate the evolution of group foraging strategies that may increase an individual's foraging efficiency. Again, these will be discussed later. Familiarity with particular roosts, habitats and roost inhabitants can in theory bring benefits through improved foraging efficiency and reduced predation.

The possible benefits of colonial living depend in part upon the stability of colonies and of the groups of individuals within them. Transient groupings can gain from reduced thermoregulatory costs or reduced predation risk, but are less likely to benefit from co-operative strategies or an improved knowledge of the roost and its surroundings. It is becoming clear that although many bats gain from quite random associations, structure and social interaction play a major role in many species and can give additional benefits.

BATS THAT FORM LARGE COLONIES

The best-known bats that form very large colonies are found amongst cave-roosting, insectivorous species and tree-roosting Old World fruit bats. The largest roosts are those of the Brazilian free-tailed bat (*Tadarida brasiliensis*, family Molossidae) in North, South and Central America. Nursery colonies in caves in Texas have been estimated to have held as many as 20 million females in the past (Davis *et al.*, 1962). Many of the males do not migrate north with the females, but stay in the southern part of their range until the females return. Mixed colonies form for mating. In common with most bats, only a single offspring is produced each year. There appears to be no structure or hierarchy within the roost and the females leave the young in enormous creches when they go out to feed. On their return, each female locates its own pup with the aid first of isolation and contact calls and then smell (McCracken, 1984; Gustin & McCracken, 1987). Each evening the females leave to forage at dusk and this mass emergence from the cave can take several hours. Predators (snakes and raptors) take advantage of this predictable event, but have minimal impact on the immense population.

Many other insectivorous bats form large colonies in caves. Schreiber's bent-winged bat (*Miniopterus schreibersii*, family Vespertilionidae, Fig. 15.1(a)) is one of the most widely distributed species, being found in temperate and tropical regions of both the Northern and Southern hemispheres. Colonies can contain tens of thousands of individuals, both male and female. Dwyer (1966) has shown that in Australian populations, colony composition can vary spatially and temporally. Nursery colonies were predominantly all female, 'adult' colonies had both sexes and some were mating roosts in which females were transient residents. Juveniles and yearlings also formed colonies. All three colony types varied seasonally in size and composition in relation to the reproductive cycle. European colonies of Schreiber's bat show similar patterns that are widespread in other insectivorous families: e.g. Rhinolophidae, Hipposideridae and the Emballonuridae, and in the Old World fruit bats (Pteropodidae).

Many fruit bats roost in camps in the canopies of trees. Despite persecution and habitat loss, some of these colonies can still number tens of thousands and sometimes over 100 000 (Kunz & Lumsden, 2003).

(a)

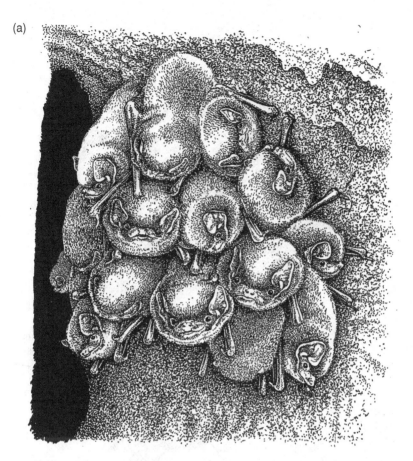

Figure 15.1   (a) Schreiber's bent-winged bat, *Miniopterus schreibersii*, a species that forms large colonies in caves in many parts of the world. (b) White tent bat, *Ectophylla alba*, which makes foliage tents for its small family-based colonies in the neotropics.

There is structure in the colonies of several of the species studied, for example in the grey-headed flying fox, *Pteropus poliocephalus* (Pteropodidae) (Nelson, 1965). This species often forms sexually segregated camps prior to parturition, with males and females occupying different trees or gathering at different levels in the canopy of a single tree. As the young are weaned males and females mix, the males form small territories within the camps (maintained by aggression and scent marking) and attract one or more females. After mating the females leave to form large single-sex groups and the males quickly establish

(b)

Figure 15.1 (*cont.*)

male-only groups. In winter, large camps may break up into smaller groups. Similar patterns have been observed in other species.

BATS THAT FORM SMALLER COLONIES

The vast majority of species form colonies of a few individuals to a few hundred. Their roosting ecology is extremely varied and has been reviewed several times (e.g. Kunz, 1982; Altringham, 1996; Kunz & Lumsden, 2003). Roosts range from the effectively permanent (caves and rock crevices) through the long-lived (tree boles and buildings) to the ephemeral (foliage). Some of the more unusual roosts include flowers, bamboo culms and the nests of birds and ants (Altringham, 1996; Kunz & Lumsden, 2003). Some species even manufacture their own roosts by modifying leaves to give greater protection from weather and predators (Altringham, 1996; Kunz & Lumsden, 2003). The distribution, density and lifespan of roost sites may have a significant influence on the distribution and density of some species, as well as influencing colony size and social structure.

In most temperate species, small colonies are made up exclusively or largely of females, the males living apart, singly or in small groups. One of the best studied examples is the greater horseshoe bat, *Rhinolophus ferrumequinum* (Rhinolophidae; Ransome, 1990). These colonies often comprise only a small number of matrilines. The brown long-eared bat, *Plecotus auritus* (Vespertilionidae), forms small colonies in which both males and females are typically present, both sexes showing marked natal philopatry (Burland *et al.*, 1999; Entwistle *et al.*, 2000; Burland *et al.*, 2001). Similarly, in Bechstein's bat, *Myotis bechsteinii* (Vespertilionidae), nursery colonies of 20–40 females are closed, with just one or two matrilines (Kerth *et al.*, 2000). In contrast to long-eared bats, male Bechstein's bats roost singly, away from the females. Colonies of females frequently fragment, with considerable mixing in the subgroups, but associations are strong between lactating females, possibly for improved social thermoregulation. Fragmentation and fusion is common to many species, but the mechanisms driving it are poorly understood. Many species also change roost frequently (O'Donnell & Sedgeley, 1999). This may be to reduce predation as described earlier, to reduce parasite burden, to improve roost microclimate (Kerth *et al.*, 2001b) or to move nearer to better foraging sites. Roost switching is not confined to bats using small or semi-permanent roosts, since many cave-roosting bats are also very mobile. In northern Greece, female long-fingered bats, *Myotis capaccinii* (Vespertilionidae), regularly fly between nursery roosts in disused mines that are 30 km apart (E. Papadatou, unpublished data).

In many tropical and subtropical species, for example in many bats of the family Phyllostomidae, a single male roosts with a harem of females. In the Jamaican fruit-eating bat (*Artibeus jamaicencis*) colonies of over 200 bats roost in caves. A dominant male defends a group of between 4 and 18 females. In groups with over 14 females there is also a subordinate male, who, although not appearing to actively defend the harem, seems to deter take-over bids from other males and will take priority as the next dominant male (Ortega & Arita, 2000; 2002). A study of a colony of greater sac-winged bat, *Saccopteryx bilineata* (Emballonuridae), in Costa Rica showed that single males defended a particular territory in the roof of a building, even if there were no females present. Harem size was between one and five females, and non-territory holding males roosted close to the harems. In this species, juvenile females disperse and the male juveniles remain with their natal colony (Voigt & Streich, 2003).

Bats that use ephemeral roosts typically form the smallest and most nomadic colonies. Good examples include the neotropical disk-winged bats (Thyropteridae) and palaeotropical sucker-footed bats (Myzopodidae). They use fleshy pads on wrists and ankles to stick to the inside of furled leaves (Kunz & Lumsden, 2003). As the leaves unfurl the bats must move on to new leaves. However, many species lacking these special adaptations make use of similar roosts in similar ways. *Cynopterus sphinx*, the short-nosed fruit bat, constructs tents in the fruit/flower clusters of the kitul palm tree. They form harems, comprising a single male with 1–37 females and dependent young. Average group size rises from 6 in the wet season to 14 in the dry season and although the overall harem structure is maintained throughout the year, individuals and groups of bats switch roosts regularly (Storz *et al.*, 2000). The yellow-eared bat, *Uroderma bilobatum*, is another tent-building bat. A study in Costa Rica focused on a colony of about 30 bats, which moved into a coconut grove prior to parturition. They divided themselves amongst an average of nine tents. Two of these tents were nursery roosts containing the majority of the females and occasionally males, but most of the males were usually found roosting alone or in small groups. There were regular movements of bats both into and out of the colony during the study (Lewis, 1992). *Ectophylla alba* is a tent-making bat that roosts in *Heliconia* leaves (Fig. 15.1(b)). Post-parturition group composition changes from mixed groups of four to eight bats, to harems and male-only roosts (Brooke, 1990). Even temperate species may roost in very small groups or singly. The females of *Lasiurus* species (Vespertilionidae) in North America frequently roost alone or with their offspring, hanging from the foliage of broad-leaved and coniferous trees (Kunz & Lumsden, 2003). Carnivorous species roost in small colonies or family groups. The Australian ghost bat, *Macroderma gigas* (Megadermatidae), is a large insectivore and carnivore that can form roosts of over 400 bats, but rather less than 100 is more typical. The carnivorous *Vampyrum spectrum* (Phyllostomidae) on the other hand typically roosts in pairs. Colony size and composition is often influenced by mating strategy, as discussed later.

FORAGING BEHAVIOUR

What do we know about the social aspects of foraging behaviour? When the large colonies of Brazilian free-tailed bats leave the roost they form dense clouds that can be picked up by radar as they fan out and forage up to 50 km away from the roost, flying at altitudes of up to 3000 m. The bats feed on aerial insect swarms, often at considerable altitude in large groups (Williams et al., 1973; Lee & McCracken, 2002). We know little about their foraging strategy or behaviour, but it is possible that group cohesion may enhance both the location of prey and foraging efficiency.

In fruit bat camps, all or most of the bats leave the roost at the same time. They typically break up into several streams of bats heading in different directions, although the numbers going in each direction vary considerably. The bats change foraging area and food plant frequently as they follow the fruiting of different trees. Again, it is not known how information about new food sources is communicated through the colony. More is known about the neotropical fruit-eater the greater spear-nosed bat, *Phyllostomus hastatus*: females emit screech calls to attract roost-mates when leaving the roost or feeding (Wilkinson, 1995). Wilkinson and Boughman (1998) were able to show that these calls facilitate co-ordinated foraging.

Bats that form small colonies show a wide diversity of foraging patterns. The earliest detailed studies were by Bradbury and Vehrencamp (1976a,b; 1977) who studied several tropical emballonurid species that showed considerable social diversity. Female proboscis bats, *Rhynchonycteris naso* (Fig. 15.2), foraged in groups in the centre of their colony territory, with several males feeding alone on the periphery, apparently defending the 1 ha territory for the small, stable colony of 5–11 bats. *Saccopteryx leptura* and *S. bilineata* also formed small colonies that defended small territories, within which individuals or small groups had favoured feeding patches. *Balantiopteryx plicata* formed larger colonies of up to 2000 females that foraged alone or in groups, but further details are lacking. The brown long-eared bat, *Plecotus auritus*, is a temperate gleaning species that forages within 1 km of the roost. Females foraged at the core of the home range, but males foraged further away, and it was suggested that this may reduce inter-sexual competition for resources rather than being for territorial defence (Entwistle et al., 1996). Bechstein's bat, *Myotis bechsteinii*, is also a temperate woodland species in which each female colony member uses a small foraging area close to the roost (Kerth et al., 2001a). However,

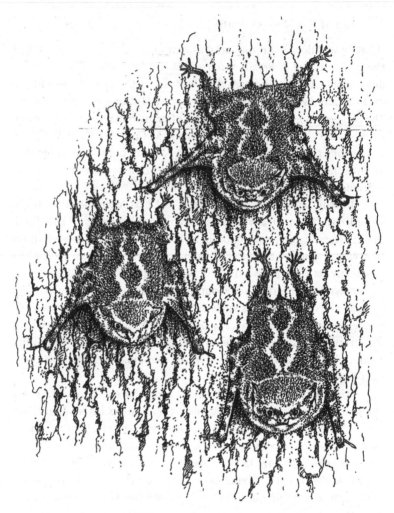

Figure 15.2   The proboscis bat, *Rhynchonycteris naso*: males forage on
the periphery of the home range used by the colony females.

no evidence for territoriality, group foraging or information transfer
amongst colony members was found. Females occupied small, largely
non-overlapping feeding sites to which they were faithful, even when
switching roost. Daughters appeared to inherit part of their mothers'
foraging sites. Males roosted separately and presumably foraged away
from the females, but this was not studied. Daubenton's bats, *Myotis
daubentonii*, typically feed low over still water, distributing themselves
along a river, using a number of roosts in trees and stone bridges.

Radio-tracking of this species (P. Senior, J. D. Altringham & R. K. Butlin, unpublished) has shown that colony males forage within the home range of the females, with both males and females returning each night to forage along the same short (<100 m) stretches of river. Bats were faithful to these feeding sites even when switching roost site and the same foraging patch could be used for several years.

There are numerous descriptions of foraging behaviour in temperate bats that comment on or describe territoriality, group foraging and information transfer, but relatively few detailed studies. Territoriality has been described in a number of species, on the basis of vocalization and chasing behaviour (Wilkinson, 1995). In many cases, the evidence is convincing, but, as Hickey and Fenton (1990) have pointed out, chasing does not necessarily imply territorial behaviour. Tandem flying is seen in many species, with one bat closely following the flight of another. There is often no evidence that it leads to the 'eviction' of any of the individuals and may serve other functions. In reviewing the literature, Wilkinson (1995) suggested that tandem and group flying may increase foraging efficiency, since group members could avoid foraging at already exploited patches and can perhaps take prey flushed out by other group members. A recent study by Reddy and Fenton (2003) on foraging red bats, *Lasiurus borealis*, provided no evidence for co-operative foraging. Information transfer in bats has also been reviewed by Wilkinson (1995), who demonstrated it very clearly in his own study of the evening bat, *Nycticeus humeralis* (Vespertilionidae) (Wilkinson, 1992). In a large nursery roost, females that returned to the roost after an unsuccessful foraging trip were shown to follow other departing females. Although the bats followed were apparently chosen at random, those following were more successful using this strategy than unsuccessful bats that did not follow others. A study of group foraging and the benefits to be had from it was made by Howell (1979) for the nectar feeder, *Leptonycteris curasoae* (Phyllostomidae) (Fig. 15.3). Small flocks fed sequentially on flowering agaves, using an optimal group foraging strategy to maximize returns on each plant and avoid returning to plants already exploited that night. In insectivorous bats, group foraging would be most beneficial to those species feeding on prey that are temporally and spatially patchy: these are primarily the aerial hawkers. Those species that feed on more evenly distributed food, such as gleaning bats, are perhaps more likely to be solitary in their foraging habits since information transfer is less beneficial. This trend is generally supported by the literature (Kerth *et al.*, 2001a), but there are exceptions and many species show very flexible foraging strategies.

Figure 15.3   The benefits of group foraging have been demonstrated in studies of the nectar-feeding bat, *Leptonycteris curasoae*.

COMMUNICATION AND BEHAVIOUR

Bats make use of vision, smell, touch and even thermoperception (Altringham & Fenton, 2003), but since most species fly in the dark and orientate using echolocation, research has naturally focused on sound as the most important communication channel. The Old World fruit bats, Pteropodidae, with the exception of some members of the genus *Rousettus*, do not echolocate and rely on vision. It has been suggested that feeding insectivorous bats attract other bats through their echolocation calls. When a bat closes on its prey its echolocation pulses show a brief, rapid increase in repetition rate, from about 10 Hz to over 100 Hz. This is known as a feeding buzz, and sends out an unambiguous signal to other bats that there is food to be had. Playback experiments, using recorded feeding buzzes have shown that in some cases, other bats are attracted by these calls (Wilkinson, 1995). A feature of the feeding buzz of many species is a terminal drop in frequency. This has no obvious function in echolocation and in theory should reduce both the resolving power and ranging acuity of the bats' calls. However, low-frequency sound travels further and it has been suggested that it may advertise the presence of food and attract conspecifics (Wilkinson, 1995). Another possibility is that physiological constraints do not allow the bats to produce high frequency calls at high repetition rates.

Bats use a range of communication or social calls and they are typically at frequencies between 5 kHz and 20 kHz, whereas echolocation calls are usually above 20 kHz. If food is abundant, the presence of other bats could reduce predation risk and calls may be used specifically to attract conspecifics. In other cases calls may be territorial, warning conspecifics to stay away. The structure and context of social calls have rarely been studied in detail and the bats may be using several types.

Many species, on returning to the roost, fly around the entrance in large groups, often emitting low-frequency social calls before finally entering the roost. Individual bats may persist in this behaviour for many minutes and the whole event may last for an hour or more (Vaughan & O'Shea, 1976). This behaviour can take place at the entrances to several roosts before the bats finally enter one of them. One suggestion is that it may help to imprint roost entrance location. It could also serve to synchronize roost switching or to introduce naïve bats to new roosts (e.g. Wilkinson, 1992; Kerth & Reckardt, 2003), since foraging bats may fly around the roost used the previous day, before moving to a new one. Whatever the reason, it is presumably of some adaptive significance, since it incurs significant costs. Considerable energy is expended in both flying and calling and this very predictable and conspicuous event is likely to increase the risk of predation by both owls and diurnal predators.

SEASONAL CHANGES IN ROOSTING BEHAVIOUR:
THE TEMPERATE CYCLE

Social structure and behaviour have evolved to meet the physiological and ecological demands acting on a species. Thus, seasonal changes in climate, food availability and reproductive status will all influence social behaviour. This is best seen, and best studied, in temperate bat species, since they experience large seasonal fluctuations in environmental conditions. As described earlier, temperate bats use migration, hibernation or even both, as mechanisms to overcome a seasonal loss or reduction in food supply, which occurs at a time when the energetic demands of homeothermy usually increase dramatically. Accompanying these shifts in behaviour there are often changes in social structure. The best way to illustrate these complexities is to describe what is known as the temperate cycle, and its variations, in some of the better-studied bats of Europe and North America. European examples come from two families, the evening bats or Vespertilionidae and the horseshoe bats, Rhinolophidae. The Vespertilionidae is the largest and most widespread family, the Rhinolophidae are confined to the Old World. Both families have temperate and tropical members. Almost all North American temperate species are members of the Vespertilionidae. The following discussion applies to both families. In mid-winter bats hibernate singly, in mixed- or single-sex groups and sometimes in mixed-species groups (Altringham, 1996). Groups can be loose or form tight clusters to regulate heat and water loss (e.g. Twente, 1955; Beer & Richards, 1956). In the spring, arousal and dispersal from hibernation sites typically leads to partial or total sexual segregation. As described earlier, females typically gather to form nursery colonies of varying size. Males most typically roost either alone or in small groups in separate roosts, but can form male-only colonies or integrate to varying extents with the nursery roosts. The females, having mated and stored sperm in the previous autumn or winter (see later) become pregnant in the spring or early summer. During pregnancy and lactation colonies may occupy several roosts, fragmenting and reforming to varying degrees as described earlier. Any males in the roost may move around with the females but play no role in any aspect of rearing the young. At the end of the summer, when the young are weaned, nursery colonies usually break up and mating occurs before hibernation in the autumn. Mating strategies, even in temperate bats, are very varied and are covered in the next section. See Lausen and Barclay (2002; 2003) for the most recent studies on the factors affecting roost selection in temperate bats.

MATING SYSTEMS

The diversity of bat mating systems across all climate zones has been described in a recent review by McCracken and Wilkinson (2000). Table 15.1 summarizes the bats reviewed, categorizing them on the basis of structural roosting and mating associations and their stability

Table 15.1 *A summary of social systems in bats, based on data collated by McCracken and Wilkinson (2000).*

Social system (Data for 67 species of approx. 1000 in 18 families)	No. (%) species	No. families in which observed
Single male/multi-female		
Year-round harem, female group stable	5 (8)	5
Year-round harem, female groups unstable	10 (15)	4
Seasonal single male/multi-female	8 (11)	2
Other single male/multi-female	12 (18)	3
Multi-male/multi-female		
Year-round, mate at roost	4 (6)	2
Year-round, mate away from roost	4 (6)	1
Seasonal	8 (11)	5
Monogamous groups	16 (24)	7

and seasonality. Although data are only available from 67 of the 1000 or more species, the variety of mating systems is striking. Even though most species are polygynous, a significant number appear to be monogamous and there are many variations on polygyny. As McCracken and Wilkinson point out, categorization is difficult and potentially misleading, since it attempts to deal with a continuum in which some species make use of more than one system, and because much of the evidence is correlative and experimental testing is frequently lacking. It is worth describing a few examples to illustrate the variety.

Vampyrum spectrum (Phyllostomidae) is a large, neotropical carnivore, feeding on birds and small land vertebrates. It is the only bat known to form stable pair bonds and to live in extended family groups, and one of only two species in which males are known to be involved in parental care (Vehrencamp *et al.*, 1977). However, other species are monogamous and they are a diverse group: *Lavia frons* (Megadermatidae) a large, sit-and-wait predator of insects, also pairs up and the single juvenile stays with the parents for up to 50 days after learning to fly (Vaughan & Vaughan, 1986; 1987). *Cardioderma cor* (Megadermatidae) is also a sit-and-wait predator, but colonial. Nevertheless, within colonies of up to 80 individuals it is monogamous (Vaughan, 1976). *Saccopteryx leptura* is a typical small, colonial insectivore, but it is frequently monogamous. However, this species is apparently not restricted to a single system – *Saccopteryx leptura* can also be polygynous and form harems (Bradbury & Vehrencamp, 1976a; 1977). The Old World fruit bat *Pteropus samoensis* is also monogamous (McCracken & Wilkinson, 2000).

(a)

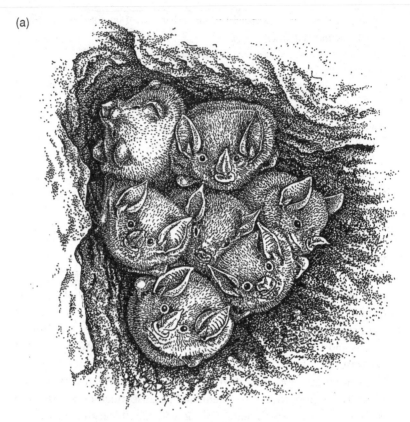

Figure 15.4   Harem of the spear-nosed bat, *Phyllostomus hastatus* (a) and
a male hammerheaded bat, *Hypsygnathus monstrosus* (b) that uses its
enlarged nasal cavities to make booming calls when lekking.

The majority of species studied show some form of polygyny,
either a single-male/multi-female grouping or multi-male/multi-female.
Year-round harems are common in tropical species, with both stable and unstable female groups (e.g. *Phyllostomus hastatus* (Fig. 15.4(a))
(McCracken & Bradbury, 1981) and *Pteropus tonganus* (Grant & Banak,
1995), respectively). Many other examples are described by McCracken
and Wilkinson (2000) and multi-male groups are equally diverse. In
temperate species the harem and multi-male groupings are seasonal.
There are a number of well-documented lekking species in the tropics,
such as the hammer-headed bat, *Hypsignathus monstrosus* (Pteropodidae,
Fig. 15.4(b)) (Bradbury, 1977), a relatively rare example of marked sexual dimorphism in bats. Finally, many temperate members of the large
genus *Myotis* are swarming bats. These insectivores usually form small

(b)

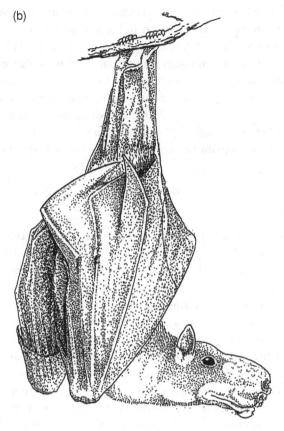

Figure 15.4 (*cont.*)

to medium-sized, unstable, mobile colonies in the summer. In late summer they disperse to swarming sites (usually underground hibernation sites) where they mate. Hundreds or thousands of bats gather at the site each night and a characteristic behaviour is persistent chasing. This suggests that there may be some element of mate choice. Mitochondrial DNA analysis shows non-random mating (Watt & Fenton, 1995), although sperm competition may underlie this observation. The sex ratio at swarming sites is frequently skewed, with males typically comprising up to 80% of the bats caught (Parsons *et al.*, 2003). This may reflect a real skew in the bats present, or the fact that the behaviour of the males makes them easier to catch. The presence of large numbers of males, possibly displaying for a smaller group of females, or one with a more rapid turnover of its members, might be indicative of a lek breeding system in swarming bats.

Climate, and in particular the length and intensity of seasonal fluctuations in climate, are major factors influencing social and mating systems. In the tropics, social organization is often stable year round. In temperate zones social groupings are invariably seasonal. This must have some impact on the mating strategies that evolve, but there are no simple patterns: many mating systems are seen in similar forms in both temperate and tropical regions. A short mating season, for example that between the weaning of young and hibernation in temperate bats, probably plays a role too, perhaps favouring brief events such as swarming.

SPATIAL SEGREGATION IN ROOSTING AND FORAGING BATS

Segregation occurs at various spatial levels, from the roost through home range to a broader geographical scale. It would be convenient to describe these in turn, but they are not easily separated. However, we can start at the smallest scale and begin by asking why do males and females so often roost apart during the nursery season? Although this question has been asked and several answers put forward, there is surprisingly little direct evidence in support of most of the explanations. In temperate bats, perhaps the most plausible reason and that supported by the greatest evidence, relates to physiological and energetic differences between the sexes. Males and females of temperate species often require a different thermal environment, forcing them to roost separately. Although there is some variation between species and during the stages of pregnancy and lactation, females remain homeothermic much of the time to maximize fetal growth rate and possibly also milk production. Males on the other hand can make use of torpor more often than females in the summer and to maximize energy savings may choose cool roosts (e.g. Hamilton & Barclay, 1994). This is discussed by Altringham (1996) and most recently by Lausen and Barclay (2002; 2003) in relation to roosting ecology.

Another reason for spatial segregation may be to reduce competition for food in the vicinity of the roost (Kunz, 1974). A large nursery colony consumes a considerable amount of food each night and females may be better able to provide for their young without competition from males. Foraging is generally more efficient and infant mortality lower if the females can feed close to the roost (Tuttle, 1979). This explanation assumes that if males and females share the same roost they will share foraging areas. Male and female *Myotis daubentonii* (Fig. 15.5) that share the same roost do indeed forage in the same home range, but

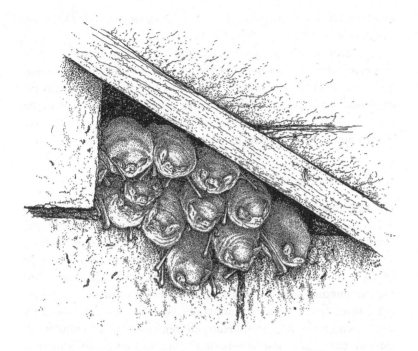

Figure 15.5 Daubenton's bat, *Myotis daubentonii*. Roosts show partial segregation. Some of the males roost with the females and forage in the same home range, but most males roost and forage away from the nursery colony.

(at least in upland areas) most of the males roost and forage separately (P. Senior, J.D. Altringham & R.K. Butlin, unpublished). Do the females (and the males that share their roost) exclude other males from their home range? We have evidence to suggest that non-colony males are feeding in less favourable habitat and get fewer opportunities to mate (P. Senior, J.D. Altringham & R.K. Butlin, unpublished). However, they may not be actively excluded, but may choose to forage away from the roost to avoid competition. *Plecotus auritus* also forms mixed-sex roosts in which the males forage on the periphery of the home range (Entwistle *et al.*, 1996). Wilkinson and Barclay (1997) showed that male big brown bats, *Eptesicus fuscus*, foraged over larger areas than females, spending more time away from the main feeding area. Again, in both cases, the males may get more food by feeding away from the core area where competition with the females would be more intense, despite the longer commuting distances. If true, how do the females exercize and main-tain this apparent dominance? These differences in foraging behaviour may be related to sex-based differences in foraging strategy or diet, but these topics need further exploration.

Segregation may also have evolved to avoid aggression, either between males or between males and the females and their offspring, but again direct evidence is lacking and there is little evidence for aggression amongst individuals in bat roosts. Finally, limitations in the size and number of suitable roosts, particularly for nurseries, may drive segregation in some species, but this too requires study.

SEGREGATION BY ALTITUDE

The first study specifically to investigate segregation on a large spatial scale was carried out by Barclay (1991). Not only do male and female little brown bats, *Myotis lucifugus*, roost apart, but their roosting and foraging areas can also be geographically discrete. In the foothills of the Canadian Rockies only male little brown bats are caught during the summer months, nursery colonies being confined to the low-lying plains away from the mountains. The bats feed on the dusk peak of small insects, predominantly chironomids. As temperatures fall through the night, these insects stop flying, so food becomes increasingly scarce, particularly at higher altitudes, where night-time temperatures are lower. Pregnant and lactating females have high energetic demands through the summer, when it is important to remain homeothermic: early weaning lowers juvenile mortality and increases their subsequent breeding success significantly (Tuttle, 1979). Males on the other hand make extensive use of torpor when prey availability is low (e.g. Hamilton & Barclay, 1994). The females therefore appear to confine themselves to lower elevations to ensure an adequate and reliable supply of food and perhaps warmer roosts. Decreasing temperature with increasing altitude clearly reduces the prey available to aerial hawkers such as the little brown bat, but what if the foraging strategy is different? In the same study, Barclay showed that both sexes of a related species, *Myotis evotis*, were found at high elevations and that females were able to rear young. The females were able to find enough food because much of their food is gleaned from surfaces: they are able to take a variety of non-volant (day-flying, resting night-flying or non-flying) prey throughout the night.

Segregation by altitude has been observed in other species too. For example, in the Black Hills of South Dakota USA, females, and particularly reproductive females, of all 11 species of bat captured declined in abundance with increasing altitude, relative to males of the same species (Cryan *et al.*, 2000). Other North American studies show similar trends (e.g. Grindal *et al.*, 1999; Storz & Williams, 1996, and see references in Cryan *et al.*, 2000). In Europe, Daubenton's bat, *Myotis daubentonii*, also segregates in the Jura mountains (Leuzinger & Brossard, 1994), Italy (Russo, 2002) and in the UK (P. Senior, J.D. Altringham & R.K. Butlin, unpublished). Studies of an Old World fruit bat, *Otopteropus cartilagonodus*, on Luzon in the Philippines (Ruedas *et al.*, 1994) showed marked sexual segregation along an altitudinal gradient and distinct differences in size and skull/jaw structure. It was suggested that this would reduce inter-sexual competition for food, but no behavioural or ecological studies have been carried out.

MIGRATION

A small but significant proportion of temperate and subtropical species are migratory (Fleming & Eby, 2003). In most cases, females are more likely to be migratory than the males. Where both sexes migrate, females typically migrate further and often start their migration earlier. This differential migration probably occurs for two reasons. Spring migration to a resource-rich summer habitat must be weighed against the high cost of migration itself. The high energy demands of pregnancy and lactation in females tip the balance in favour of migration. This applies to both the insectivorous microbats and the small number of New World nectar feeders that have been studied. Females may also migrate to find nursery roosts with more favourable microclimates: usually warmer sites that reduce thermoregulatory costs. Males can remain in, or migrate to, less favoured habitats since they require less food and cooler roosts due to their more frequent use of torpor during the summer months. The males of some temperate species (e.g. *Nyctalus noctula* and *Pipistrellus nathusii*) set up mating roosts or territories on the autumn migration routes of the females. A small number of tropical species also migrate. The reasons for migration are again related to variations in food supply, but are more varied, and female-biased migration is less common (Fleming & Eby, 2003).

Segregation also occurs within a sex: male Daubenton's bats segregate, some residing with the females, others occupying non-overlapping roosting and foraging areas (P. Senior, J.D. Altringham & R.K. Butlin, unpublished). This segregation persists over several seasons and perhaps throughout life. Ongoing research is investigating the mechanisms that drive this segregation and the consequences to mating patterns and population structure.

CONCLUSION

There is considerable diversity in social patterns within and between taxonomic groups and ecological guilds. Bats are likely to be a fruitful hunting ground in attempts to understand mammalian sexual segregation, but we have barely begun to exploit this resource. Few of the hypotheses put forward to explain sexual segregation in bats have been formally advanced and tested. What determines whether males and females use separate roosts, or segregate within the roost? The most likely explanations are reduced competition for food and, in temperate species, differences in roost microclimate requirements. There is evidence to support the latter argument, but the first remains untested. What determines whether males and females forage separately? Males of some species do have different foraging habits to females and may even have separate foraging areas. Energetic constraints on the females can explain why there might be differences, but what mechanisms underlie the separation and in some cases apparently allow the females to exploit the better habitat? Are there differences in diet between males and females that are related to segregation? Juveniles and adults of some species have different diets, but we know little about intersexual differences that might be related to foraging behaviour. Finally, studying the consequences of segregation to mating success and population structure will give a new angle on the evolution of mating systems.

ACKNOWLEDGEMENTS

Thanks to Tom McOwat for the drawings that illustrate this chapter. Thanks also to the editors, Kathreen E. Ruckstuhl and Peter Neuhaus, for the invitation to contribute this chapter and to Brock Fenton, Robert Barclay and a third, anonymous, reviewer, for their helpful comments on the manuscript.

16

Sociality and ecology of the odontocetes

OVERVIEW

Several years ago, I visited colleagues studying guanacos, *Lama guanicoe*, small South American camelids, in southern Chile. In just a few days, I was able to witness nursing bouts, mating attempts and births after which we could catch and tag the newborn. In my 20 year's studying beluga whales, *Delphinapterus leucas*, in the St Lawrence Estuary in eastern Canada, I have observed thousands of surface events but only five mating attempts, one possible birth, no nursing and we are still speculating on what belugas feed on! This might explain why whales and dolphins rarely feature in the primary literature on behavioural ecology. The most obvious reason is the difficulty facing a terrestrial mammal observing an aquatic one (Connor *et al.*, 1998). Nonetheless, a few long-term studies initiated in the early 1980s are starting to change this. These studies, based on individual identification and genetic profiling have overcome some of the difficulties and revealed an impressive diversity among cetacean societies (see recent reviews in Mann *et al.*, 2000b). They have paved the way for comparisons of cetacean sociality with that of their terrestrial relatives from which stem most theoretical work on the evolution of mammalian social systems (Norris, 1994; Clapham, 1996; Weilgart *et al.*, 1996; Connor *et al.*, 1998). In this chapter, I examine a particular aspect of odontocetes' or toothed whales' social life: sexual segregation. Odontocetes or toothed whales are a suborder of cetaceans, which includes porpoises, dolphins, beaked whales and sperm whales.

Sexual Segregation in Vertebrates: Ecology of the Two Sexes, eds. K. E. Ruckstuhl and P. Neuhaus. Published by Cambridge University Press. © Cambridge University Press 2005.

SEXUAL SEGREGATION IN WHALES

Whereas sexual segregation is common among group-living mammals (primates: Chapman, 1990; ungulates: Main & Coblentz, 1990; macropodes: Jarman, 1991, see other reviews in this book) it has been most studied in sexually size-dimorphic social ungulates, among which sexual segregation seems to be the rule (Clutton-Brock *et al.*, 1987a; Main & Coblentz, 1990; Mysterud, 2000; Ruckstuhl & Neuhaus, 2000; 2002). Sexual segregation is best defined as a behavioural pattern in which individuals tend to segregate, outside breeding periods, with individuals of their own sex (Main *et al.*, 1996; see also Chapter 2 for the different definitions). It involves sex differences in grouping patterns (social segregation) and ranging patterns (spatial and or habitat segregation). As such, it represents an important component of the sociality and ecology of a species.

Current explanations for when and why sexes should segregate in ungulates invoke sex differences in predator avoidance, energy requirements, foraging behaviour, social motives, as well as intraspecific competition (Beier, 1987; Clutton-Brock *et al.*, 1987a; Verme, 1988; Main *et al.*, 1996; Conradt, 1998a; Ruckstuhl, 1998; Barboza & Bowyer, 2000). Whether there is one all-encompassing explanation for sexual segregation in ungulates or a combination of several factors favouring it is still a matter of debate. Ruckstuhl and Neuhaus (2000) pointed out that several factors invoked in these hypotheses are largely influenced by body size. In order to avoid this confounding effect, they proposed a predictive framework for sexually size-dimorphic and monomorphic species. Using this approach, Ruckstuhl and Neuhaus (2002) performed a comparative analysis for a wide variety of social ruminants. They concluded that sexual differences in activity budgets, which impose costs on activity synchronization, is most likely the central factor driving sexual segregation and that other factors are simply additive. In order to better understand when and why sexes should segregate, they emphasized the need to examine this question in other mammalian groups with various degrees of sexual segregation and sexual size-dimorphism.

Only a few of the 70 known species of odontocetes have been studied in moderate detail, yet extensive variations have been found in the expression of both social and spatial patterns of sexual segregation. Non-breeding sexually mature males sperm whales, *Physeter macrocephalus*, for example, feed in high latitudes, some 5000 km away from females who live in tropical waters year-round (Best, 1979; Whitehead *et al.*, 1991) (see Fig. 16.1). Male and female killer whales, *Orcinus orca*,

Figure 16.1 Sperm whales provide an extreme case of both social and spatial sexual segregation. Females live in tropical waters year-round in fairly stable social units some 5000 km away from non-breeding sexually mature males feeding in high latitude. Courtesy of Flip Nicklin, Minden Pictures.

from the so-called 'resident' fish-eating population of the eastern North Pacific, on the other hand, live their entire lives with their mothers (Bigg *et al.*, 1990). Important variations are also found in body size and sexual size-dimorphism. The largest of the odontocetes, the male sperm whale, reaches 18 m in length and weighs up to 49 tons, more than twice as much as the female sperm whale (Rice, 1989), and 750 times the size of the harbour porpoise, *Phocoena phocoena* (Read & Tolley, 1997), in which females are bigger than males. Odontocetes therefore provide a useful group to test Ruckstuhl and Neuhaus' (2002) model.

The limited amount of quantitative data on most factors invoked to explain sexual segregation, as well as the limited number of species for which this level of information is available, do not allow for taxon-wide formal hypothesis testing. Instead, I selected a few of the best-studied species (or populations) to illustrate the potential role of each factor and their possible interactions. I begin this analysis by examining the underlying assumptions behind the main hypotheses included in Ruckstuhl and Neuhaus' (2000) predictive framework (see Box 16.1). I then compare the segregation patterns observed in the selected species with the different sets of predictions. Whenever relevant, examples are

Box 16.1 When and why sexes should segregate?

Summary of assumptions and predictions for *monomorphic* and *sexually dimorphic* species derived from five hypotheses proposed to explain sexual segregation (adapted from Ruckstuhl and Neuhaus, 2000).

The **predation risk hypothesis** is based on the assumptions that vulnerability to a predator will vary with body size and presence or absence of an accompanying calf. Therefore, if difference in predation risk is the primary factor responsible for segregation, then:

In monomorphic species, females with young should segregate from males and non-reproducing females (social segregation) and seek safer habitat (spatial segregation).

In sexually dimorphic species, all females should segregate from males (social segregation) and seek safer habitat (spatial segregation). Females with young should also segregate from non-reproducing females.

The **forage selection hypothesis** is based on the assumption that metabolic requirements will vary with body size and reproductive status. Therefore, if difference in forage selection is the primary factor responsible for segregation, then:

In monomorphic species, reproducing females should segregate from males and other females (social segregation) and seek different food sources (spatial segregation, in same or different habitat). Non-reproducing females should be found with males feeding on the same food source.

In sexually dimorphic species, all females should segregate from males (social segregation) and seek different food sources (spatial segregation, in same or different habitat). Females with young should also segregate from other females and seek different food sources (in same or different habitat).

The **scramble competition hypothesis** is based on the assumption that under low food availability, intense feeding pressure by females could exclude larger males from preferred areas. Therefore, if scramble is the primary factor responsible for segregation, then:

Under high food availability, females and males from monomorphic species and sexually dimorphic species should form mixed-sex groups.

Under low food availability, females and males from monomorphic species should form mixed-sex groups and males from sexually dimorphic species should be excluded from highest quality habitat by female intensive feeding pressure.

The **activity budget hypothesis** is based on the assumption that difference in body size or in reproductive condition will lead to differences in foraging behaviour. These differences will in turn impose costs to activity synchronization. Therefore, if difference in activity budget is the primary factor responsible for segregation, then:

In monomorphic species, reproducing females should segregate from males and other females (social segregation) and adopt a different activity schedule.

In sexually dimorphic species, all females should segregate from males (social segregation) and adopt a different activity schedule. Females with young should also segregate from other females.

The activity budget hypothesis further predicts that the degree of segregation will increase with the degree of sexual size dimorphism.

The **social preference hypothesis** is based the assumption that sex-specific preferences and aggressive behaviour could lead to social and spatial segregation. Therefore, if difference in social preferences is the primary factor responsible for segregation, then:

In monomorphic species and sexually dimorphic species, independent of their respective size or reproductive status, females and males should segregate into single-sex groups.

drawn from different populations (intraspecific variation) or related species. I close this chapter with a discussion concerning the usefulness of Ruckstuhl and Neuhaus' (2002) model to explain sexual segregation in odontocetes.

METHODS

To examine the extent and the possible causes of sexual segregation in odontocetes, I used one species per family (I used recent classification by Rice (1998)), except for the very large Delphinidae family, for which I included four species from four different genera: three coastal species with wide variation in body size and one pelagic species (Table 16.1). Delphinidae represents almost half the species (35 species/17 genera)

Table 16.1 *List of selected species and some aspects of their ranging, grouping, social bonding and biological parameters (F = females, M = males, J = juveniles, RF = reproducing females, NRF = non-reproducing females).*

Species (family) population	Habitat	Ranging pattern	Grouping pattern	Social bonding strategies	Sexual size dimorphism[a]	Inter-birth interval (year)	Breeding seasonality (peak calving season in months)	Source[b]
Sexually dimorphic species								
Sperm whale, *Physeter macrocephalus* (Physeteridae)	Pelagic	F live between the tropics (range span 1450 km); M feed in high latitude; older adult M migrate between high latitude and tropical breeding ground.	F live in small stable units (~12 related and unrelated ind.), which temporally join in small groups (24 ind.); JM form small bachelor groups (1–5 ind.); older M rove alone.	FF: long-term social bonds between related and unrelated females. MM: no long-term social bond between males.	2.41 c		Seasonal (3)	1, 2, 3, 4, 5, 6
Boto, *Inia geoffrensis* (Iniidae)	Freshwater river, lake and flooded forest	Undefended home range; slight seasonal migration related to annual flooding cycle; during high water, F prefer flooded forest, M prefer rivers.	Mainly solitary; most groups of two are cow-calf pairs; occasional loose feeding aggregation.	No social bond.	1.57b	?	Seasonal (3)	10, 11
Killer whale, *Orcinus orca* (Delphinidae) Eastern North Pacific 'resident'	Coastal	Population divided into community with extended summer range; migration not well known.	Matrilineal groups (2–9 ind.) join in subpods (2–11 groups) and in pods (3–59 individuals)	MF: lifetime family bonds.	1.46 c	5	Year-round (6)	7, 8, 9, 32

Beluga. *Delphinapterus leucas* (Monodontidae)	Pelagic, coastal and estuarine	Seasonal migration between coastal summer range and pelagic winter range; matrilineally directed geographical philopatry (summer range); F use estuaries and shallow areas more than M.	Gregarious: variable herd size and composition (fission-fusion); F with and without dependent calf form large herds; J occasionally form distinct herds: all M herds are smaller, mixed herds can be huge (up to one thousand)	FF: loose network of females. MM: long-term social bonds (alliance) between males.	1.41 c	3	Seasonal (4)	12, 13, 14, 15, 16
Bottlenose dolphin. *Tursiops truncates* (Delphinidae) Sarasota, Florida	Coastal	Population divided into communities each with limited annual range and limited immigration/ emigration; M and F ranges widely overlap; F occupy smaller core area and use shallow and sheltered bay more than M.	Variable group size (2–15) and composition (fission-fusion).	FF: loose network of associates (Band); associations modulated by reproductive condition. MM: long-term social bonds (alliance).	1.33 a	3	Diffusely seasonal (4)	17, 18, 19, 20, 29
Harbour porpoise, *Phocoena phocoena* (Phocoenidae) Bay of Fundy, Canada	Coastal	Range in coastal areas during summer; winter movement not well known; extremely mobile; cover extended range (>1000 km^2); M and F have different habitat preferences.	Cow-calf pairs and small groups; sometimes form large feeding aggregations (10s–100s)	Long-term social bonds have not been documented.	0.76 a, d	1	Seasonal (1)	24, 25, 26, 27, 28

(cont.)

Table 16.1 (cont.)

Species (family) population	Habitat	Ranging pattern	Grouping pattern	Social bonding strategies	Sexual size dimorphism[a]	Inter-birth interval (year)	Breeding seasonality (peak calving season in months)	Source[b]
Sexually monomorphic species								
Dusky dolphin, *Lagenorhynchus obscurus* (Delphinidae)	Coastal and shelf	Small discontinuous populations; migratory pattern not well understood.	Variable school size (2-500) with smaller stable subgroups (~15 ind.) (fission-fusion).	No sex-specific stable associations documented but possible long-term individual associations.	1.09 c	3	Seasonal (3)	21, 22, 23
Striped dolphin, *Stenella coeruleoalba* (Delphinidae)	Pelagic	In some regions, seasonal movements associated with warm oceanic currents.	Travel in highly cohesive schools (up to 500); three main types of schools are recognized: juvenile, adult and mixed.	Long-term social bonds have not been documented.	1.08 c	4	Diffusely seasonal (5)	30, 31

a. Sexual size dimorphism ratios are calculated as the male body weight over female body weight using (a) asymptotic, (b) mean or (c) maximum adult weights.

b. 1, (Rice, 1989); 2, (Whitehead et al., 1991); 3, (Christal et al., 1998); 4, (Letterall et al., 2002); 5, (Whitehead, 1993); 6, (Whitehead, 2003); 7, (Bigg et al., 1990); 8, (Dahlheim and Heyning, 1999); 9, (Baird, 2000); 10, (Martin & da Silva, 2004); 11, (Best & da Silva, 1993); 12, (Kleinenberg et al., 1964); 13, (Smith et al., 1994); 14, (Sergeant, 1973); 15, (Michaud, 1999); 16, (Brodie, 1971); 17, (Wells et al., 1987); 18, (Scott et al., 1990); 19, (Wells, 2003); 20, (Wells & Scott, 1999); 21, (Würsig & Würsig, 1980); 22, (Würsig & Würsig, 1980); 23, (Brownell Jr. & Cipriano, 1989); 24, (Smith & Gaskin, 1983); 25, (Read, 1990); 26, (Read & Tolley, 1997); 27, (Read, 1999); 28, (Hoek, 1992); 29, (Read et al., 1993); 30, (Miyazaki & Nishiwaki, 1978); 31, (Perrin et al., 1994); 32, (Olesiuk et al., 1990).

of this suborder. Because of the limited amount of detailed information available, none of the Kogiidae and Ziphiidae species were included and only one of the four families of freshwater dolphins (Iniidae) was included. The eight species retained for this review exhibit sufficient variation in their habitat, ranging and grouping patterns, social organization and extent of sexual size dimorphism to provide natural field tests of the above hypotheses (see Box 16.1).

Data used in this review come from a variety of sources including whaling records, strandings, sightings or long-term studies based on individually recognizable individuals. Much of the information was taken from species accounts (Ridgway & Harrison, 1989; 1994; 1999; Mann et al., 2000b). When available, more recent data have been used. Information on sexual size dimorphism and other biological parameters were taken from recently published summary tables (Connor et al., 2000; Whitehead & Mann, 2000; Boness et al., 2002).

Selected species have been classified as monomorphic or dimorphic following the 20% difference in body size criteria used by Ruckstuhl and Neuhaus (2002). There is no strict basis for setting this criterion at 20% in odontocetes. However, as several life-history features are allometrically scaled on body mass, I assumed that a 20% difference has an effect on some of the factors invoked to explain sexual segregation. One of the species selected displays a strong reversed dimorphism, females being 30% heavier than males; it has been included here among the dimorphic species (Table 16.1).

RESULTS

Predation risk hypothesis

According to the predation risk hypothesis, when male reproductive success is influenced by physical condition, selective pressure will favour behaviours that maximize energy gain even if they increase predation risk. Conversely, when female lifetime reproductive success is related to offspring survival, selective pressures should favour behaviours minimizing predation risk, at the cost of foraging in poor quality habitat. According to this view, social and spatial segregation of the sexes could be predicted on the basis of sex-specific vulnerability to predation or tolerance to predation risk (see also Chapter 9 by Main & du Toit).

Although predation is believed to be one of the main selective pressures acting on odontocetes (Norris & Dohl, 1980), there is no direct evidence for sex- or size-specific vulnerability. Nonetheless, predation seems to be exerting considerable pressure on young animals and smaller species. In Shark Bay, one-third of bottlenose dolphin calves, *Tursiops truncatus*, bore shark bites (Mann & Barnett, 1999), and over 40% are estimated to die before weaning (Mann *et al.*, 2000a). Predation may be a weaker selection pressure in large-bodied sperm whales capable of communal defence (Arnbom *et al.*, 1987). However, sperm whales do bear tooth rake marks on their flukes, which are suggestive of harassment, probably from killer whales (Dufault & Whitehead, 1995a) and a successful attack on a group of females and young was recently witnessed off the coast of California (Pitman *et al.*, 2001).

Accuracy of the predictions for size-dimorphic species

With the exception of killer whales in which there is no sexual segregation, the general patterns of spatial segregation observed among selected size-dimorphic species are consistent with the predictions of the predation-risk hypothesis (summary in Table 16.2). The year-round residence of female sperm whales in tropical waters could be a strategy to reduce killer whale predation, which is most prevalent in high latitudes (Dahlheim & Heyning, 1999). The same argument has been recently invoked to explain why baleen whales migrate to lower latitudes (Corkeron & Connor, 1999).

The functions attributed to summer aggregations of female belugas in coastal waters include thermal advantages for calves, feeding, moulting and protection from predators (Sergeant & Brodie, 1969; Sergeant, 1973; Fraker *et al.*, 1979; St Aubin *et al.*, 1990) (see Fig. 16.2). While the first has been discounted (Doidge, 1990), feeding does not seem to be an important activity in the High Arctic estuaries (Smith *et al.*, 1994), and moulting has not been observed as a highly synchronized phenomenon in the St Lawrence Estuary where belugas also display sexual segregation, females being found in the shallower and more protected waters of the upper estuary (Pippard, 1985; Michaud, 1993). Whereas Arctic female belugas may not be totally efficient in preventing successful attacks by killer whales in coastal habitats (see Frost *et al.*, 1992), where they are less easy to detect, negative reactions to the approach of killer whales have been documented on a few occasions, with belugas crowding close to shore and slowly following the coastline (Lowry *et al.*, 1987). Female belugas' preference for shallow coastal habitats could also allow them to forage without leaving their calves unattended for long periods, thereby minimizing risk of predation on their calves.

Table 16.2 *Summary of the accuracy of predictions under the predation risk hypothesis (F = females, M = males, RF = reproducing females, NRF = non-reproducing females).*

Dimorphic species	F segregate from M	F are found in safer habitat	RF segregate from other F	Sources[a]
Sperm whale	Yes	Possibly	No	1, 2
Boto	Yes	Yes	?	3
Killer whale Eastern North Pacific 'resident'	No	No	No	4
Beluga	Yes	Yes	No	5, 6, 7, 8
Bottlenose dolphin Sarasota, Florida	Yes	Yes	Partly[b]	9, 10
Harbour porpoise Bay of Fundy	Yes	No	NA[c]	11

Monomorphic species	RF segregate from M	RF are found in safer habitat	NRF stay with M	Sources
Dusky dolphin	Yes (temporarily)	Yes	Yes	12
Striped dolphin	Yes	Possibly (dilution)	Yes	13

a. Sources: 1, (Best, 1979); 2, (Whitehead *et al.*, 1991); 3, (Martin & da Silva, 2004); 4, (Bigg *et al.*, 1990); 5, (Kleinenberg *et al.*, 1964); 6, (Pippard, 1985); 7, (Michaud, 1993); 8, (Smith *et al.*, 1994); 9, (Wells *et al.*, 1987), 10, (Scott *et al.*, 1990); 11, (Smith & Gaskin, 1983); 12, (Würsig & Würsig, 1980); 13, (Miyazaki & Nishiwaki, 1978).
b. Female bottlenose dolphins form networks of preferred associates (bands) but individual association within these networks are modulated by reproductive condition (Wells *et al.*, 1987).
c. Not applicable: as the birth-interval in harbour porpoise is one year, there are very few non-reproductive adult females in the population (Read & Hohn, 1995).

It has also been suggested that the tendency of female bottlenose dolphins in the Sarasota community to use shallow and sheltered bays more than males, is a response to predation pressure (Wells *et al.*, 1980; Scott *et al.*, 1990). In Shark Bay, Mann *et al.* (2000a) found that calf mortality is significantly lower for females spending more time in shallow waters than for females ranging further offshore. The resulting higher reproductive success strongly suggests that intense selection pressure favours movement of females with calves to safer habitats.

Figure 16.2   Belugas are highly gregarious. During summer, calves gather in freshwater estuaries while males spend more time in open deep waters. Courtesy of Flip Nicklin, Minden Pictures.

As botos, *Inia geoffrensis*, have no known predator (Best & da Silva, 1993), the anti-predator value of female botos' marked preference for shallow protected waters in flooded forest is not obvious (Martin & da Silva, 2004). Martin and da Silva (2004) proposed that female avoidance of male harassment is a probable cause of the segregation. However, before rejecting the anti-predation explanation, it might be worthwhile considering the possibility that this preference for a 'safer habitat' has been maintained well beyond the disappearance of an ancient predator, or a switch of habitat, a phenomenon to which Byers (1997) referred to as 'ghosts of predators past'.

The situation of the harbour porpoise, in which the size dimorphism is reversed, is particular. In the Bay of Fundy (Atlantic Canada) males, as predicted by their smaller size, are found in the most protected waters (Smith & Gaskin, 1983). However, in this population, female harbour porpoises become pregnant every year and nurse their calves for 8 to 12 months (Read, 1990; Read & Hohn, 1995). Adult females are therefore always simultaneously pregnant and lactating. According to the predation risk hypothesis, females should also seek safe habitat. Instead, female harbour porpoises with calves do segregate from males, but use riskier waters where their two major predators, white sharks, *Carcharodon carcharias*, and killer whales (Arnold, 1972; Jefferson et al., 1991), can be found. Smith and Gaskin (1983) argued that female habitat selection corresponds to different nutritional needs (see later).

The predation risk hypothesis also predicts that reproducing and non-reproducing females should be found in separate groups. While this may be the case with bottlenose dolphins, in which female–female associations are modulated by reproductive condition (Wells et al., 1987), it is not the case for female sperm whales living in fairly stable units (Whitehead et al., 1991; Christal et al., 1998). Observations on female beluga herds also failed to find female segregation based on reproductive condition (Kleinenberg et al., 1964; Sergeant, 1973).

Accuracy of the predictions for monomorphic species

Grouping patterns observed in dusky dolphins, *Lagenorhyncus obscurus*, and striped dolphins, *Stenella coeruleoalba*, also closely follow the predictions of the predation risk hypothesis (Table 16.2). Here again, variation in vulnerability to predation could explain the observed patterns. Würsig and Würsig (1980) speculated that the formation of dusky dolphin nursery groups during surface-active feeding bouts, could constitute an anti-predator strategy, as killer whales and sharks may be attracted to the feeding sites. In this case females would seek safer locations for their calves at the cost of losing a good feeding opportunity. While the movements of males between mating and non-mating schools of striped dolphins (described by Miyasaki and Nishiwaki, 1978) apparently result in the segregation of reproductive females from males and non-reproductive females, there is no direct evidence that females with calves segregate to safer habitats. However, the concept of a safe habitat may not apply directly in the open sea. There, formation of groups may be the primary anti-predator strategy (Norris & Schilt, 1988). Nursery groups may be an efficient strategy to increase calves' safety through the dilution effect.

Forage selection hypothesis

In ruminants, for which the forage selection hypothesis was proposed in the first place, digestive capabilities are related to body size (Demment & van Soest, 1985). According to this hypothesis, females of smaller size would compensate for a lower digestive efficiency by selecting high quality forage, while the males' higher digestive efficiency would allow them to select habitat on the basis of food biomass availability and not only on food quality. These differences would lead to differences in forage selection, hence to habitat segregation (see du Toit, Chapter 3). While cetaceans have retained a compartmented stomach from their closest terrestrial relative, the artiodactyls (Nikaido et al., 1999;

Geisler & Uhen, 2003), they are not ruminants. However, a less restric-
tive version of the forage selection hypothesis assumption could be
that energy requirements are affected by body size and reproductive
condition.

Although few studies have examined sex differences in the diets
of odontocetes, their results are consistent with this assumption. While
in size-dimorphic species the bigger sex tends to eat larger and different
prey than the smaller one (beluga: Vladykov, 1946; Lowry et al., 1985;
sperm whale: Rice, 1989; bottlenose dolphin: Barros & Odell, 1990; Cockcroft & Ross, 1990), in monomorphic species their appears to be no difference between males and non-reproducing females (pantropical spotted dolphin, *Stenella attenuata*: Bernard & Hohn, 1989; dusky dolphin:
McKinnon, 1994). Furthermore, female diets do vary with reproductive
condition in several species (pantropical spotted dolphin: Bernard &
Hohn, 1989; harbour porpoise: Recchia & Read, 1989; common dolphin:
Young & Cockcroft, 1994).

Accuracy of the predictions for size-dimorphic species

Patterns of segregation predicted by the forage selection hypothesis for
size-dimorphic species are essentially the same as those predicted by the
predation risk hypothesis. As mentioned above, the observed patterns
among the examined size-dimorphic species are consistent with the
predicted patterns (summary in Table 16.3). There is, however, not much
evidence that forage selection is the driving force behind the observed
patterns. In sperm whales and belugas, where males and females are
segregated into quite different habitats, it is difficult to discriminate
whether sex differences in feeding habit reflect local availability of prey
and behavioural constraints imposed by the presence of a calf, rather
than sex-specific requirements.

The same caution is required when interpreting the differences
for species in which spatial segregation is less obvious. Small sex-related
differences in prey preference reported in bottlenose dolphins could
also reflect sex-related difference in habitat preference (Barros & Odell,
1990). Similarly, it is not clear whether differences in the diet of lactating and that of resting or pregnant female Indian Ocean bottlenose
dolphins, *Tursiops aduncus*, correspond to forage-selection or result from
selecting safer habitat (Cockcroft & Ross, 1990).

On the other hand, the sexual segregation pattern observed in
the harbour porpoise falls perfectly within the predictions under the
forage selection hypothesis. In the Bay of Fundy, females differ from

Table 16.3 Summary of the accuracy of predictions under the forage selection hypothesis (F = females, M = males, RF = reproducing females, NRF = non-reproducing females).

Dimorphic species	F segregate from M and differ in diet	RF segregate from other F and differ in diet	F are not found in safer habitat	Sources[a]
Sperm whale	Yes/Yes	No/?	Possibly	1, 2, 3
Boto	Yes/?	?	Yes	4
Killer whale Eastern North Pacific 'resident'	No/?	No/?	No	5
Beluga	Yes/Yes	No/?	Yes	6, 7, 8, 9, 10
Bottlenose dolphin Sarasota, Florida	Yes/Yes	Yes/Yes	Yes	11, 12, 13
Harbour porpoise Bay of Fundy	Yes/Yes	NA[b]	No	14, 15

Monomorphic species	RF segregate from M and differ in diet	RF segregate from other F and differ in diet	NRF stay with M and have the same diet	RF are found in safer habitat	Sources
Dusky dolphin	Yes (temporarily)/No	Yes/No	Yes/Yes	Yes	16, 17
Striped dolphin	Yes/?	Yes/?	Yes/?	Possibly (dilution)	18

a. Sources: 1, (Best, 1979); 2, (Whitehead, 1993); 3, (Rice, 1989); 4, (Martin & da Silva, 2004); 5, (Bigg et al., 1990); 6, (Kleinenberg et al., 1964); 7, (Pippard, 1985); 8, (Michaud, 1993); 9, (Smith et al., 1994); 10, (Vladykov, 1946); 11, (Wells et al., 1987); 12, (Barros & Odell, 1990); 13, (Cockcroft & Ross, 1990); 14, (Smith & Gaskin, 1983); 15, (Recchia & Read, 1989); 16, (Würsig & Würsig, 1980); 17, (McKinnon, 1994); 18, (Miyazaki & Nishiwaki, 1978).

b. Not applicable: as the birth-interval in harbour porpoise in one year, there are very few non-reproductive adult females in the population (Read & Hohn, 1995).

males in both feeding and habitat preferences but are not found in safer habitat (Smith & Gaskin, 1983). Sex differences in forage selection may be responsible for the segregation of the sexes in the harbour porpoise.

Accuracy of the predictions for monomorphic species

Sex-specific diet has not yet been examined in striped dolphins. However, a detailed study of the closely related pantropical spotted dolphins has found, in concordance with the assumptions underlying the forage selection hypothesis, that there are no differences between male and non-reproductive female diets, but that significant variation exists between pregnant and lactating female diets (Bernard & Hohn, 1989). Interestingly, Bernard and Hohn (1989) suggested that by feeding at the surface on flying fish, as opposed to the deeper squid, preferred by others, females avoid leaving their vulnerable calf unattended. Their shift in preferred prey could be a predator avoidance strategy.

Conversely, no age or sex difference was found in the dusky dolphin diet (McKinnon, 1994). McKinnon (1994) suggested that all classes of reproductive females, males, as well as juveniles are able to satisfy their energy and water requirements with the abundant anchoveta.

Scramble competition hypothesis

The scramble competition hypothesis was proposed to explain situations where males occupy habitat of poorer quality than females (Clutton-Brock et al., 1987a, see also Chapter 3 by du Toit). This hypothesis predicts that under low food availability, intense feeding pressure by females could exclude larger males from preferred areas. Passive inter-sexual competition for resources will only occur under a special combination of environmental conditions. It is unlikely to occur when predation-risk forces females into safe but poor forage quality habitat or when high-quality food is abundant (Main, 1998).

Because we have limited information on the food habits of odontocetes and even less on the abundance of their prey, the scramble competition hypothesis may not presently be testable. However, there are some indications that spatial segregation could result from inter-sexual or scramble competition in one species. Whitehead and Weilgart (2000) reported an unusual arrival of male sperm whale aggregations in the Galapagos waters, soon after these waters had been deserted by females (Christal & Whitehead, 1997). Whitehead (2003: 347) suggested that this is indicative that males are out-competed by females.

He proposed that scramble competition might best explain why males spend the non-breeding season in high latitudes, in female-free waters.

Main (1998) suggested that the higher mobility of males and the absence of parental duty predisposed them to avoid overexploited areas. Their higher rate of movement could increase the rate at which they encounter new patches of food. Main (1998) also argued that the female tendency for strong site fidelity and a relatively smaller home range concentrates feeding pressure on limited areas and therefore contributes to limiting food availability for the other sex. All of these arguments may well apply to the case of sperm whales as reported earlier. Although a female range of approximately 1000 km may seem huge on the terrestrial scale, it is relatively small on the ocean basin scale (Dufault & Whitehead, 1995b).

There is, however, no definite evidence that tropical waters would indeed be preferred habitats for male sperm whales. Best (1979) suggested for instance that larger food items found near the poles could be more valuable for larger males. This would support the forage-selection hypothesis. There are two other alternatives to the impressive spatial segregation in sperm whales: Best (1979) also suggested that the limited range of females could be dictated by thermal requirements for calves. This argument has since been discounted by recent thermoregulation modelling (Watts et al., 1993; but see also Corkeron & Connor, 1999 for cautionary). Finally, the safety of offspring could also be at stake as discussed earlier.

Activity budget hypothesis

The activity budget hypothesis is based on the assumption that sex differences in body size will lead to differences in foraging behaviour and movement rates (Conradt, 1998a; Ruckstuhl, 1998; see also Chapter 10). According to this hypothesis, these differences would lead to a high fission rate of mixed-sex groups, which would in turn impose costs on activity synchronization and hence prompt segregation among animals of differing body size or reproductive conditions. This hypothesis further predicts that the level of segregation will be proportional to the degree of sexual size dimorphism.

In ungulates, sex differences in activity budgets are believed to be driven by sex differences in the proportion of time allocated to grazing, or by different rhythms of foraging imposed by different digestive efficiencies. While there are no data available with which to test this yet, sex and age differences in diving and swimming capability, as well as behavioural constraints associated with the presence of a calf, could impose costs on activity synchronization and hence promote segregation among odontocetes. Using interspecific comparisons, Schreer and Kovacs (1997) found strong allometric relationships between maximum diving depth ($r = 0.75$) and duration ($r = 0.84$) and body size. These suggest that even a moderate body-size dimorphism could impose some cost on activity synchronization. An allometric relationship has also been found between the cost of transportation (COT) and body size of marine mammals ($COT = 7.94\,mass^{-0.28}$: Williams, 1999). Relative to standard metabolic rate ($SMR = 10.1\,mass^{0.75}$: Kleiber, 1975), the cost of transport is expected to be inversely proportional to mass. Here again, theoretically, the cost of transportation could differ between the sexes in size-dimorphic species. These differences may not be meaningful for daily activities, but could be more significant during long-range movements such as northbound migrations and could explain at least in part the differences in ranging patterns of male and female sperm whales.

Behavioural constraints associated with the presence of a calf could also impose additional costs to activity synchronization. In pinnipeds, several studies have shown that diving performance of pups increases with age (Lyderson et al., 1994; Le Boeuf et al., 1996). If the diving capabilities of odontocete calves are limited, this could impose limits on their mothers' foraging behaviour (Papastavrou et al., 1989). Observations on captive bottlenose dolphins have shown that by midterm, as pregnant females' swimming activity decreased, they tended to segregate from their groups (McBride & Kritzler, 1951).

Higher fission rates or costs associated with activity synchronization cannot be invoked to explain the spatial segregation found in nonsocial species like the boto. However energetic constraints associated with smaller size or with the presence of a calf could still explain why female botos stay away from the strong river currents preferred by males (Martin & da Silva, 2004).

While patterns of segregation in most species examined were consistent with the patterns predicted by the activity budget hypothesis, it is not possible, with the available information, to evaluate whether there are sex-related differences in activity budgets and whether these could have led to the levels of segregation observed. Nonetheless, as predicted by the hypothesis, the segregation patterns are more pronounced in species with a larger sexual size dimorphism, such as the sperm whale or the beluga, than in species with a less pronounced dimorphism. This finding is consistent with an increasing cost of male–female activity synchronization as body size differences increase (Ruckstuhl & Neuhaus, 2000).

Social preference hypothesis

The social preference hypothesis predicts that sex-specific preferences or aggressive behaviour of females toward males could lead to social and spatial segregation. However, Main et al. (1996) noted that social preference is not likely to cause spatial segregation unless one sex actively excludes the other. While strong social bonds between individuals of the same sex do exist in several odontocetes species (sperm whale: Whitehead et al., 1991; bottlenose dolphin, T. aduncus: Connor et al., 1992; beluga: Michaud, 1999; Connor et al., 2000; Whitehead & Mann, 2000; bottlenose whale: Gowans et al., 2001), there are few observations suggesting that females could effectively keep males away. Connor et al. (1992) report that in Shark Bay, female bottlenose dolphins (T. aduncus) sometimes associate, and can be successful in driving away male alliances. However, these interactions are related to mating attempts and therefore would not help to explain why males and females live separately outside the breeding season.

DISCUSSION

Comparing the patterns of segregation in 35 species of ungulates with the predictions of three of the hypotheses examined herein, Ruckstuhl and Neuhaus (2002) proposed a mechanistic model according to which sex differences in activity budget are the proximate cause of segregation while sex differences in predation risk and forage selection are additive factors explaining, for example, the range of spatial segregation. The patterns of sexual segregation found among the odontocetes examined in this review generally fit this model. However, evidence is still lacking that body-size differences between sexes could make it sub-optimal for odontocetes of different sexes to synchronize their activity. Furthermore, few interesting departures from the predictions have been found and will be discussed later.

As predicted by the activity budget hypothesis, segregation patterns found among size-dimorphic species are more pronounced than those observed in monomorphic species. Furthermore, the degree of segregation tends to increase with increasing sexual size dimorphism. With the small number of species examined here it is obviously not possible to measure the relationship between size dimorphism and the degree of segregation as it has been done for large herbivores (Mysterud, 2000). However, the absence of segregation in one of the most dimorphic species, the killer whale, and the difference between the extent of spatial segregation in belugas and Sarasota bottlenose dolphins, which have a sexual size dimorphism ratio close to that of belugas, clearly suggest it is not a universal nor a simple relationship. The lack of pronounced segregation in Sarasota bottlenose dolphins could be explained by the females' reproductive schedule. The seasonally diffused breeding period in this species, as well as multiple oestrous cycles (Connor *et al.*, 1996) allowing females to conceive rapidly after losing a calf (Mann *et al.*, 2000a), are all selecting for year-round male residency.

Clinal geographic variation in bottlenose dolphin body size (Wells & Scott, 1999) could provide a perfect testing ground to further examine the relationship between extent of sexual segregation and the degree of size dimorphism. Tolley *et al.* (1995) already suggested that variation in the level of male bonding found in bottlenose dolphins may be inversely correlated to the body size or the degree of size dimorphism. Here again, body size variation may not be the sole factor at play. Wilson (1995) suggested that the deeper waters of Moray Firth may render female consorting more difficult, hence reducing the adaptive advantage of alliances found in the smaller Sarasota bottlenose dolphins (Wells *et al.*, 1987). Furthermore, lower predation pressure found in Moray Firth could also lower the advantage of roving in groups (Wilson, 1995). In order to test for a possible correlation between size dimorphism and the degree of sexual segregation among bottlenose dolphins, more comprehensive measurements of both size dimorphism and segregation are needed. While body length measures are available for several populations (Wells & Scott, 1999), reliable measures of adult male and female body weights, which would be biologically more significant (Read *et al.*, 1993), are still missing for most populations. Ideally, for such comparative analysis, a quantitative measure of segregation of the type proposed by Conradt (1998b) would be needed.

The absence of dispersal from the natal group by either sex in fish-eating killer whales in the eastern North Pacific, is an unusual case of social bonds. The functional advantages of the male killer whale natal philopatry are not well understood. Connor *et al.* (2000) proposed it was made possible by the wide distribution of resources in the aquatic environment, absence of breeding site and territoriality and the low costs of locomotion. These characteristics result in large day and home ranges, allowing males to encounter potential mates while travelling with their own mother. Baird and Whitehead (2000) also proposed that the absence of male dispersal in fish-eating killer whales is facilitated by the absence of a relationship between group size and food intake. Although there is no social segregation in these killer whales, there are some indications that males could segregate ecologically, either by diving deeper (diving capabilities correlates with size (Schreer & Kovacs, 1997) but yet time–depth recorder failed to find differences (Baird, 2000)) or through limited access to shallows where smaller females chase fish (see discussion in Baird, 2000).

While sex differences in activity budget, or by extension sex-related behavioural constraints or constraints imposed by reproductive condition, may be a proximate cause (mechanism) for social segregation, they are not sufficient to explain the diversity of sexual segregation patterns found in odontocetes. As discussed by Ruckstuhl and Neuhaus (2000, 2002) in their reviews of sexual segregation in ungulates, most other factors invoked under the different hypotheses appear to influence, to some extent, the patterns of segregation described in this brief review.

This review suggests that differential vulnerability or tolerance to predation risk plays an important role in driving males and females to adopt different social strategies. There are, however, interesting exceptions. Because of their small size and relative vulnerability to predators, harbour porpoises and particularly females with calves would be expected to seek safe habitat or adopt another predator avoidance strategy, but apparently do not, at least in the Bay of Fundy (Smith & Gaskin, 1983). I speculate that female harbour porpoises' preference for highly productive but not necessarily protected habitat may be a trade-off to satisfy the exceptional energetic requirements associated with their unusual annual reproductive cycle, referred to as 'life in the fast lane' by Read and Hohn (1995). Examination of male and female ranging patterns in Californian waters, where most females appeared to be on a two-year reproductive schedule (Hohn & Brownell, 1990) would provide a useful test of this hypothesis.

As noted in several ungulate studies, it is difficult to discriminate between forage selection and other factors potentially affecting sex-specific habitat selection (Clutton-Brock et al., 1987a; Main et al., 1996; Gross, 1998; Main, 1998). For example, reduced predation pressure may constitute an adaptive advantage to female sperm whales roaming between the tropics. Subtle differences between male and female bottlenose dolphins' diet may simply reflect prey availability in protected areas preferred by females. However, forage selection or especially intraspecific or scramble competition may be more important to explain the impressive spatial segregation of this large-bodied species whose populations are more likely regulated by food supply than by predation (Whitehead & Weilgart, 2000).

It may not be possible to demonstrate the action of forage selection or scramble competition without setting up experiments, which are, for obvious reasons, not applicable to the study of free-ranging odontocetes. Recent developments in fatty acid and stable isotope signature analysis or the use of crittercams (see review in: Bowen *et al.*, 2002) may eventually facilitate this kind of investigation. Further studies of dolphin feeding habits would provide useful insights into the interaction between forage selection and predator avoidance strategies (Scott *et al.*, 1990; Mann *et al.*, 2000a). Lactation lasts for more than one year in several species of odontocetes (Perrin & Reilly, 1984). It would therefore be possible to test forage selectivity by looking at prey consumed by males and females during seasons when sexual segregation is less pronounced.

While sex-specific social bonds or aggressive behaviour of one sex toward the other may not be sufficient to lead to sexual segregation as predicted by the social preference hypothesis, benefits of sex-specific social bonds may explain why females of a few species of odontocetes stay with other females of different reproductive condition. Ruckstuhl and Neuhaus (2002) also noted that in ungulates, females with and without young are often found together. They speculate that these female associations may form because there are simply not enough females to form a predator safe group or because they have the same diet. For odontocetes, I suggest that this variance from the predictions of the effects of reproductive status under most of the preceding hypotheses indicates that benefits from stable female groups or networks outweigh the costs potentially imposed by differences in optimal activity rhythm or forage preferences.

Communal care for calves (sperm whale: Weilgart & Whitehead, 1986; beluga: Béland *et al.*, 1990), babysitting (sperm whale: Whitehead, 1996) and defence against predators (sperm whale: Arnbom *et al.*, 1987) have been found to be some of the possible benefits of communal life in odontocetes. Female alliance could also help to avoid or reduce male harassment (bottlenose dolphin: Connor *et al.*, 1992; Richards, 1996) or infanticide (bottlenose dolphin: Patterson *et al.*, 1998; Dunn *et al.*, 2002). In sperm whales and bottlenose dolphins, the degree of relatedness within these female groups or networks was found to be higher than between them (Richard *et al.*, 1996; Mesnick, 2001; Krützen *et al.*, 2004). An increasing number of long-term studies unravelling the complex social life of odontocetes also suggest that stable social units could facilitate the cultural transmission of learned information (whales and dolphins: Rendell & Whitehead, 2001; killer whale: Yurk *et al.*, 2002; bottlenose dolphin: Wells, 2003).

SUMMARY AND CONCLUSIONS

It is clear from this review that sexual segregation in odontocetes, just as other components of a population's social organization, is a dynamic and complex behavioural phenomenon that probably cannot be explained by any one single factor. While sex differences in activity budget, or by extension sex-related behavioural constraints or constraints imposed by reproductive condition, may be proximate causes (mechanisms) for social segregation, they are not sufficient to explain the diversity of sexual segregation patterns found in odontocetes. Each of the four other factors invoked under the different hypotheses appear to influence, to some extent, the patterns of segregation described in this brief review.

Acknowledging the weakness of a unitary explanation, Ruckstuhl and Neuhaus (2000) emphasized the heuristic value of their approach. Whereas their predictive framework has proven very useful to evaluate the role of the different factors invoked in the most common hypotheses formulated to explain sexual segregation among the species examined in this review, the mechanistic model they proposed (Ruckstuhl & Neuhaus, 2002) may not be sufficient to account for the overall diversity of forms and functions of sexual segregation. I believe that this can be best explained by the general model of sociality according to which the grouping and ranging patterns of males and their mating strategies depend primarily on the distribution of females, which is determined primarily by the distribution of resources and threats from their predators (Trivers, 1972; Bradbury & Vehrencamp, 1977; Emlen & Oring, 1977; Clutton-Brock, 1989). This model emphasizes that different sexes use different cues and have different glues (Wrangham & Rubenstein, 1986).

ACKNOWLEDGEMENTS

An earlier version of this essay had been submitted in my preliminary exam in partial fulfilment of the requirements of my Ph.D. at Dalhousie University. I am grateful to my committee members, Don Bowen, Marty Lennard, Shannon Gowans, and my supervisor, Hal Whitehead, whose comments and suggestions helped me to prepare this chapter. I would like to thank Kathreen Ruckstuhl for the source of inspiration and for organizing the workshops that have nourished this new version. Useful information has been generously provided by Tony Martin, William Perrin, Ben Wilson, Bob Reid, Richard Connor and Wayne Perryman. Pierre Béland, Sarah Mesnick and Robin Baird greatly improved the manuscript through their suggestions. 'Merci beaucoup' to Shawn Thompson and Véronick de la Chenelière for reviewing my English. And a special thank you to Cyrille Barrette who first encouraged me to look at terrestrial mammals to understand the elusive whales and dolphins.

DAVID P. WATTS

17
Sexual segregation in non-human primates

OVERVIEW

Sexual segregation is fairly common in non-human primates, usually in the form of social segregation (Box 17.1). Spider monkeys (*Ateles* spp.) and chimpanzees (*Pan troglodytes*) have fission-fusion social systems in which individuals form temporary subgroups (parties) within socially bounded communities; social segregation is not complete, but single-sex parties are common, and males are more gregarious than females and associate predominantly with each other. Some nocturnal lemurs (e.g. grey mouse lemurs, *Microcebus murinus*: Radespiel et al., 2001a,b) and bushbabies (*Galago* spp., *Galagoides* spp., *Otolemur* spp.) forage solitarily, but form sleeping associations that consist mostly of females. Macaques (*Macaca* spp.) form cohesive mixed-sex groups, but maturing males in some species spend time alone or in peripheral all-male groups before joining mixed-sex groups. Habitat segregation is rare, although males may use larger home ranges than females (e.g. orangutans, *Pongo pygmaeus*: Singleton & van Schaik, 2001; chimpanzees: Hasegawa, 1990) or expand their ranges during mating seasons (e.g. grey mouse lemurs: Eberle & Kappeler, 2002).

However, most diurnal primates, even those that breed seasonally, form stable, cohesive groups in which males and females associate permanently. Even when some males are socially peripheral (e.g. squirrel monkeys, *Saimiri* spp.; see later), females associate permanently with others. Stable female groups without permanently associated males are known only in mandrills (*Mandrillus sphinx*; Abernethy et al., 2002). Sexual segregation has received much less attention than questions about why stable mixed-sex groups are so common, about costs and

Sexual Segregation in Vertebrates: Ecology of the Two Sexes, eds. K. E. Ruckstuhl and P. Neuhaus. Published by Cambridge University Press. © Cambridge University Press 2005.

Box 17.1 Examples of sexual segregation in
non-human primates

(1) **Individually based fission-fusion social system with
socially-bounded communities**
Individuals belong to communities within which individuals
have relatively affiliative social relationships and between
which relations are antagonistic. No cohesive groups;
temporary parties vary in size, composition, and duration.
Males more gregarious than females; social segregation,
variable habitat segregation (males use larger ranges than at
least females with dependent young). Examples: Spider
monkeys, chimpanzees.

(2) **Individually based fission-fusion social system without
socially bounded communities**
Adults and adolescents of both sexes mostly forage alone; no
groups or closed communities, but individuals sometimes
congregate at food sources and sometimes travel together.
Adult females are more gregarious than adult males and use
smaller ranges, although females become less gregarious
when the energy demands of lactation peak. Social
segregation. Example: Orangutans.

(3) **Gregarious resting, dispersed or solitary foraging**
Nocturnal species in which individuals form daytime sleeping
associations that vary in stability and that may be unisexual,
and in which multiple individuals use a common home range
or have overlapping home ranges but mostly forage alone.
Social and temporal segregation. Examples: Grey mouse
lemurs other cheirogaleids?; galagos.

(4) **Stable mixed-sex groups; some or all sexually mature males
socially and spatially peripheral to groups**
(a) Adult males mostly on periphery of cohesive groups: e.g.
Peruvian squirrel monkeys; (b) adolescent males temporarily
forage alone within their natal groups' ranges: e.g. Japanese
macaques, long-tailed macaques; (c) some adult and
adolescent males form unstable all-male groups on the
periphery of mixed-sex groups: e.g. Japanese macaques.

(5) **Hierarchical fission-fusion societies with unstable all-male
groups**
Stable, cohesive single-male (sometimes multi-male) breeding
units congregate into larger groupings during foraging; males

not in breeding units form unstable groups that use the same foraging area but maintain social segregation from breeding units. Example: Geladas.

(6) **'Hordes' with permanent female membership and only temporary male membership**
Females and immature offspring permanently in large groups ('hordes') that appear to have stable membership and to forage cohesively; sexually mature males associate with hordes mostly during mating seasons and are otherwise solitary. Example: Mandrills.

benefits of group living, and about ecological influences on variation in social relationships within groups.

In this chapter, I summarize the current 'socio-ecological model' for explaining group living and variation in social relationships in non-human primates, review some data on sex differences in foraging behaviour and activity budgets, and assess how well hypotheses invoked to explain sexual segregation in ungulates apply to non-human primates. Of those that Ruckstuhl & Neuhaus (2000) consider, the predation risk hypothesis (see Chapter 9) receives little support because most primates are in mixed-sex groups for defence against predators, although it can apply to small-scale spatial segregation within groups. The forage selection hypothesis (Chapter 3) also has little support. Segregation in some species may minimize feeding competition, but competition with females is probably not unusually costly for males, contrary to the scramble competition hypothesis (Chapter 3). However, scramble competition probably constrains female gregariousness and contributes to segregation in some species with fission-fusion social systems. Available data give little support to the activity budget hypothesis (see Chapter 10), but it merits further consideration. The 'social preferences hypothesis' (Chapter 11) probably helps to explain social and spatial segregation in chimpanzees and spider monkeys and may account for some social segregation within cohesive groups. I conclude by focusing on grey mouse lemurs, squirrel monkeys, spider monkeys and chimpanzees, and by considering how research on sexual segregation in primates could proceed.

THE SOCIO-ECOLOGICAL MODEL

Explanations of variation in primate grouping patterns and social relationships (e.g. van Schaik, 1983, 1989; Janson, 1992; Sterck et al., 1997)

generally start from two premises. Firstly, females need to maximize foraging efficiency while promoting their own survival and that of their
dependent offspring, while the main strategic goal of males is to maximize mating opportunities (Wrangham, 1979). Secondly, the main cause
of group living is predation: group members benefit from the dilution
effect, enhanced vigilance and, sometimes, direct defence (van Schaik,
1983). The balance of predation risk and feeding competition determines group size, and food distribution determines whether feeding
competition is mainly within-group scramble or contest competition,
or a mix of one or both of these with between-group contest competition. The competitive regime determines whether females typically disperse and influences the quality of social relationships among females.
For example, strong within-group contest competition without strong
between-group contest competition leads to strict female dominance
hierarchies (van Schaik, 1989; Sterck et al., 1997).

   Not all primates use grouping or large group size for protection
against predators during activity periods. Some small-bodied species
form small groups and rely more on being inconspicuous (Janson, 1998).
Most nocturnal species forage solitarily; this may minimize the risk
from predators that locate prey visually and/or auditorily, although it
may result from a high potential for feeding competition (Eberle &
Kappeler, 2002). Even so, some nocturnal species (bushbabies, mouse
lemurs and dwarf lemurs (*Cheirogaleus* spp.)) form daytime sleeping associations that may have anti-predation functions (Box 17.1; Bearder, 1987;
Radespiel et al., 2001b; see later).

   Females in species that obtain anti-predation benefits from grouping could presumably get these from all-female groups. Nevertheless,
their groups almost always include males. In species without seasonal
breeding, permanent association with females may be the best male
mate search tactic. However, permanent male–female association also
characterizes many species that breed seasonally (e.g. rhesus macaques,
Macaca mulatta; redtail monkeys, *Cercopithecus ascanius*). Even when male
transfer during breeding seasons is common (e.g. Verraux's sifakas,
Propithecus verrauxi: Richard et al., 1993) or groups receive male influxes
during mating seasons, as in some redtail monkey (Cords, 1984) and
patas monkey (*Erythrocebus patas*: Chism & Rowell, 1986) populations,
some or all males reside in stable bisexual groups during the rest of
the year.

   Unlike some ungulate species in which males maximize foraging
efficiency by using the best quality foraging areas (Main & Coblentz,
1990; Chapter 9; but see Ruckstuhl & Neuhaus, 2002), primate males

in group-living species typically 'go where the females are' (Altmann, 1990). Variation in the number of males per group depends mostly on variation in female group size and reproductive synchrony (and perhaps on predation risk; van Schaik & van Noordwijk, 1989); these determine operational sex ratios and influence the form and intensity of male mating competition (Mitani *et al.*, 1996; Nunn, 1999; Eberle & Kappeler, 2002). In turn, operational sex ratios can influence male dispersal decisions (e.g. yellow baboons, *Papio cynocephalus*; Alberts & Altmann, 1995).

Males may gain protection against predation by associating with females, especially in species with low size dimorphism (e.g. vervets, *Chlorocebus aethiops*: Isbell *et al.*, 2002). They could presumably gain similar benefits from all-male groups, and predation risk may explain formation of all-male groups in species like grey langurs (*Semnopithecus entellus*; Rajporohit *et al.*, 1995) and Thomas' langurs (*Presbytis thomasi*; Steenbeek *et al.*, 2000). But in these and some other cases (e.g. mountain gorillas, *Gorilla gorilla beringei*: Yamagiwa, 1987), all-male groups comprise either males evicted from bisexual groups by other males, or males from previously mixed-sex groups that have lost their females through emigration or death. Females continue to reside with other males. Immigrating males must fight resident males to gain access to groups that contain females; therefore segregation is a forced outcome of male–male mating competition, not part of a strategy to minimize predation risk or to improve competitive ability by maximizing foraging efficiency.

Females can gain anti-predation benefits from male presence if males defend them and their young. For example, male red colobus (*Procolobus badius*) defend their groups against chimpanzees (personal observation). High costs may limit active defence, but males may use alarm calls as less costly deterrents of pursuit by predators that use stealth (e.g. black-and-white colobus, *Colobus polykomos*, and Diana monkeys, *Cercopithecus diana*; Zuberbühler *et al.*, 1999). They may also provide disproportionate anti-predator benefits if they are more vigilant than females, as in brown capuchins (*Cebus apella*) and white-fronted capuchins (*C. albifrons*; van Schaik & van Noordwijk, 1989). Even males may gain survival benefits from greater vigilance that outweigh the reproductive benefits of excluding other males from female groups (van Schaik & van Noordwijk, 1989).

However, a complete explanation of permanent male–female association in primates requires consideration of mate guarding (Palombit, 1999) and, especially, infanticide by males, which is widespread (Sterck *et al.*, 1997; van Schaik & Kappeler, 1997). Infanticide is associated with generally low female reproductive rates and with lactation that lasts longer than gestation (van Schaik, 2000a). Much evidence (reviewed in van Schaik, 2000b) supports the sexual selection hypothesis: by killing unrelated infants, males cause females to become fertile again sooner than otherwise and can then increase their reproductive success if they mate with the females. Females use various counter-strategies, including individual defence against attacking males, but often rely crucially on potential fathers to defend infants, especially when males are considerably larger than females (van Schaik, 2000b; Palombit et al., 2000). Lactation periods that exceed the length of gestation are rare in ungulates; for this reason and because many breed seasonally and annually, infanticide would usually not change the time of female receptiveness (van Schaik, 2000a). Infanticide risk may crucially alter the relative costs and benefits of segregation for both sexes and do much to explain its rarity in primates. Correspondingly, male provision of anti-predator and anti-infanticide services may help to explain mixed-sex groups in group-living equids and a few other ungulates (Cameron et al., 2003).

THE ECOLOGY OF TWO SEXES IN PRIMATES

Interspecific variation in diet quality in primates is generally consistent with the Jarman–Bell principle (Bell, 1971; Jarman, 1974): smaller species typically eat higher-quality diets than larger species (Sailer et al., 1985). At the same time, sexual dimorphism in body size varies greatly across species (reviewed by Smith & Jungers, 1997). Clutton-Brock (1977) proposed that high sexual dimorphism in body mass could lead to sex differences in foraging strategies like those predicted by the forage selection hypothesis (e.g. Short, 1963; Beier, 1987): relatively lower male basal metabolic rates in highly dimorphic species allow them to eat diets that provide lower net energy gain rates than females require. Clutton-Brock (1977) also noted that size dimorphism could lead to sex differences in positional behaviour and substrate use in highly or totally arboreal species. His review and several later studies showed some evidence of size-related sex differences in diet and foraging substrate use, but available sex-specific data, while still sparse, are not entirely consistent with expectations. Gautier-Hion (1980) found that males of three guenon species (*Cercopithecus nictitans, C. cephus* and *C. pogonias*) ate diets more similar to each other's than to those of conspecific females during some seasons. However, contrary to expectations, folivory did not increase with body size, nor was insectivory inversely related to body size, in comparisons among species and between the sexes. Female Costa Rican squirrel monkeys (*Saimiri oerstedi*) foraged more, ate more and rested less than males (which are 30% heavier), but male and female foraging techniques and food preferences were similar (Boinski, 1988). Male white-fronted capuchins (*Cebus capucinus*), which are about 40% larger than females, ate more vertebrates and large invertebrates, spent more time on large supports and broke branches and stripped bark more than females, but sex differences in diet were otherwise slight (Rose, 1994). In another capuchin (*C. olivaceous*), males harvested fruit and animal prey more efficiently, and used substrates slightly differently than females, but the differences were not tightly linked to body size and both sexes were 'catholic' in diet and foraging behaviour (Fragaszy & Boinski, 1995). Male western lowland gorillas (*Gorilla gorilla gorilla*) at Bai Hokou ate more high-fibre food and less fruit than females (Remis, 1999), but sex differences in diet are absent in the highly folivorous Virunga mountain gorillas (*G. g. beringei*) (Watts, 1984). These species all form permanent mixed-sex groups.

PRIMATES AND HYPOTHESES APPLIED TO UNGULATES

Sexual dimorphism and sexual segregation

Sexual dimorphism in body size plays a major role in explanations of sexual segregation in ungulates (Main *et al.*, 1996; Ruckstuhl & Neuhaus, 2000). The forage selection, activity budget and scramble competition hypotheses predict segregation between all adult females and adult males when dimorphism is high. The forage selection hypothesis also predicts that females eat better quality diets in highly dimorphic species. The activity budget hypothesis predicts that segregation is absent in monomorphic species unless sex differences in predation risk and anti-predation strategy exist, and that sex differences in time active increase with increasing dimorphism (Ruckstuhl & Neuhaus, 2000). The predation risk hypothesis predicts that females with dependent young use safer habitat than males, but that females without young form mixed-sex groups in non-dimorphic species (Ruckstuhl & Neuhaus, 2000). These hypotheses are not mutually exclusive, and all have some support. For example, Festa-Bianchet (1988) found that female bighorn sheep (*Ovis canadensis*) with dependent young used safer habitat than males, while significant sex differences in activity budgets occurred in bighorn sheep (Ruckstuhl, 1998) and in dimorphic Alpine ibex (*Capra ibex*), but not in monomorphic zebra (*Equus burchelli*) or oryx (*Oryx gazelle*; Ruckstuhl & Neuhaus, 2001, 2002). However, comparative analysis strongly supports the activity budget hypothesis, but not the predation risk hypothesis (Ruckstuhl & Neuhaus, 2002).

Hypotheses proposed for ungulate sexual segregation clearly do not consistently apply to primates, given that sexual segregation and formation of stable mixed-sex groups occur in taxa that span the range of variation in body size dimorphism. Social segregation occurs in some cheirogaleids and galagines (Box 17.1), which are monomorphic or have low dimorphism; in Peruvian squirrel monkeys (*Saimiri boliviensis*) and chimpanzees, in which males are 30% heavier than females (Smith & Jungers, 1997); and in orangutans and mandrills, in which males are twice or more the size of females (Smith & Jungers, 1997). However, all macaques (male body mass 30% to 70% higher than female) and baboons (males 70% to 80% larger; Smith & Jungers, 1997) form stable mixed-sex groups that usually contain multiple males and females.

Scramble competition and forage selection

Limited data on sex differences in diet in primates do not show that females necessarily eat higher quality diets in dimorphic species (see earlier), and habitat segregation associated with variation in the quality of available food is rare or absent. Scramble competition for food has a fundamental role in primate socio-ecology (van Schaik, 1989), but, contrary to the 'scramble competition' hypothesis, it probably affects females more than males because of the high costs of reproduction for females and because no obvious morphological or behavioural factor like the allometry of ungulate bite size (Main & Coblentz, 1990) would lead females in dimorphic species to reduce food availability for males disproportionately. Comparison of two of the most highly dimorphic primates illustrates the importance of scramble competition. Male mountain gorillas feed about 12% longer per day than females on the same range food of (Watts, 1988). Mountain gorillas fit the socio-ecological model well: their diet is highly folivorous, feeding competition is low and females reside permanently in mixed-sex groups and depend on male protection against infanticide by extra-group males (Watts, 1991; Sterck et al., 1997). In contrast, orangutans are highly frugivorous. Their large size and mostly arboreal lifestyle means that predation risk is low, but preferred food (ripe fruit) is often scarce relative to metabolic needs and foraging in stable groups would be too costly (Rodman & Mitani, 1987). Females have extremely low reproductive rates, and those with medium-sized infants and thus peak energy demands of lactation, are less gregarious than those with younger or older infants or without infants, as expected if they need to minimize feeding competition (van Schaik, 1999).

Predation risk

Habitat segregation is rare in primates, although within-group segregation related to variation in exposure to predators occurs. For example, juveniles and females with dependent infants are more likely than adult males and other adult females to be in the centre of brown capuchin (*Cebus apella*) groups, although animals to the front and sides have higher foraging efficiency (Janson, 1990). Similar spatial patterns occur in baboons (Busse, 1984; Rhine *et al.*, 1985) and grey-cheeked mangabeys (*Lophocebus albigena*; Waser, 1985). Selfish herd effects may sometimes produce within-group social segregation independently of sex. For example, high-ranking male and female long-tailed macaques (*Macaca fascicularis*) are more often in their group's 'main party', where they enjoy priority of access to food patches too small to feed all group members, while low-ranking adults of both sexes forage on the group's periphery, where they face less feeding competition but are more exposed to predators (van Schaik & van Noordwijk, 1989). Busse (1984) found similar rank-related differences for female chacma baboons.

Some habitat segregation may exist between all-male groups and stable bisexual groups, but in the one species for which data on habitat quality are available (red howlers, *Alouatta seniculus*: Pope, 2000a), all-male groups use poorer foraging habitat than mixed-sex groups. All-male groups comprise males evicted from breeding groups. By forming 'coalitions', evicted males improve their chances of successfully challenging males resident in female groups and gaining breeding opportunities (Pope, 1990). Likewise, all-male groups in grey langurs may use poor quality habitat, but, beyond gaining protection against predators, some group members may benefit by collaboratively challenging males resident in mixed-sex groups (Rajporohit *et al.*, 1995). Female red howlers also co-operate to evict other females from groups when these become too large and to prevent female immigration. Like extra-group males, evicted females use poorer habitat and have poorer quality diets than females in established groups. Some join extra-group males and perhaps other evicted females, but apparently they breed successfully in these new groups only if they establish feeding territories (Pope, 2000a,b).

Body size influences mating success in male baboons, and priority of access to food gained via dominance over females and smaller peers should help to maximize growth rates, particularly during adolescent growth spurts (Pereira, 1988; Pereira & Leigh, 2003). Male rhesus macaques on Cayo Santiago gain weight before mating seasons; those who sire offspring are significantly fatter at the start of the season than non-sires, presumably because they do better in endurance competition (Bercovitch, 1997). Sexual segregation does not occur in these species, but segregation to gain foraging benefits may occur in two other macaque species. Adolescent male long-tailed macaques become largely solitary within the home ranges of their natal groups for several months during periods of high fruit abundance. Solitariness increases predation risk, but apparently enables males to maximize their foraging efficiency and to grow rapidly. Male peers in groups do not show such growth spurts. After attaining full growth, solitary males either rejoin natal groups or immigrate into new groups. They then challenge resident alpha males, often successfully, whereas other male immigrants do not make such challenges (van Nordwijk & van Schaik, 2002). Adolescent male Japanese macaques (*Macaca fuscata*) on Koshima Islet also become solitary within their natal group's ranges, then rejoin groups when at or near full adult body mass (Mori & Watanabe, 2003). Solitariness may promote maximum growth by minimizing feeding competition, especially given that coalitionary support from fully grown males allows adult females to win many feeding contests with adolescent males. Many males in other Japanese macaque populations spend substantial periods alone or in all-male groups and may join mixed groups during mating seasons (Sprague *et al.*, 1998). Jack and Pavelka (1997) claimed that spatially distinct subgroups of 'peripheral' males usually occupied 'sub-optimal' portions of the ranges of associated mixed-sex groups, implying that peripheralization did not improve foraging efficiency, and argued that excluding young males benefits females and fully adult males. However, neither they nor Sprague *et al.* (1998) provided detailed data on activity budgets or foraging behaviour.

Males in all-male Thomas' langur groups have lower foraging efficiency than those in mixed-sex groups. However, the difference may occur because males in all-male groups are frequently attacked by those in mixed-sex groups and consequent high levels of vigilance compromise foraging efficiency, not because they suffer more feeding competition or use poorer habitat (Steenbeek *et al.*, 2000). In contrast to grey langurs and red howlers, members of all-male groups do not cooperatively take over breeding groups (Steenbeek *et al.*, 2000).

Mandrills (*Mandrillus sphinx*) have a unique social system among primates: females and dependent young form extremely large and apparently stable groups ('hordes'), while males rarely associate with females and are mostly solitary, outside mating seasons (Abernethy *et al.*, 2002). Mandrills are among the most highly size-dimorphic primates and males have testosterone-mediated ornaments that appear to be honest signals of mate quality (Wickings & Dixon, 1992). Foraging efficiency may impact both body size and ornament development, but data to test whether segregation maximizes foraging efficiency and/or results from activity budget incompatibility are not available.

Activity budgets

In a comparative analysis of ungulate data, Ruckstuhl and Neuhaus (2002) found a significant positive relationship between the per-cent difference in female and male time active and the per-cent difference in male and female body mass, as the activity budget hypothesis predicts. Most published studies of primate activity budgets give only values for females or group means, but Table 17.1 gives data on sex differences for 24 species of monkeys and apes representing 19 genera. Contrary to the activity budget hypothesis, the ratio of female to male feeding time decreases as dimorphism increases. With data on the per-cent time spent feeding $\log_{10}$ transformed (cf. Ross, 1992), mean generic values show a significant inverse correlation between the ratio of female to male feeding time and the ratio of male to female body mass ($r = -0.55$, d.f. $= 17$, $p < 0.05$). Use of untransformed data (as in Ruckstuhl & Neuhaus, 2002) yields a correlation that approaches significance ($r = -0.44$, d.f. $= 17$, $p = 0.06$; Fig. 17.1). This result should be interpreted cautiously because the analysis does not correct for phylogeny, but correction would probably not produce a significant positive relationship. A negative relationship is expected if males and females eat similar quality diets, but the nutritional demands of reproduction lead to higher female intake in less dimorphic species, while male intake is relatively greater in highly dimorphic species (Hanley, 1982). However, feeding time may not represent food intake accurately. For example, male siamang (*Hylobates symphalangus*) are 10% larger than females and feed about 15% longer, but with intake rates only half those of females (Chivers, 1974).

Both sexes may compromise their activity budgets to obtain benefits of grouping. Females often increase feeding time during gestation and, especially, lactation (e.g. yellow baboons: Altmann, 1980; geladas, *Theropithecus gelada*: Dunbar & Dunbar, 1988; mountain gorillas: Watts, 1988; white-fronted capuchins: Rose, 1994), and reproductive costs can

Table 17.1 *Activity budgets and sexual dimorphism in body mass in non-human primates. Data are from 24 species representing 19 genera of haplorhine primates (Old and New World monkeys; apes). Body mass data from Smith and Jungers (1997) unless otherwise noted. % Difference =* [(♂ *body mass*/♀ *body mass*) × 100] − 100.

Species	% Time feeding ♂	% Time feeding ♀	Body mass (kg) ♂	Body mass (kg) ♀	% Difference	Reference
Alouatta palliata	14.0	18.0	7.15	5.35	33.6	Smith, 1977
Ateles paniscus	22.5	30.7	9.11	8.44	7.9	McFarland-Symington, 1988
Callicebus torquatus	25.0	25.0	1.28	1.21	5.8	Kinzey, 1977
Cebus albifrons	35.9	51.4	3.18	2.29	38.9	van Schaik & van Nordwijk, 1989
Cebus apella	41.7	51.5	3.65	2.52	44.8	van Schaik & van Nordwijk, 1989
Cebus capucinus	47.0	53.0	3.30	2.30	43.5	Rose, 1994
Cebus olivaceous	19.0	22.5	3.29	2.52	30.1	Fragaszy & Boinski, 1995
Chlorocebus aethiops	18.7	19.6	4.26	2.98	43.0	Baldellou & Aden, 1997
Colobus guereza	21.0	26.5	13.5	9.2	46.7	Fashing, 2001
Erythrocebus patas	30.0	30.0	12.4	6.5	90.8	Nakagawa, 1989
Gorilla gorilla	67.4	59.9	163.0	98.0	66.3	Watts, 1988
Hylobates syndactylus	59.9	51.9	11.9	10.7	11.2	Chivers, 1974
Lagothrix lagotricha[a]	16.3	17.7	7.1	4.5	57.8	DiFiore & Rodman, 2001; A. DiFiore, pers. com.
Lophocebus albigena	41.0	48.9	8.3	6.0	38.3	Waser, 1977
Macaca fascicularis	15.0	13.5	5.4	3.6	50.0	van Nordwijk et al., 1993
Macaca fuscata	15.9	24.8	11.0	8.0	37.5	Maruhashi, 1981
Pan troglodytes	50.0	60.0	42.7	33.7	26.7	Wrangham & Smuts, 1980
Papio cynocephalus	52.9	52.8	21.8	12.3	77.2	Post et al., 1981
Papio anubis	42.0	51.0	25.1	13.3	88.7	Bercovitch, 1983
Pongo pygmaeus	44.4	41.3	78.5	35.8	119.3	Rodman, 1988[b]
Presbytis thomasi	24.5	35.0	6.29	6.17	1.9	Steenbeek et al., 2000
Procolobus badius	28.3	34.6	9.67	7.21	34.1	Marsh, 1981
Saimiri oerstedi	11.7	16.4	0.88	0.68	29.4	Boinski, 1988
Theropithecus gelada[c]		42.0	45.4	20.3	14.8	Dunbar, 1977

[a] Body mass data for *Lagothrix* from DiFiore & Rodman, 2001.

[b] Mean values from five studies.

[c] Body mass data for *Theropithecus* from Stammbach, 1987.

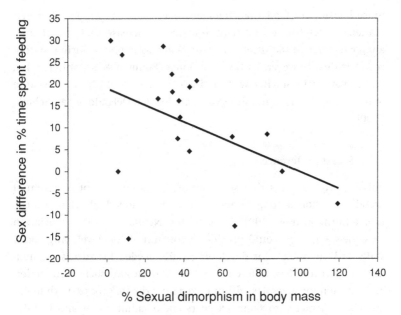

Figure 17.1   Relationship between sexual dimorphism in body mass and sex differences in time spent feeding in haplorhine primates (monkeys and apes). Data points are means for 19 genera; Table 17.1 gives raw values. The relationship is non-significant ($r = -0.44$, d.f. = 17, $p = 0.06$). With feeding time values $\log_{10}$ transformed, the relationship is significantly negative ($r = -0.55$, d.f. = 17, $p < 0.05$).

eventually constrain their activity budgets seriously (Altmann, 1980; Dunbar & Dunbar, 1988). Altmann (1980) noted that female baboons could benefit from sometimes foraging apart from their groups, but could not afford to forsake the anti-predation benefits of grouping; similar trade-offs are probably widespread. Conversely, males may need to guard offspring against infanticide and/or to guard mates, even if sex differences in optimal activity budgets exist. Alternatively, sex differences in food intake rates may compensate for the costs of permanent association between the sexes. Elevations of female feeding rates during lactation, as in yellow baboons (Silk, 1987), can provide at least partial compensation (cf. Ruckstuhl & Neuhaus, 2002, for zebra).

In principle, the activity budget and forage selection hypotheses could apply to orangutans. Data averaged across five studies (Rodman, 1988) showed that male orangutans feed only slightly longer than females. However, males at Kutai had longer patch residence times, shorter day journeys and less diverse diets than females (Rodman, 1979), and males in some populations (e.g. Tanjung Puting: Galdikas, 1988) eat more low-quality food than females. Rodman (1979) argued that permanent association would require males to increase their travel costs and/or females to sacrifice dietary diversity, as happens during mating consortships. However, as he later noted (Rodman & Mitani, 1987), a male's pay-off in staying with a single female would be low because female reproductive rates are extremely low (cf. Delgado & van Schaik, 2000).

Social preference

This hypothesis holds that segregation occurs because of preferential social attraction among same-sex immature individuals that persists into adulthood (Bon, 1991; Ruckstuhl & Neuhaus, 2000). Association in same-sex groups could provide opportunities for developing social skills and, for males, for developing skills needed for success in mating competition. Sex differences in association and social interaction are common for juveniles within mixed-sex primate groups, with males especially attracted to same-sex peers and/or adult males and females attracted to female peers and adult females. Juvenile male baboons associate preferentially with adult males, although association time drops at adolescence as adult male aggression rates rise and remains low in adulthood. Juvenile females associate preferentially with adult females, especially those with infants (Pereira, 1989). Juvenile male long-tailed macaques play and groom mostly with same-sexed peers with whom they may later emigrate, while juvenile females concentrate on developing female social networks that later include adult males (van Noordwijk et al., 1993). Juvenile vervet monkeys show similar preferences for same-sex peers and adults, with males especially attracted to those adult males born in the juveniles' groups who have not yet dispersed (particularly older brothers) and to alpha males (Fairbanks, 1993). In these and other cases (reviewed in Fairbanks, 1993), juveniles obtain opportunities to learn valuable sex-specific social and reproductive skills. For example, juvenile male baboons develop fighting skills through play and can learn how to assess competitors (Pereira, 1989; cf. Bon et al., 2001, for Alpine ibex). When members of one sex typically either do not disperse from their natal groups, disperse with same-sex peers or join groups into which familiar same sex individuals have immigrated, these sex biases also facilitate the development of valuable long-term social relationships (Pereira, 1989; Fairbanks, 1993; van Noordwijk et al., 1993; Mitchell, 1994).

Males of macaque species in which unstable, peripheral all-male groups occur may also benefit from opportunities to assess competitors and to develop alliances within these groups. Similar benefits may accrue to males in the loose all-male groups that form within gelada herds, but breeding males exclude these males from mixed-sex groups (Dunbar, 1984), so their formation does not clearly support the social preference hypothesis, despite the probable importance of fighting skills developed in all-male groups for male success at gaining and maintaining breeding positions.

CASE STUDIES

Grey mouse lemurs

Mouse and dwarf lemurs are small-bodied, mostly monomorphic, nocturnal Malagasy strepsirhine primates. Mouse lemurs have 'dispersed promiscuous' mating systems with high male scramble and sperm competition (Radespiel et al., 2001a; Eberle & Kappeler, 2002). Grey mouse lemurs (Microcebus murinus) at Kirindy show seasonal body mass dimorphism, with males significantly larger than females shortly before and during the mating season, but females otherwise sometimes significantly heavier than males (Schmid & Kappeler, 1998). Females hibernate for six months per year, but only some males hibernate, and for shorter periods than females. Males become active earlier than females prior to the mating season. Food abundance is low and predation risk high at this time, but resumption of foraging helps males to gain body mass in preparation for the mating season, when male body size probably influences reproductive success (Schmid & Kappeler, 1998).

Individuals of both sexes forage alone, probably because of the high potential for competition over the arthropods, plant gum and fruit that compose their diet (Atsalis, 1999; Eberle & Kappeler, 2002). Mouse lemurs sleep in nest holes in tree trunks during the day and hibernate in these holes. Many females form stable sleeping associations with matrilineal relatives while females with no nearby relatives sleep alone. Females may co-operatively defend access to good nest holes, especially against males; these provide energy savings and anti-predator benefits and are in limited supply (Perret, 1998; Radespiel et al., 1998; Radespiel et al., 2001b; Schmid & Kappeler, 1998). Males mostly sleep alone, but may nest with up to four other males, most of them close relatives (Eberle & Kappeler, 2002) and sometimes sleep with close female kin (Radespiel et al., 2001b). They also sometimes inspect groups of hibernating females, which may give them information about mate distribution in the next mating season (Eberle & Kappeler, 2002).

As Radespiel *et al.* (2001b) note, social, but sexually segregated, resting and solitary foraging is unusual among primates but occurs in some other nocturnal strepsirhine (e.g. some bushbabies) and a few other mammals, such as European badgers (*Meles meles*) and Bechstein's bats (*Myotis bechsteinii*).

Martin (1972) reported that mouse lemurs at Mandena formed discrete population nuclei, with females, young and dominant males in the centre and other males on the periphery. However, male home range overlap was high and habitat segregation absent at other sites (Fietz, 1999; Radespiel, 2000; Radespiel *et al.*, 2001a; Eberle & Kappeler, 2002). Apparent habitat segregation at Mandena might have been an artefact of less systematic sampling, although regional differences in ecology could lead to variation in social systems and the Mandena population may even be a different species (Radespiel *et al.*, 2001a).

Mouse and dwarf lemurs (and other nocturnal strepsirhine primates) show remarkable diversity in social and mating systems, and not all show sexual segregation. For example, like grey mouse lemurs, golden brown mouse lemurs (*Microcebus ravelobensis*) have a dispersed, multi-male, multi-female social system with promiscuous mating and social sleeping, but sleeping groups occur year-round and most are mixed-sex (Weidt *et al.*, 2004). Also, opposite-sex sleeping group members encounter each other and interact affiliatively at relatively high rates while foraging. Whether segregation occurs seems to depend partly on the spatial distribution of females, but also on the distribution of sleeping sites and the possible costs and benefits of sharing these. Competition for sparsely distributed safe sleeping sites (tree holes) is high in grey mouse lemurs. Small groups of related females can gain mutual anti-predator and thermoregulatory benefits by sharing these and, given little or no sexual dimorphism in body mass, can collaboratively exclude males (Radespiel *et al.*, 2001b). Golden brown mouse lemurs mostly sleep in nests constructed in tree crowns and presumably face higher thermoregulatory costs, and sharing nests with males may allow females to meet these costs (Weidt *et al.*, 2004). Further research on the costs and benefits of sleeping in groups should help to explain variation in segregation.

Squirrel monkeys

Variation in the behavioural ecology of three squirrel monkey species is consistent with the socio-ecological model (Boinski, 1999; Boinski *et al.*, 2002). Squirrel monkeys face high predation pressure, and all species form large, stable, mixed-sex groups. Females are philopatric in *S. boliviensis*; males are philopatric and females transfer in the other two species. *S. sciureus* and *S. boliviensis* rely heavily on food that comes in defendable clumps; feeding contests are common and females form dominance hierarchies. Female coalitions are common in *S. boliviensis*, whose rich food patches often can support multiple individuals, but not entire groups; they are absent in *S. sciureus*, for which monopolizable patches often support only single animals (Boinski et al., 2002). Contest feeding competition is rare in *S. oerstedi* and females do not have dominance relationships.

Interspecific variation in sexual segregation occurs in squirrel monkeys, in association with variation in agonistic relationships between males and females. As in grey mouse lemurs, sexual dimorphism in body size and sexual disparities in power are relatively low. Neither sex is unequivocally dominant to the other in *S. oerstedi*, in which social segregation within mixed-sex groups occurs (Boinski et al., 2002). Also, females with infants associate preferentially with each other, which increases overall anti-predator vigilance (Boinski, 1987). Sexual segregation is absent in *S. sciureus*, in which males are dominant (Boinski, 1994, 1999; Boinski et al., 2002). Female coalitions win contests against males in *S. boliviensis* and males tend to stay on group peripheries (Mitchell, 1994). Mitchell (1994) describes this social segregation as the result of preferential female–female association and affiliation that help females maintain alliances advantageous in within- and between-sex feeding competition; low male–male association in mixed-sex groups; and female aggression against males.

The social preferences hypothesis apparently applies to *S. boliviensis* males, although without male exclusion of other males from mixed-sex groups. Immature males form same-sex peer play groups in their natal groups, and maturing males disperse to bachelor groups, where they may remain for several years before immigrating into mixed-sex groups. Pairs of males often immigrate together in 'alliances' and co-operatively challenge resident males. *S. boliviensis* breeds seasonally; males gain weight before breeding seasons, and many try to transfer from bachelor groups into mixed-sex groups during breeding seasons (Mitchell, 1994). Bachelor groups probably do not use higher quality habitat than mixed-sex groups, contrary to predation risk hypothesis, and a modified scramble competition hypothesis could better explain their formation: segregated males may forage more efficiently because they avoid the costs of competing with females, who can get access to food at male expense.

Chimpanzees and spider monkeys

Relatively large size and highly frugivorous diets, hence low predation risk but high potential for feeding competition, underlie convergence in chimpanzee and spider monkey social systems. In both, males are philopatric and females transfer between communities as adolescents. Males are more gregarious than females and associate with each other more than with females, although in chimpanzees, the extent of these sex differences varies across populations and over time within them (Nishida, 1968; Goodall, 1986; McFarland-Symington, 1988; Chapman, 1990; Pepper et al., 1999; Boesch & Boesch-Achermann, 2000; Shimoka, 2003).

The high potential for scramble competition over fruit especially constrains female gregariousness, in contrast to the ungulate version of the scramble competition hypothesis. Males generally use more of a community's range than pregnant and lactating females, although sexually cycling females range more widely (chimpanzees: Wrangham, 1979; Wrangham & Smuts, 1980; Hasegawa, 1990; Chapman & Wrangham, 1993; but see Boesch & Boesch-Achermann, 2000; spider monkeys: McFarland-Symington, 1990). Party size often varies positively with ripe fruit abundance, although independent effect of sexually receptive females on male aggregation in chimpanzees can obscure this relationship (chimpanzees: Wrangham et al., 1996; Newton-Fisher et al., 2000; Mitani et al., 2002; Hashimoto et al., 2001; spider monkeys: MacFarland-Symington, 1988; Chapman, 1990; Shimoka, 2003). Female chimpanzees may join males at large food sources but then not travel with the males (Wrangham & Smuts, 1980; Wrangham, 2000; personal observation), and at Gombe, they feed less when with males for extended periods than otherwise (Wrangham & Smuts, 1980). At least at Gombe and perhaps elsewhere, females compete over access to 'core areas' within community ranges, and variation in core area quality influences reproductive success (Pusey et al., 1997).

Thus, social segregation is partly a by-product of greater male tolerance for gregarious foraging (cf. Ruckstuhl & Neuhaus, 2000, for ungulates), but scramble competition is not a complete explanation. Conspecifics present a threat like that in the predation risk hypothesis: both sexes face aggression from coalitions of extra-community males near territory boundaries. In chimpanzees, this aggression can be lethal for adult males and for lactating females and their infant and juvenile offspring (Goodall, 1986; Wrangham, 1999; Wilson & Wrangham, 2003; Wilson et al., 2004). Female chimpanzees may tend to avoid boundary areas, especially when low fruit abundance limits party sizes, unless they are sexually receptive or have mated with males from both communities (Herbinger *et al.*, 2001; Williams *et al.*, 2002; Williams *et al.*, 2004). Relatively lower costs of feeding competition make it easier for males to stay in parties large enough to provide safety. Similar factors may operate in spider monkeys, in which males also engage in inter-group aggression and male coalitions may be an infanticide threat (Chapman *et al.*, 1995), although lethal attacks are unknown.

Association among male chimpanzees facilitates territorial defence and can bring benefits from offensive aggression against neighbours during 'border patrols' and incursions into neighbouring territories. The main benefits are probably protection of community members (including direct protection of the males' offspring against infanticide) and territory expansion, which increases food availability and can improve survivorship of females and offspring and increase female reproductive rates (Wrangham, 1999; Watts & Mitani, 2001; Williams *et al.*, 2004). Male spider monkeys also engage in patrols and incursions, but their long-term consequences are unknown (Chapman *et al.*, 1995).

Male–male association has other potential benefits. Male alliances influence the outcome of male–male competition for status, and thus for mating opportunities, within chimpanzee communities (Goodall, 1986; Nishida & Hosaka, 1996). Allies spend much time together, groom extensively, preferentially share meat and often accompany each other on patrols (Watts & Mitani, 2001; Mitani & Watts, 2002). Male spider monkeys form coalitions that enable them to win feeding contests against females, although their role in male–male competition is unclear. In chimpanzees, success during group hunts of prey (notably red colobus monkeys, *Procolobus badius*) varies positively with the number of males present (Mitani & Watts, 2002), and males who participate in hunting groups can signal their quality as allies in other contexts.

At least in chimpanzees, these potential benefits from male cooperation make the social attraction hypothesis relevant. Sexual segregation starts at adolescence, as males spend increasing amounts of time away from their mothers with adult males despite frequent aggression from adult males (Pusey, 1990; Mitani & Amsler, 2003). They are gradually integrated into adult male social networks (e.g. dyadic grooming is mostly unidirectional from adolescents to adults at first, but eventually becomes bidirectional; Pusey, 1990; Watts, unpublished data) and gradually take greater roles in community defence (Watts & Mitani, 2001). Analyses of data from Ngogo (Mitani *et al.*, 2002) show additional age-related segregation among males that presumably reflect common interests (e.g. males close in age may tend to be each other's best potential allies), although the age categories (e.g. young prime vs. old adults) were much broader than those associated with segregation in ungulates like Alpine ibex (Bon *et al.*, 2001) and bighorn sheep (Ruckstuhl & Fiesta-Bianchet, 2001). The development of sexual segregation has received little attention in spider monkeys, but may be similar.

Shared interests, like protection against infanticide and opportunities for infant and juvenile offspring to play with peers, presumably explain chimpanzee 'nursery parties' that consist of adult females and immature offspring (Goodall, 1986). At Ngogo, for example, females with dependent offspring associate with each other more often than expected by chance (Pepper *et al.*, 1999).

FUTURE DIRECTIONS

In reviewing the primate literature, I encountered the term 'sexual segregation' only once (Radespiel *et al.*, 2001b), although primatologists have often described the phenomenon without using the term. The literature on sexual segregation in non-primates has implications for understanding sex-related variation in spatial positions within groups, variation in dispersal strategies (e.g. whether males emigrate voluntarily from their natal groups and whether they form all-male groups) and other aspects of primate behavioural ecology. By giving more attention to this literature, primatologists could also contribute to a better understanding of sexual segregation in non-primates. I conclude by noting three relevant directions for future work.

(1)   Inter-sexual activity budget conflicts, like those associated with lactation in baboons (Altmann, 1980), may commonly be another cost of group living in primates. How often females resolve these by increasing feeding rates is unknown; relevant data are sparse and mostly address questions about feeding competition among females. Too few published studies give activity budget data broken down by sex and age, and analysis of activity synchrony using Ruckstuhl's (1999) and Ruckstuhl and Neuhaus' (2001) synchrony index would be valuable, especially for species with high body size dimorphism. More data on sex-specific activity budgets in colobine monkeys, which have ruminant-like complex stomachs and are foregut fermenters, could provide good comparisons to ungulates.

(2)   The few studies that have assessed sex differences in diet show wide variation (cf. van Noordwijk *et al.*, 1993; Rose, 1994). More are needed to test hypotheses about feeding competition and about the effects of sexual dimorphism generally, because such tests require comparative analyses (Ruckstuhl & Neuhaus, 2002). More detailed data on feeding rates and diet quality in mixed-sex and all-male groups and for solitary males would help to illuminate the relationship between feeding competition and sexual segregation.

(3)   Many species of higher primates have complex social relationships, and social complexity is probably linked to the evolution of complex cognitive abilities. Primates are thus excellent candidates for the evaluation of the 'social attraction' hypothesis, but research on the ontogeny of relationships and of social strategies has languished (Pereira & Fairbanks, 2002). More attention to the development of sex differences in behaviour would contribute to comparative research on sexual segregation.

Part VII   Implications for conservation

Part VII Implications for conservation

18

Sexual segregation in birds: patterns, processes and implications for conservation

OVERVIEW

It has been suggested recently that sexual segregation of male and female birds outside the breeding season, or away from their nest-sites, could be a general ecological trait of many migratory species (e.g. Lopez Ornat & Greenberg, 1990; Cristol et al., 1999), and one that may have important implications for population dynamics and conservation (Weimerskirch et al., 1997a; Croxall et al., 1998; Marra et al., 1998; Marra & Holmes, 2001). Segregation by sex can occur at varying spatial scales, ranging from broad geographical differences in distribution, to local differences in habitat or even microhabitat use. Segregation is relatively independent of mating systems or sex role specialization during reproduction. In many species of crane (Gruidae), for example, adult males and females, once paired, generally stay within each other's sight throughout the annual cycle, and even migrate together to reach their winter quarters (Archibald & Meine, 1996). In the great albatrosses, *Diomedea* spp., on the other hand, not only do members of a pair meet only briefly at the nest in between their long foraging trips (a direct consequence of their lifestyle), but they also largely forage in different geographical areas throughout both the breeding and non-breeding seasons (Weimerskirch et al., 1993; Prince et al., 1998; Weimerskirch & Wilson, 2000; see also Chapter 6). That both cranes and albatrosses are strictly monogamous, highly faithful to their mates during their long lives, display little differentiation in reproductive sex-roles and are highly migratory, does not prevent them from having such

Sexual Segregation in Vertebrates: Ecology of the Two Sexes, eds. K. E. Ruckstuhl and P. Neuhaus.
Published by Cambridge University Press. © Cambridge University Press 2005.

a contrasting pattern of sexual segregation in geographic and habitat distribution.

Two principal hypotheses seek to explain the general patterns of sexual segregation in birds. The social dominance hypothesis suggests that segregation arises from the despotic exclusion of subordinate individuals from favoured areas by dominant conspecifics (e.g. Gauthreaux, 1978; Greenberg, 1986). The specialization hypothesis, on the other hand, proposes that males and females are found in different areas because of sex-specific habitat preferences, differential tolerance to ecological factors such as extreme temperatures, or sex-specific constraints arising from role specialization and competitive pressures in reproduction (e.g. Selander, 1966; Ketterson & Nolan, 1983; Morton, 1990). These hypotheses deal with proximal causes, and unfortunately tell us little about the evolution of differences between the sexes in these related characteristics, nor about the ecological factors responsible for the development and maintenance of the observed patterns of spatial segregation. However, relatively little conclusive work has been done relating segregation of the sexes to ultimate causes such as the evolution of sexual dimorphism (a complex and controversial subject on its own) or the relative importance of intraspecific and interspecific competition for food.

In this chapter our main aim is therefore to provide a broad, but brief, overview of the patterns of sexual segregation found at various spatial scales, review available evidence for each of the processes that potentially underlie such patterns and discuss the possible consequences for population dynamics and conservation.

DIFFERENTIAL MIGRATION

Differential distance migration

In many, perhaps most, migratory bird species, one of the sexes moves further from the breeding areas to reach more distant wintering grounds (termed 'differential distance migration by sex'). Differential distance migration usually results in a certain degree of mostly latitudinal segregation of the sexes during the non-breeding season, although an extensive overlap in male and female distribution always remains. Cristol and colleagues (1999) recently presented an excellent review of this subject, including some new ideas, and much of what follows is a summary of their findings.

In most species proven to be differential migrants by sex, it is the females that move further to reach their winter quarters. Amongst the species studied so far, this is true for most grouse, most owls, all the ducks and all the passerines. However, there are numerous exceptions to this general pattern, particularly amongst diurnal birds of prey and waders (Cristol et al., 1999). Coincidentally, in most, but far from all, species there is also a pattern of differential migration by age, with juveniles migrating further than adults (Cristol et al., 1999). Three main hypotheses have been proposed to account for such patterns (e.g. Myers, 1981; Ketterson & Nolan, 1983).

(1)   The dominance hypothesis suggests that dominant birds, which in some cases remain sedentary, force subordinate individuals to move to areas further from the breeding grounds.

(2)   The body size hypothesis proposes that winter segregation is linked to the relative degree of cold-weather resistance. Briefly, smaller individuals are constrained to winter in milder climates, whereas larger birds can remain in harsher environments closer to the breeding grounds.

(3)   The arrival time hypothesis states that individuals under greater pressure to arrive at the breeding grounds in advance of their potential competitors benefit more from wintering closer to nesting areas.

These three hypotheses are not mutually exclusive, which makes them particularly hard to test and refute. It should also be noted here that hypotheses (2) and (3) can be viewed as particular cases of the general specialization hypothesis.

In their extensive review, Cristol et al. (1999) were able to classify 53 species as differential migrants (by sex and/or age). They evaluated the three hypotheses described earlier, and found each to be consistent with the observed migration pattern in more species than expected by chance. However, there were also obvious examples that contradicted each hypothesis. For example, although many duck species are differential migrants by sex (e.g. Owen & Dix, 1986), pairs usually form away from the breeding grounds, and males and females arrive in nesting areas at the same time, contrary to the arrival time hypothesis (Carbone & Owen, 1995).

Unfortunately, in evaluating the relative merits of each hypothesis, it is difficult to overcome the problem that body size, dominance and arrival time are closely correlated with each other and with sex, and sex is related to migration distance. The analysis of Cristol et al. (1999) therefore remained inconclusive, and ended with a plea for more empirical data and research. They also included suggestions for two potentially revealing approaches, which were to model the migration distance of each sex and age class as a multifactor optimality problem, to be solved class by class, and/or to focus on comparisons of closely related pairs of species, of which only one was a differential migrant, in order to identify any important life-history distinctions. In fact, it seems likely that differential migration did not evolve in relation to any one selective pressure, and that multifactor explanations are required.

Other forms of differential migration

Besides differential distance migration, variation in the timing of migratory movements also gives rise to sexual segregation at certain times of the year. In most species of birds, particularly amongst passerines, males move ahead of females during spring migration (Gauthreaux, 1982; Kissner et al., 2003). In autumn, the patterns are more varied within and between groups; males can precede females, females precede males, or there can be no difference between the sexes (Gauthreaux, 1982). Generally, such sexual differences are not pronounced, varying from a few days to a few weeks (e.g. Kissner et al., 2003).

Sexual segregation at a large geographical scale also occurs as a result of moult migration. In many dabbling and diving ducks (Anatidae), males leave the females incubating or looking after the brood and gather together in (often massive) moulting aggregations, usually on large bodies of water. Moult migration is often, but not always, to a site halfway to the wintering grounds. Immature and failed females generally join the males at such moulting sites, whereas females with chicks usually moult at the breeding areas (Owen & Black, 1990). Moult migration leads to some of the most locally or regionally biased sex ratios known in birds.

In comparison with differential distance migration, sex differences in timing of migration can be explained much more readily by the independent reproductive roles of males and females, and by variation in the intensity of intra-sex competition resulting from such specialization.

SEGREGATION BY HABITAT AND MICROHABITAT

Within one single geographic region, males and females often show segregation by habitat to a greater or lesser extent. Sometimes this leads to a strong separation of the sexes, with some habitat types harbouring mostly females and others dominated by males (Sherry & Holmes,

1996). More commonly the segregation is incomplete, and differences in sex ratios between habitats, although statistically significant, are relatively slight (e.g. Owen & Dix, 1986). In many species, differences in habitat use between the sexes are so small that males and females still travel together in mixed-sex flocks, but maintain a slight difference in microhabitat distribution (foraging height in trees, for example; Desrochers, 1989). In practice, these scenarios can be seen as variations on a theme, attributable to broadly underlying processes such as the despotic exclusion of subdominants by dominant conspecifics, or the differential habitat selection of males and females.

Patterns of microhabitat segregation

Sex differences in microhabitat use are probably a widespread characteristic of birds (e.g. Selander, 1966; Durell, 2000). Woodpeckers (Picidae), for example, are often sexually dimorphic in plumage, making them a convenient group to study differences in male and female foraging behaviour. Many studies have found that male and female woodpeckers occupy slightly different niches, often foraging at different heights and branch diameters in trees, and sometimes differing in tree species preferences or in the use of live and dead substrates (Morrison & With, 1987; Grubb & Woodrey, 1990; Pasinelli, 2000). The same type of segregation has been described for other arboreal and mostly insectivorous birds of diverse families (e.g. Holmes, 1986; Grubb & Woodrey, 1990; Suhonen & Kuitunen, 1991; Recher & Holmes, 2000).

   Microhabitat differences in foraging locations of the sexes also occur in estuarine or marine habitats. On mudflats, bar-tailed godwits, *Limosa lapponica*, segregate on the basis of water depth, permanently adjusting their positions as tides rise and fall (Smith & Evans, 1973; Zwarts, 1988). In some species of cormorant, *Phalacrocorax* spp., dives by males are usually deeper and/or of longer duration (Wanless *et al.*, 1995; Kato *et al.*, 2000; but see Wanless *et al.*, 1993), and in comparison with those of females, are much more likely to take place in the afternoon than in the morning (Bernstein & Maxson, 1984; Wanless *et al.*, 1995; Kato *et al.*, 2000). These cormorants feed benthically, and one suggestion is that males maximize potential dive depths by co-ordinating their feeding rhythm to the optimum period in the diel light cycle when ambient illumination and light penetration through the water column is greatest (Wanless *et al.*, 1999). In several other seabird taxa (particularly penguins), males and females apparently differ in species or size class of prey (Volkman *et al.*, 1980; Pierotti & Annett, 1991; Quinn, 1990; Williams, 1991; Wagner, 1997) or in stable isotope signatures (see Chapter 6) (Forero *et al.*, 2002; Nisbet *et al.*, 2002). Although other factors, such as prey-size selection, could be responsible, these findings suggest that variation in microhabitat selection by seabirds possibly is more widespread than previously thought.

Microhabitat segregation by sex (and age class) is also apparent at communal roost sites, where such phenomena are easier to study. Studies of common starlings, *Sturnus vulgaris*, in England, and of bramblings, *Fringilla montifringilla*, in Switzerland, for example, reported that males are more often found in the central areas of the roosts and females at the periphery (Summers *et al.*, 1987; Jenni, 1993). Such minor differences in habitat use do not necessarily signal a segregation of the sexes at other spatial scales, and may have limited relevance in terms of habitat management and conservation. However, understanding the mechanisms of segregation in a context of proximity, where males and females are interacting frequently in a common general environment, can provide useful insights and potential models for the study of segregation at larger spatial scales.

Processes in microhabitat segregation

Differential niche utilization of foraging males and females is often linked with a dimorphism in structures associated with feeding and locomotion, which suggests a certain degree of specialization by the sexes. For example, in the Hispaniolan woodpecker, *Centurus striatus*, males and females are only slightly dimorphic in body mass and wing length but females have a much (21%) shorter and shallower bill and the highly specialized tongue is even more dimorphic. Not surprisingly, males and females show important differences in foraging techniques and also minor differences in microhabitat use (Selander, 1966). In the famous and extinct New Zealand Huia, *Heterolocha acutirostris*, male and female bill size and shape were so different that each sex was initially described as a separate species in spite of the similarity in plumage and general body size. Such differences also correlated with different foraging techniques and substrate use (e.g. Oliver, 1955). Many other examples could be given of bird species as diverse as grebes, pelicans, waders, woodpeckers and passerines, where sexual dimorphism in feeding apparatus is much more pronounced than in body size (Selander, 1966), which is almost certainly indicative of a degree of niche divergence (Shine, 1989a). Dimorphism in body size, independent of shape, can also have consequences for microhabitat utilization. It has been suggested, for example, that sexual segregation by water depth in diving birds, such as cormorants or penguins, is related to size dimorphism (Williams, 1991; Kato *et al.*, 2000), given the positive relationship between maximum dive depth and body mass (Burger, 1991).

Physical characteristics are not always fixed, and some may change even in individuals that have attained adult age and completed growth. Such remarkable phenotypic flexibility can be found, for example, in the size and shape of the bills of great tits, *Parus major*, in England (Gosler, 1987). The strong sexual dimorphism in the proportions of great tit bills is evident in some seasons and years, coinciding with a clear sexual segregation by microhabitat. In years when the main food (beech-mast) was very abundant and within easy access to all sex and age classes, there was a low variability in bill proportions in the great tit population, and no sexual segregation by habitat. The correlation between bill size and handling time varied with different food types, and this functional relationship was presumed partly to underlie the sexual dimorphism in bill form and niche segregation. Male dominance therefore appeared to be responsible for niche shifts and consequent adjustment of bill proportions, rather than the variation in bill size predisposing males and females to exploit different foraging niches (Gosler, 1987).

Experimental studies provide compelling evidence for the role of social dominance in shaping the differential use of space by each sex. Male and female downy woodpeckers, *Picoides pubescens*, studied in Ohio, USA, differ in vertical distribution in trees and in typical substrates, even though they are morphologically very similar. When Peters and Grubb (1983) removed males from feeding territories, females selected branch diameters, foraging heights and substrate angles normally used only by males. The opposite did not occur when females were removed, indicating that the pattern observed was not the result of a symmetrical release in exploitation competition. Male supplanting attacks on females were observed on several occasions, further supporting the argument that male dominance was important (see also Matthysen *et al.*, 1991).

In a similar study with winter flocks of black-capped chickadees, *Parus atricapillus*, in Canada, Desrochers (1989) showed that the removal of males resulted in a niche shift of females. In the absence of males, females foraged closer to the trunk and lower down, the positions generally occupied by the larger sex. When the males were released, females reverted to their original foraging sites. During the removal period, females did not use all the typical foraging locations exploited when both males and females were present, which would have been expected if female shifts were simply due to a release in exploitation competition. This implies greater profitability or increased safety for chickadees of either sex when feeding lower in the woods, and that it was male social dominance (well established in this species) that prevented females from occupying preferred male microhabitats (Desrochers, 1989).

Sexual segregation within roosts also seems to be driven by asymmetric contests between males and females. In both starling and brambling roosts, there are many aggressive interactions, suggesting competition for roosting sites. It is difficult to imagine that females would prefer to sleep in the periphery where exposure to cold weather and predators is greater than in the centre, if they were not displaced from the core by males. Experiments with captive starlings confirm the role of social exclusion in shaping female distribution in roosts (Feare et al., 1995). When isolated, all birds showed similar preferences in roosting positions, and when together, individuals fought for preferred positions. Such contests were generally won by dominant individuals. The removal of dominants immediately led to shifts in roosting sites of subdominant starlings (Feare et al., 1995).

Patterns of broad habitat segregation

Sexual segregation at large spatial scales is a relatively poorly documented phenomenon in birds. We found robust evidence for this type of segregation in only 37 species (Table 18.1), although of course, our review was far from exhaustive. It is unlikely that this reflects the rarity of such patterns, but rather the difficulty of distinguishing the sexes using external morphology. Recent advances in molecular sexing partially overcome this problem, but only if birds can be captured to obtain tissue samples. In addition, sexual segregation is often weak or incomplete at large spatial scales, which means that differences in sex ratios may not be immediately apparent upon a superficial examination. This is not to say that sexual segregation is a general characteristic of most migratory birds. Of two recent studies on monochromatic species sexed from DNA, weak sexual segregation by habitat was found in a study on European robins, Erithacus rubecula (Catry et al., 2004), but not in hermit thrushes, Catharus guttatus (Brown et al., 2002).

There are two families for which species coverage is probably sufficient to make reasonable generalizations concerning the incidence of large-scale sexual segregation. One includes the albatrosses (Diomedeidae), in which satellite-tracking studies have revealed sexual segregation in foraging areas during at least part of the breeding season in several

Table 18.1 *Examples of large-scale habitat segregation by sex in birds. All the examples refer to pairs of habitat types within the same general geographical region (differences do not result from differential migration). Microhabitat segregation (e.g. use of different substrates or parts of the same tree or dives at different water depths) is not included here.*

Species	Region	Habitat with relatively more females	Habitat with relatively more males	References
Breeding season				
Adélie Penguin *Pygoscelis adeliae*	Antarctic continent	Continental shelf break (>1000 m depth)	Inshore (<20 km), shallow waters	Clarke et al., 1998
Wandering Albatross *Diomedea exulans*	South-west Indian Ocean	Subtropical waters	Antarctic and sub-Antarctic waters, shelf edge	Weimerskirch et al., 1993; Weimerskirch, 1995; Nel et al., 2002
Black-browed Albatross *Thalassarche melanophrys*	South Atlantic South Atlantic	Continental shelf Less windy, mixed subtropical–sub-Antarctic waters (during incubation only)	Antarctic waters Sub-Antarctic and Antarctic waters (during incubation only)	Prince et al., 1998 Phillips et al., 2004
Grey-headed Albatross *Thalassarche chrysostoma*	South Atlantic	More northerly sub-Antarctic and Antarctic waters (during incubation only)	Windier Antarctic waters (during incubation only)	Phillips et al., 2004
Buller's Albatross *Thalassarche bulleri*	New Zealand plateau and Tasman Sea	Further north and over deep water	Further south and over shelf	Stahl & Sagar 2000
Southern Giant Petrel *Macronectes giganteus*	South Atlantic	At sea, including distant continental shelf	Coastal seal and penguin colonies	González-Solís et al., 2000b

(cont.)

Table 18.1 (cont.)

Species	Region	Habitat with relatively more females	Habitat with relatively more males	References
Northern Giant Petrel *Macronectes halli*	South Atlantic	At sea, including distant continental shelf	Coastal seal and penguin colonies and close shelf	González-Solís et al., 2000b
Brown Booby *Sula leucogaster*		Offshore (>90 km)	Inshore (<20 km)	Gilardi, 1992
Non-breeding season				
Wandering Albatross *Diomedea exulans*	Southern Ocean	Tropical and subtropical waters	Antarctic and sub-Antarctic waters	Weimerskirch & Wilson, 2000; Nel et al., 2002
Harlequin Duck *Histrionicus histrionicus*	British Columbia, Canada	Further inshore coastal waters	Further offshore	Rodway et al., 2003
Goldeneye *Bucephala clangula*	South Sweden	Shallower coastal Baltic Sea	Deeper coastal Baltic sea, inland lakes	Nilsson, 1970
Canvasback *Aythya valisineria*	Chesapeake Bay area, USA	Smaller bodies of water	Large open bodies of water (main estuary and tributaries)	Nichols & Aramis, 1980
	Louisiana, USA	Coastal areas	Inland sites	Woolington, 1993
Pochard *Aythya ferina*	United Kingdom	Reservoirs and gravel pits	Rivers, marshes and natural lakes	Owen & Dix, 1986
	Manchester, England	Aquatic vegetation-rich habitat	Invertebrate-rich habitat	Marsden & Sullivan, 2000
Lesser Scaup *Aythya affinis*	South Carolina, USA	Areas in reservoir with floating-leaved vegetation	Areas of open water in reservoir	Bergan & Smith, 1989

Species	Location			Reference
Ring-necked Duck *Aythya collaris*	South Carolina, USA	Areas in reservoir with emergent vegetation	Areas of open water in reservoir	Bergan & Smith, 1989
Tufted Duck *Aythya fuligula*	Tay estuary, UK	Freshwater habitats	Saltwater habitats	Boase, 1926
	South Sweden	Smaller ponds, more vegetation	Larger ponds, less vegetation	Nilsson, 1970
Hen Harrier *Circus cyaneus*	Arkansas, USA	Tall dense vegetation (e.g. tall corn)	Short, sparse vegetation (e.g. stubble)	Preston, 1990
	Scotland	Upland heather moor	Coastal including wetland margins	Marquiss, 1980
European Sparrowhawk *Accipiter nisus*	United Kingdom	Farmland, open country	Forest	Newton, 1986
American Kestrel *Falco sparverius*	USA	Open habitats, more pastures	Semi-open habitats, more woodlots	Ardia & Bildstein, 1997, 2001 and references included
Capercaille *Tetrao urogallus*	Norway	Pure pine forests, higher tree density	Forests with spruce trees, lower tree density	Gjerde, 1991
Rock Ptarmigan *Lagopus mutus*	North America	Shrubby openings in the boreal forest	Above timberline	Weeden, 1964
Willow Grouse *Lagopus lagopus*	Alaska British Columbia, Canada	Lower altitude, boreal forest habitats	Higher altitude, timberline and tundra areas	Weeden, 1964; Gruys, 1993
Oystercatcher *Haematopus ostralegus*	Exe estuary, England	Mudflats and adjacent fields	Mussel beds	Durrel & Goss Custard, 1996
Curlew *Numenius arquata*	Tees estuary, England	Intertidal areas	Fields	Townshend, 1981

(cont.)

Table 18.1 (*cont.*)

Species	Region	Habitat with relatively more females	Habitat with relatively more males	References
Bar-tailed Godwit *Limosa lapponica*	Coastal Guinea-Bissau	Intertidal areas with fiddler crabs	Intertidal areas without fiddler crabs	Zwarts, 1988
Red-backed Shrike *Lanius collurio*	Queensland, Australia	Unvegetated sandflats	*Zostera* seagrass beds	Zharikov & Skilleter, 2002
	Kalahari basin, Botswana	Denser, taller and more lush bush or woodland	More exposed habitats, in open, shrubbier bushveld	Herremans, 1997
Robin *Erithacus rubecula*	Southern Portugal	Scrubland	Woodland	Catry *et al.*, 2004
Cetti's Warbler *Cettia cetti*	Portugal	Pure reedbeds	Riparian vegetation with trees and bushes	Bibby & Thomas, 1984
Eastern Great Reed Warbler *Acrocephalus orientalis*	Malaysia	Reedbeds	Scrub habitats	Nisbet & Medway, 1972
Hooded Warbler *Wilsonia citrina*	Yucatan Peninsula, Mexico	Brushy successional woods	Tall closed canopy forest	Lynch *et al.*, 1985; Lopez Ornat & Greenberg, 1990
Kentucky Warbler *Oporornis formosus*	Belize	Early successional forest	Mature forest	Conway *et al.*, 1995
Cape May Warbler *Dendroica tigrina*	Dominican Republic	Desert and dry forest	Pine forests	Latta & Faaborg, 2002

Species	Location			Reference
Black-throated Blue Warbler *Dendroica caerulescens*	Yucatan Peninsula, Mexico; Puerto Rico	Shrubby second growth	Tall mature forest	Lopez Ornat & Greenberg, 1990; Wunderle, 1995
Prairie Warbler *Dendroica discolor*	San Salvador, Bahamas	Scrub and early successional habitats	Forest	Murphy et al., 2001
Magnolia Warbler *Dendroica magnolia*	Yucatan Peninsula, Mexico	Scrub, open fields and pastures	Tall mature forest or old second growth	Lopez Ornat & Greenberg, 1990
Northern Parula *Parula americana*	Yucatan Peninsula, Mexico	Scrub, open fields and pastures	Tall mature forest or old second growth	Lopez Ornat & Greenberg, 1990
Common Yellowthroat *Geothlypis trichas*	Yucatan Peninsula, Mexico	Scrub, open fields and pastures	Tall mature forest or old second growth	Lopez Ornat & Greenberg, 1990
American Redstart *Setophaga ruticilla*	Yucatan Pen., Mexico Jamaica	Open habitats, scrub Coastal scrub forest, dry limestone forest, logwood thickets	Taller forest habitats Mangrove, wet limestone forest	Lopez Ornat & Greenberg, 1990 Parrish & Sherry 1994; Sherry & Holmes, 1996
Ruby-crowned Kinglet *Regulus calendula*	California, USA	Coastal scrub	Riparian forest, mixed evergreen forest	Humple et al., 2001
Snow Bunting *Plectrophenax nivalis*	Scotland	Lower altitude	Higher altitude in mountains	Smith et al., 1993

(see Table 18.1), but by no means all, taxa (see e.g. Anderson *et al.*, 1998; Prince *et al.*, 1999; Hedd *et al.*, 2001; Phillips *et al.*, 2004; see also Chapter 5). In other species, overall foraging ranges of males and females overlap a great deal, and although there may be differences in core areas (e.g. Hyrenbach *et al.*, 2002), these could conceivably result from a combination of individual preferences unrelated to sex, and small samples sizes. The wandering albatross, *Diomedea exulans*, however, is a classic example in which sexual segregation in foraging zones was first suggested by latitudinal variation in at-sea distributions of birds sexed from plumage (Weimerskirch & Jouventin, 1987) and subsequently confirmed for breeding birds at several sites using satellite transmitters (Chapter 6). Recent data on their distribution during the non-breeding season obtained using novel Global Location Sensing (GLS) loggers indicates that these habitat preferences are maintained throughout the year (Table 18.1).

In New World warblers (Parulidae) wintering in Middle America, sexual segregation by habitat occurs in many species. Studies to date indicate that, in general, males are associated with tall mature forest, while females are more frequent in scrub habitats or early secondary growth (Table 18.1).

Processes in broad habitat segregation

As with sexual segregation by microhabitat, there is strong evidence that large-scale habitat segregation is driven by specialization in some species or systems, and by social dominance in others.

The hooded warbler, *Wilsonia citrina*, is a medium- to long-distance migrant nesting in eastern North America and wintering in Mesoamerica, where it generally defends exclusive feeding territories. Detailed studies in the Yucatan Peninsula, Mexico, showed that wintering males defend territories within relatively tall, closed-canopy forest, whereas the majority of female territories are located in shorter, more open successional scrub and deciduous woodland, with very little overlap (Lynch *et al.*, 1985). In border areas between the two habitat types, males and females sometimes defend territories with a common boundary. When males from border territories were experimentally removed, females from adjacent habitats did not move into the vacated sites, which instead either remained empty for the rest of the season or were taken up by new males (Morton *et al.*, 1987). The conclusion was that direct male exclusion of females from forest habitat was not a sufficient proximal explanation for the observed sexual segregation, and that males and females actively selected different habitats (Morton *et al.*, 1987). Further studies with captive naïve hand-raised hooded warblers showed that such preferences have an innate basis; males chose tall habitats with separated stems over short habitats with dense stems, whereas females showed no preferences (Morton, 1990). Furthermore, females had an innate attraction to environments with many oblique elements, while males preferred vertical ones (Morton, 1990; see also Morton *et al.*, 1993). Limited evidence for male and female specializations in habitat and foraging techniques has also been found in another New World warbler, for example the black-throated blue warbler, *Dendroica caerulescens* (Wunderle, 1995).

As with microhabitat segregation, broad sex-specific habitat preferences have also been linked to morphological differentiation and niche divergence of males and females. For example, female curlews, *Numenius arquata*, have longer bills than males and are thought to be better adapted for reaching locally abundant fiddler crabs, *Uca tangeri*, hiding in burrows in the sediment. This probably explains why female curlews were more abundant in areas with fiddler crabs, while males predominated in fiddler crab free areas, in coastal Guinea-Bissau (Zwarts, 1988; see also Townshend, 1981).

Large-scale sexual segregation in foraging areas has also been linked to breeding role specialization. In the marginally dimorphic Adélie penguin, *Pygoscelis adeliae*, there may be an advantage for the more aggressive males to spend longer with the chick during the guard period, and consequently they have less time than females to travel to the most productive areas at sea (Clarke *et al.*, 1998). Another example of a slightly different selection pressure related to reproductive roles has been suggested for brown boobies, *Sula leucogaster*. In this species, males feed closer to the colony, possibly in order to maximize time spent defending the territory or to increase the number of opportunities for extra-pair copulations (Gilardi, 1992).

In contrast to these examples, studies on another New World warbler, the American redstart, *Setophaga ruticilla*, has provided strong evidence for the role of social dominance and despotic exclusion in causing broad habitat segregation of males and females at the wintering grounds in Jamaica (e.g. Sherry & Holmes, 1996). By studying patterns of settlement in autumn, aggressive interactions, territorial displacements and replacements after experimental removals, Marra (2000) showed convincingly that the predominance of males in mangrove habitats, as opposed to adjacent scrub, was the result of despotic exclusion of most females by dominant males (see also Marra *et al.*, 1993).

There was also independent evidence that the habitat type occupied by males was of a higher quality, with higher and more stable insect biomass, including groups important in redstart diet (Parrish & Sherry, 1994) and that birds in mangroves attained better body condition and general fitness (Marra & Holberton, 1998; Marra *et al.*, 1998; Marra & Holmes, 2001).

Removal experiments are particularly informative when testing competing hypotheses explaining sexual segregation. Another such study, supporting the social dominance hypothesis, was carried out with American kestrels, *Falco sparverius*, in Pennsylvania, USA. In this case, males quickly occupied habitat patches from where females (the larger sex in this species) had been removed (Ardia & Bildstein, 1997).

Several researchers working with waterfowl have found evidence for the social dominance hypothesis including:

(1)   In winter flocks, males (the dominant sex), direct their agonistic interactions mostly towards females (Hepp & Hair, 1984; Choudhury & Black, 1991).
(2)   Male dominance influences the timing of foraging by females (Choudhury & Black, 1991).
(3)   Males are more successful in defending small feeding territories (Saylor & Afton, 1981).
(4)   Females tend to occur more frequently on the periphery of rafts (Nichols & Aramis, 1980) and in small flocks in low density areas (Nilsson, 1970).
(5)   Males were more numerous in what were considered to be superior habitats (Owen & Dix, 1986).
(6)   After accounting for sexual size dimorphism, birds in male-biased habitats were heavier (Nichols & Aramis, 1980) and body mass seemed to affect survival (Haramis *et al.*, 1986).

In spite of all this compelling evidence, alternative explanations for such patterns have been suggested (e.g. Marsden & Sullivan, 2000).

It has been suggested that sexual segregation in some albatrosses (Table 18.1), which is particularly obvious during the brood guarding period when adults are constrained to forage close to the colony, is explained by the competitive exclusion of females (Weimerskirch *et al.*, 1993; Stahl & Sagar, 2000). However, female albatrosses typically weigh considerably less than males, and show lower wing loading, which probably makes them better able to exploit more northerly waters where winds are generally lighter (Shaffer et al., 2001; Phillips et al., 2004; Chapter 5). A difference in flight performance might therefore be a proximate factor explaining the sexual segregation in foraging ranges evident in some species during at least part of the year. However, it could also have developed secondarily, with the evolutionary basis of the latitudinal segregation related to sexual selection for larger males and, in the case of wandering albatrosses but not the smaller mollymawks, their dominance of feeding opportunities in nearby shelf waters.

Several studies have found that the smaller, subordinate sex, often the female, occurs more frequently in microhabitats that also hold more juveniles and/or more birds in apparently poorer body condition (e.g. Nichols & Aramis, 1980; Summers et al., 1987; Jenni, 1993; Smith et al., 1993; Marra & Holmes, 2001; Latta & Faaborg, 2002). This correlation supports the hypothesis that social dominance is a key factor underlying segregation by sex and age class. However, other explanations are possible, and dominance and habitat specialization can interact in complex ways, as follows.

Oystercatchers, Haematopus ostralegus, show a remarkable correspondence between bill morphology and dietary specialization. Birds with pointed bill tips (usually females) feed mostly on worms and clams on mudflats, whereas those with blunt bills (generally males) mostly forage on mussel beds. In birds with chisel-shaped bill tips, the sex ratio is even (Swennen et al., 1983; Hulscher, 1985; Durell et al., 1993). Differences in bill shape are thought to influence or constrain the feeding specialization developed by each individual; long, slender bills are better suited for probing for intertidal invertebrates, while short, thick bills seem to present advantages when hammering mussels (Durell et al., 1993). These specializations lead to sexual segregation in estuaries, with males found more frequently on mussel beds, and females on mudflats and adjacent fields (Swennen et al., 1983; Durell & Goss Custard, 1996), and these differences then lead to segregation at roost sites which are usually located in the proximity of feeding areas (Swennen, 1984).

Studies conducted at oystercatcher roosts in the Dutch Wadden Sea and in South Wales showed interesting similarities (Swennen, 1984; Durell et al., 1996). In both areas there was clear age and sexual segregation between roosts. Juveniles were found in larger numbers in roosts where a greater proportion of the adults had anatomical defects of the eyes, bill or legs (such defects are fairly common in oystercatchers). This suggests that juveniles and handicapped birds were relegated to sub-optimal habitats. Interestingly, the studies found opposite patterns in relation to the distribution of the sexes. In Wales, females predominated in poor quality roosts where juveniles were found, whereas in the Wadden Sea it was the males that were more frequent in areas with many juveniles and handicapped individuals (Swennen, 1984; Durell et al., 1996). These differences may be explained if the availability of female and male-preferred habitats differs in the two geographical regions. In this case, differences between Wales and the Wadden Sea would result in contrasting levels of interference competition in male- and female-dominated habitats, resulting in birds of lower status making different choices in each region (Durell et al., 1996). According to this explanation, sexual segregation in oystercatchers is mostly driven by specialization, while age segregation results mainly from social interactions and dominance.

Beyond single-factor hypotheses: discussing ultimate causes of sexual segregation

The social dominance and ecological specialization hypotheses are clearly not mutually exclusive, and in fact both processes can co-occur and interact in complex ways, as suggested by the observations on the ecology and behaviour of the oystercatcher, detailed earlier. Although some studies, such as the one with the hooded warbler, strongly point to the role of specialization as a proximate cause for habitat segregation by sex in winter, they leave open the possibility that such specialization evolved (and is maintained) as a result of asymmetric competition driven by male social dominance (Morton, 1990).

Examination of the ultimate causes of sexual segregation leads immediately to the important issue of the evolution of sexual size dimorphism and the selective pressures responsible for its maintenance. This is a complex and controversial subject. Selective pressures linked to (i) sexual selection, (ii) differentiation of reproductive roles and (iii) food competition between the sexes, are suspected to interact to produce and maintain current patterns of sexual size dimorphism (e.g. Hedrick & Temeles, 1989). However, the relative importance of these factors in different species is largely unknown. An eloquent illustration of the level of our ignorance is provided by the long lasting, but strikingly inconclusive, debate concerning the evolution of reversed sexual size dimorphism in species belonging to many avian families (e.g. Mueller, 1990; Catry et al., 1999).

Because both dominance and specialization-based niche segregation are often modulated by differences in body size, the morphological divergence of males and females can be both a cause and a consequence of such processes (Shine, 1989a). For example, male–male competition and sexual selection resulting in larger male size (Bennet & Owens, 2002) would give males an advantage in social contests, leading to a sexual segregation scenario where subdominant females are forced into an alternative niche. Sexual size dimorphism might then be amplified in response to selection by ecological factors, promoting further segregation of males and females in breeding habitats, or in non-breeding ranges, as a result of increased asymmetry in social contests and of increased specialization of the sexes.

Another interesting issue when discussing the (ultimate) causes of sexual segregation, is the influence of interspecific competition on the degree of competitive interaction, and hence potential niche divergence, between males and females within a species. Studies on the *Parus* tit community at Wytham Woods, England, suggest that sexual dimorphism in bill shape and niche segregation is not apparent in tit species low in the interspecific dominance hierarchy, for which interspecific competition is relatively important, (Gosler, 1987, 1990). By contrast, in Ireland where the diminutive coal tit, *Parus ater*, is numerically dominant and apparently faces reduced competition, the species is both sexually dimorphic and shows sexual segregation in microhabitat use (Gosler & Carruthers, 1994). These studies support the idea that interspecific competition can play an important role constraining the differentiation of male and female niches, at least at the microhabitat-use level.

Comparisons between socially monogamous species living on islands and continental forms suggest that the former are more dimorphic in body size (Selander, 1966; Bennett & Owens, 2002). This has been interpreted as an effect of reduced interspecific competition faced by birds living on oceanic islands, allowing greater niche divergence between males and females (Selander, 1966; Bennett & Owens, 2002). This suggests that the patterns of niche divergence detailed in the coal tit study could apply more generally.

Such hypotheses linked to the origin and evolution of the patterns of sexual segregation in birds cannot be tested using the type of experimental approaches so successful in pinpointing the proximal causes of current patterns. On the other hand, much more empirical data on habitat segregation is needed before suitable comparative analyses can be carried out.

SEGREGATION IN FLOCKS WITHIN THE SAME AREAS

In comparison with some mammalian species (e.g. Ruckstuhl & Neuhaus, 2000), it is uncommon for birds to live in sexually segregated groups within the same general area. In some lekking bird species, males congregate at specific locations throughout the breeding season. However, such aggregations can hardly be described as flocks, and they serve a specific reproductive purpose that will not be considered further here. Outside the breeding season, sexual segregation in flocks (including the existence of flocks partly or entirely dominated by one sex) within the same general area can be found in several species of game birds, including grouse (Tetraonidae), specifically capercaille, *Tetrao urogallus*, black grouse, *Lyrurus tetrix* (Koskimies, 1957), willow and rock ptarmigans, *Lagopus lagopus* and *L. mutus* (Weeden, 1964), curassows (Cracidae; del Hoyo, 1994), wild turkeys, *Meleagris gallopavo* (Porter, 1994), and several large bustard (Otididae) species (e.g. Ena *et al.*, 1985; Rahmani, 1989; Marchant & Higgins, 1993; Collar, 1996). As this type of segregation behaviour is present in some (but not all) species from several families in at least two different orders, it must have evolved (or been lost) more than once.

Sexual segregation in flocks has also been reported in curlew sandpipers, *Calidris ferruginea*, wintering in South Africa (Puttick, 1981), but it is not known if this was a result of social preferences for same-sex groups or was the by-product of microhabitat selection. Sexual segregation in flocks arriving and departing from a single roost in Louisiana, USA, has been reported for a strongly dimorphic passerine, the boat-tailed grackle, *Quiscalus mexicanus* (McIlhenny, 1937), but it is not known whether males and females shared the same foraging areas.

It is somewhat surprising that we could find no published hypotheses attempting to explain these occasionally very strong biases in flock sex ratios, perhaps because in most cases differences in (micro) habitat were assumed to be responsible. One possibility is that segregation between flocks is somehow linked to reproductive behaviour (male–male competition and establishment of social hierarchies) outside the breeding season. For example, several grouse species gather at leks and display in autumn and winter, when no reproduction is taking place (Rintamäki *et al.*, 1999).

Three main hypotheses have been proposed to explain sexual segregation in ungulates: the predation risk hypothesis, the forage selection hypothesis and the activity budget hypothesis (Ruckstuhl & Neuhaus, 2002). Only the third hypothesis does not imply the existence of sexual segregation by habitat or microhabitat. This hypothesis also seems to be the one that best explains sexual segregation in ungulates (Ruckstuhl & Neuhaus, 2002), and it would be interesting to evaluate whether it also applies to large game birds. In common with the ungulates, they also exhibit considerable sexual size dimorphism, are mostly herbivorous and probably show much variation in sex-specific foraging efficiency and digesta retention times (Moss, 1983).

CONSEQUENCES OF SEXUAL SEGREGATION

Fitness implications

The social dominance hypothesis predicts that the subdominant sex should pay a price for its inability to secure a position in the most favourable habitats or areas (e.g. Marra & Holmes, 2001). Such a cost could result from the intrinsically lower quality of the habitat, higher levels of competition or increased mortality associated with longer migrations between breeding and wintering sites. However, virtually nothing is known about the consequences of migrating different distances, although having to travel further can probably be assumed to incur considerable additional cost, because mortality in many species is known to be high during migration (Owen & Black, 1991; Sillett & Holmes, 2002).

As described earlier, several studies have found differences in body mass of birds wintering in different habitats. Unfortunately, body mass per se can be a poor, and even misleading, indicator of condition or fitness, as it potentially reflects not only nutritional stress but also individual choices in the strategic regulation of fat and protein reserves as well as of organ size (e.g. Witter & Cuthill, 1993; Gosler, 1996; Landys-Ciannelli et al., 2003). Few studies have progressed beyond a straightforward comparison of mass when comparing birds from different habitats. However, muscle scores are known to be good indicators of fitness or condition in some passerines (Gosler, 1991, 1996; Gosler & Carruthers, 1999) and some studies have found, after controlling for sex, higher muscle scores in birds coming from male-biased habitats (Latta & Faaborg, 2002; Catry et al., 2004).

Baseline corticosterone levels are thought to provide a good indication of physiological condition, and differ between male- and female-biased American redstart habitats (Marra & Holberton, 1998). Furthermore, spring departure dates of either sex were later from female- than from male-biased habitats (Marra et al., 1998). By using stable isotope analysis, the timing of arrival on breeding territories was linked to winter habitat type, and birds wintering in habitats dominated by females arrived later at breeding sites than those arriving from male-dominated sites (Marra et al., 1998). Such a delay is known to affect reproductive performance of several species. This evidence is unique in the way it links winter sexual segregation to events on the breeding grounds, and it fits well with the data showing that social dominance and despotic interactions determine the winter habitat distribution of this species.

Winter site persistence and between-year return rates have also been suggested as possible indicators of fitness, as they could correlate with true survival rates. Some studies of New World warblers have found higher local survival rates in male-biased habitats (e.g. Wunderle, 1995; Marra & Holmes, 2001; Latta & Faaborg, 2002; but see Conway et al., 1995). Unfortunately, mortality often cannot be distinguished from emigration, making it difficult to relate local survival to true fitness. Furthermore, some of the studies mentioned earlier concentrate on territorial birds and ignore floaters, that can represent a large part (sometimes the majority) of the wintering population (e.g. Cuadrado 1997) and whose fates are also of relevance.

Alternative behavioural evidence for variation in fitness of birds occupying different habitats comes from the studies on oystercatchers in the Exe estuary. In winter, individuals (mostly females) specializing on worms and clams on mudflats have to forage for longer when compared to birds foraging on mussel-beds, and also supplement their diet by feeding in fields at high water (Durrell, 2000). This suggests that they are having greater difficulty in maintaining adequate intake rates at low tide. Furthermore, supplementary feeding in the fields is not always an option because the ground freezes during cold weather, and has the disadvantage of increased exposure to parasites and predators (Durell, 2000). As a result, oystercatchers (mostly females) that specialize on worms and clams generally have lower body condition and poorer survival (Durell & Goss-Custard, 1996; Durell et al., 2001a).

The few case studies described earlier are rare examples of convincing evidence for fitness consequences of sexual segregation, and there is clearly much scope for further research. However, the following should also be kept in mind. In a world increasingly influenced by human activities and undergoing rapid environmental change, we might expect any form of habitat or geographic sexual segregation to leave males and females exposed to different threats. Even if and when males and females adopted distinct strategies with equal success in a natural scenario, abrupt changes brought about by human interference could now result in fitness costs for one or the other sex. Segregation of the sexes can result, for example, in differential exposure to, and accumulation of, pollutants, as applies to male and female giant petrels, *Macronectes* spp., although with no reported fitness consequences (González-Solís *et al.*, 2002b; Chapter 6). In Europe, many duck species are differential distance migrants, with disproportionately large numbers of females wintering in the Mediterranean region (e.g. Campredon, 1983; Owen & Dix, 1986). In this area, hunting seasons are traditionally longer and hunting pressures higher, which could lead to differential sex mortality, particularly as this is one of the most important sources of mortality in these populations (Tamisier, 1985). Given the widespread and accelerating level of human intervention in all areas of the planet, it may become increasingly difficult to perform unbiased comparisons of fitness for individuals wintering in widely separated geographic areas, or indeed at smaller spatial scales.

Evidence for differential mortality

Many (probably most) migratory birds studied in any detail have proven to a greater or lesser extent to be differential migrants (Cristol *et al.*, 1999). Of course, this view is strongly biased by the disproportionately high number of detailed bird studies that have taken place in temperate or polar regions, and little can be said about the large majority of bird species that live in the tropics. Nonetheless, if sexual segregation in space and (or) time often leads to differential mortality, then it could have general implications for the population dynamics of many species.

Whatever the causes, sex differences in mortality rates are often found in studies of avian population dynamics. The adult sex ratios of many populations are believed to be biased, although this is usually very difficult to quantify accurately. In passerines, for example, experimental removal studies and observations of the pairing success of both sexes suggest that males outnumber females in most populations (Breitwisch, 1989; Villard *et al.*, 1993; Marra & Holmes, 1997), although the confounding effect of occasional polygyny on apparent sex ratios is hard to evaluate (Vidal *et al.*, 1994). In ducks, direct counts at breeding and wintering grounds suggest a strong male bias in adult populations (e.g. Owen & Dix, 1986; Baldassarre & Bolen, 1994). Such imbalance in sex ratios is likely to be caused by sex-specific mortality rates of independent juveniles or adults (Owen & Dix, 1986). Direct assessments of survival rates have confirmed a general tendency for higher female mortality (Promislow *et al.*, 1992; Owens & Bennett, 1994). Could it be that the (generally) larger males, by their despotic behaviour, relegate females to less favourable environments, reducing their survival probability? Does the reverse take place in taxa with reversed size dimorphism, leading to sex-ratio biases in the opposite direction? More detailed data and analyses are needed to investigate this. Some comparative studies have concluded that there is not an increase in mortality associated with higher degrees of sexual size dimorphism (Owens & Bennett, 1994), but this is not the same as proving no cost of behavioural dominance for subordinate birds relegated to poor environments.

There is another aspect that should be considered when discussing fitness consequences of habitat segregation. If males are typically larger in most species as a consequence of sexual selection, it is possible that females might have a body size better adjusted for the ecology of the species and better adapted for survival in the non-breeding season (e.g. Lack, 1968; Searcy & Yasukawa, 1981; but see Kissner et al., 2003). In this case, females need not necessarily suffer higher over-winter mortalities than males even if they were relegated to lower-quality microhabitats, nor experience poorer survival because they migrate longer distances. The fact that females appear to show poorer survival in many species could, however, also simply result from their generally greater investment in parental care (Bennett & Owens, 2002).

SEXUAL SEGREGATION AND POPULATION DYNAMICS

The segregation of males and females can have obvious consequences for the management and conservation of natural populations. If males and females are found in different places, or at the same place but at different times, then they will often be subject to alternative sources of mortality, leading to an imbalance in the sex ratio and consequences at the population level (Durell et al., 2001b). Biased adult sex ratios probably are a natural characteristic of most bird populations, but a particularly strong imbalance could arise if a new anthropogenic source of mortality acted to disrupt dynamic equilibria in population processes.

Differential sex-related mortality will have a variable impact on population dynamics, depending on the circumstances. Increased mortality in a non-limiting sex (e.g. males in a species with a polygynous mating system) has a lower potential to affect population size than increased mortality of the scarcer sex. Alternatively, increased mortality of the most abundant sex could in some situations have a positive impact on population growth rates, if density-dependent processes are regulating fecundity.

The question of whether sexual segregation occurs because of niche specialization or the despotic exclusion of subordinates is also relevant when discussing implications for population dynamics. In theory, if each sex has its own habitat specialization, then the complete destruction of the normally female-biased habitat could ultimately lead to the extinction of the population, no matter how much of the other habitat was left. If, on the other hand, habitat segregation was the result of despotic male behaviour, and preferences do not differ between the sexes, then complete destruction of the female-biased habitat would lead to a decline in the overall population, due to reduced breeding output, but only to the point where the reduction in male numbers would allow females to invade the higher quality areas formerly occupied by males, and a new equilibrium would be reached through site-dependent population-regulation mechanisms (e.g. Rodenhouse et al., 1997; Newton, 1998). Increased mortality or excessive harvesting in the female-biased habitat would have similar consequences. The consequences of habitat loss and sex-related mortality in a species with sex-specific habitat specialization, the oystercatcher, have been modelled by Durell et al. (2001b).

Any readjustment to reach a new equilibrium, in the scenario of site-dependent regulation, will also depend on species-specific factors such as life-history characteristics and behavioural flexibility (e.g. Bowers, 1994). For example, in long-lived birds it might take many years until the dominant sex became scarce enough for the effects of reduced competition in the favoured habitats to be felt, in which case the sex-ratio imbalance could become very pronounced and population growth rates severely depressed. This would be exacerbated if subordinate individuals remained faithful to their traditional areas even when higher-quality habitat is vacated. In short-lived species, the potential for readjustment is much greater, because a decrease in reproductive output at the population level will be followed by a rapid population decline and a reduction in competition for high-quality sites or habitats, favouring the subordinate sex.

APPLICATION TO CONSERVATION

There are surprisingly few studies proving definitively that sexual segregation has consequences for the population dynamics of birds. This probably results from the difficulty of studying such issues, particularly

in migratory birds, and the lack until recently of much research interest. The problem of following individual migratory birds through different life-history stages (see Marra *et al.*, 1998), the difficulty of isolating
the effects of segregation from many other population processes and
the scope for compensatory processes to counterbalance any localized
habitat effects all combine to make the demonstration of an influence
of sexual segregation in population dynamics very difficult indeed.

Nonetheless, current knowledge is sufficient to suggest that sexual segregation can, and should, be taken into account by researchers,
managers and decision-makers. A paradigmatic example is the wandering albatross. Along with many other species of albatross and petrel,
their breeding populations in the Southern Ocean have declined dramatically as a consequence of incidental mortality in longline and,
to a lesser extent, trawl fisheries (Weimerskirch & Jouventin, 1987;
Croxall *et al.*, 1998; Nel *et al.*, 2002, Chapters 5 & 6). Longline fleets
initially targeted tuna, *Thunnus* spp., in subtropical, and later in temperate, waters, expanding rapidly in the 1960s and peaking in the late
1980s, at which point exploitation of Patagonian toothfish, *Dissostichus
eleginoides*, stocks began in waters south of the Antarctic Polar Front
(Tuck *et al.*, 2003). Longline fishing was therefore concentrated for many
years in areas visited predominantly by female wandering albatrosses
(Table 18.1). Long-term monitoring at two separate sites shows that periods of population decline coincide with a reduction in survival rates
of females compared to males, whereas mortality rates were equal during periods of stability (Croxall *et al.*, 1998; Weimerskirch *et al.*, 1997a).
That this was a consequence of incidental mortality was underscored
by analysis of ring recoveries of birds from South Georgia, the majority
of which were females killed during fishing operations on vessels in
the northern Patagonian shelf (Croxall & Prince, 1990). As a result of
differential mortality, the adult sex ratio of the breeding populations
of wandering albatrosses became biased, which resulted in a disproportionate reduction in active breeding pairs (Weimerskirch *et al.*, 1997a).

Other studies have also shown significant age and sex biases in
by-catch rates of seabirds, frequently, but not always, towards males
(Bartle, 1990; Murray *et al.*, 1993; Gales *et al.*, 1998; Ryan & Boix-Hinzen,
1999). This could explain why male black-browed albatrosses, *Thalassarche melanophrys*, have a much lower (by 2%) survival rate than females
in the currently declining population at South Georgia (Croxall *et al.*,
1998). The link to fisheries was not made in that study, perhaps because
there was little evidence for sex differences in foraging areas during
chick-rearing (Prince *et al.*, 1999). However, black-browed albatrosses

were taken as by-catch in high numbers during summer in the South Georgia area (Dalziell & de Poorter, 1993) and there are substantial long-line fisheries operating in regions where these birds winter (Prince *et al.*, 1998; Tuck *et al.*, 2003). Sexual segregation could therefore be implicated in the decline of this species, although whether this might result from large-scale differences in pre-incubation, incubation or winter foraging areas, or from 'microhabitat' differences such as the competitive exclusion of females from around fishing vessels (see e.g. Ryan & Boix-Hinzen, 1999), requires further investigation.

These are examples where consideration of the importance of sexual segregation in determining distribution patterns can help focus conservation initiatives. As regards albatrosses and petrels, this would involve the forced introduction of mitigation or better policing of vessels operating in particular areas, measures which can be extremely effective in reducing the level of incidental mortality (Murray *et al.*, 1993; Brothers *et al.*, 1999). In a different arena, research on sexual segregation driven by social dominance could also help in identifying the most important, high quality habitats for species of conservation interest. North American migratory songbirds, for example, are suffering important declines, a topic that has been of considerable recent conservation concern (e.g. Terborgh, 1990). Research in the winter quarters in Mesoamerica has demonstrated convincingly that dominant birds (males), for some species at least, generally prefer mature tropical forests and mangrove habitats over successional scrub and agricultural landscapes, and that such preferences have clear fitness implications, highlighting the importance of conserving those fast-disappearing habitats (e.g. Sherry & Holmes, 1996).

Sexual segregation of harvested populations also can and should be taken into account by managers. Male and female ducks are often subject to different hunting pressures, either because they occupy different geographical areas or habitats (see earlier). In North America, some hunting regulations already take into account biased sex ratios of duck populations. For example, harvesting by point systems has included regulations intended to redistribute the harvest pressure towards male and away from female mallards, *Anas platyrhynchos* (Rexstad & Anderson, 1988; Johnson & Moore, 1996).

In general, the recognition that males and females (as well as other social groups, such as birds of different ages) can differ in their distributions and/or habitat preferences, should lead to a more careful and thorough evaluation of the distribution, habitat requirements and potential threats affecting species of conservation concern or of economic value. Data on sex ratios of samples taken away from breeding sites should be obtained, and threats should be evaluated on a sex-specific basis where segregation is found to be sufficiently strong to justify such an approach. Population modelling should also take sex differences in distribution and mortality into account (see e.g. Durell *et al.*, 2001b).

ACKNOWLEDGEMENTS

We thank Robert Moss for a useful discussion concerning sexual segregation in grouse. Andy Gosler and Richard Holmes made helpful comments on an earlier draft. During part of the research leading to this review, Paulo Catry was supported by Fundacão para a Ciência e Tecnologia (FCT-Portugal) as part of the Programa Plurianual (UI and D 331/94) and by a fellowship (Praxis XXI BPD/16304/98 and BPD/11631/02).

19

Sexual segregation: a necessary consideration in wildlife conservation

OVERVIEW

The importance of animal behaviour to the development of conservation strategies has received increased recognition in recent years. Behaviourists realize now that evolutionary studies rely on a diversity of species occurring in their natural habitats, whereas conservationists now recognize that animal behaviour plays a large role in ecological processes and, therefore, can have great implications for conservation. Several recent texts (Clemmons & Buchholz, 1997; Caro, 1998; Gosling & Sutherland, 2000; Festa-Bianchet & Apollonio, 2003) have been dedicated to this important union. Nonetheless, much remains to be learned about the specific ways by which animal behaviour influences populations, and how that knowledge can best be incorporated into effective conservation strategies (Shumway, 1999).

Sexual segregation is a behavioural and ecological phenomenon that can have great implications for wildlife conservation. A number of hypotheses have been proposed for explaining sexual segregation (Main et al., 1996; Ruckstuhl & Neuhaus, 2002; see general overview in Chapter 2). Understanding the causes of sexual segregation will provide further insight on the selective forces shaping animal behaviour and will lead to improved conservation strategies. Regardless of the mechanism(s) leading to sexual segregation, the outcome itself is an important issue that should be considered in conservation programmes. Indeed, the temporal and spatial groupings of males and females have implications for habitat management, population monitoring, research and management.

Sexual Segregation in Vertebrates: Ecology of the Two Sexes, eds. K. E. Ruckstuhl and P. Neuhaus. Published by Cambridge University Press. © Cambridge University Press 2005.

In this chapter, we review ways in which sexual segregation can influence effectiveness of conservation strategies. The type of sexual segregation (habitat/spatial versus social) can clearly influence the implications for conservation and management. For example, spatial segregation probably will have greater implications for habitat management. Although our chapter focuses on habitat segregation, we also present examples of situations in which social segregation may have conservation implications. We focus many of our examples on bighorn sheep, *Ovis canadensis*, because this species has been the subject of research related to sexual segregation and is a species of conservation concern. Our hope is that this review will prompt others to consider the many ways in which a particular behavioural phenomenon can influence conservation strategies. In addition, we suggest that an increased understanding of sexual segregation will help us to understand better its influences on population dynamics and ecosystem functions, from both a theoretical and an applied view.

HABITAT PROTECTION AND MANAGEMENT

Habitat loss

Habitat loss is a major cause of species endangerment worldwide and, therefore, habitat protection and management play critical roles in conservation. Habitat protection, however, is a costly strategy that requires careful planning. In single-species conservation, delineation of protected areas depends on suitable patches of habitat and the ability of a species to move from one patch to another (Hanski & Gilpin, 1991). The designation of protected areas often is influenced by land ownership, land availability and the current distribution of target species. Because financial and land resources often are limited, compromises are made with only 'key' areas being provided protection. If males and females use habitats differently, even for part of the year, loss or modification of a particular habitat may impact one sex, with subsequent impacts to the population (Kie & Bowyer, 1999; Bowyer et al., 2001). For example, biologists and managers have long recognized the need to protect steep and rugged mountainous habitat for bighorn sheep (Fig. 19.1) but adjoining areas of more gentle terrain often have not received similar protection (Schwartz et al., 1986; Bleich et al., 1990, 1996). Bleich et al. (1997) reported that male and female bighorn sheep in their study area used habitat differently during periods of sexual segregation, with females using more steep and rugged areas and males using gentle, and

Figure 19.1   Bighorn sheep, *Ovis canadensis*, are sexually dimorphic and exhibit pronounced periods of sexual segregation throughout their range. Photographs © Benjamin R. Miller.

even flat, terrain for foraging. Loss of the latter habitat would mean potential loss of important foraging areas for males, as well as loss of movement corridors, and could have implications for the entire population (Bleich *et al.*, 1990).

Many species, including ungulates (Bleich *et al.*, 1997), cetaceans (Stevick *et al.*, 2003), bats (Russo, 2002), fish (Sims *et al.*, 2001a) and birds (Rodway *et al.*, 2003) exhibit sex-based differences in habitat use and selection (also see other chapters in this book). Conservation of all these species relies on protection and management of habitat. Habitat needs of both sexes must be evaluated and incorporated into management plans (Bleich *et al.*, 1997); in the absence of such consideration, losses or alteration of habitat used differentially by the sexes may have subtle, but important, implications for conservation.

The sexes may also differ in their temporal use of habitat. Festa-Bianchet (1988) reported that pregnant bighorn sheep females moved from winter to summer ranges earlier than males and non-pregnant females. Similarly, Stevick *et al.* (2003) reported that male humpback whales, *Megaptera novaeangliae*, moved to breeding grounds earlier than did females. Such temporal periods of segregation and differences in habitat use may have implications for management decisions related to harvest programmes, recreational activities or construction projects.

Habitat fragmentation

Males and females also can differ in their vulnerability to habitat fragmentation due to differences in habitat use. For example, the distribution of bighorn sheep is driven largely by their association with mountainous areas. To a certain extent, bighorn sheep habitat may be considered 'naturally' fragmented (Bleich et al., 1990). Development of roads and other anthropogenic features, however, has resulted in additional fragmentation (Bleich et al., 1996; Rubin et al., 1998), which ultimately will affect both sexes, but possibly in different ways. Males have larger home ranges and are more likely to cross roads than are females (Rubin et al., 1998), which exhibit a higher degree of philopatry (Geist, 1971). Females, therefore, may suffer fewer mortalities because of vehicle collisions, but the distribution of females may be more immediately affected by barriers, with potential demographic implications as subpopulations become fragmented (Rubin et al., 1998). Males may suffer greater mortality from vehicle collisions and, when movement by males is compromised, individual populations and the entire metapopulation will be influenced genetically, as flow of nuclear genes mediated by male movement is reduced (Schwartz et al., 1986; Bleich et al., 1990, 1996).

The influence of ranging behaviour on susceptibility to fragmentation was also suggested by Woodroffe and Ginsberg (2000), who argued that wide-ranging carnivores are more susceptible to edge effects that result from habitat fragmentation. For example, if two species with different ranging behaviour (one with large home ranges and one with small home ranges) were to inhabit the same size reserve, a greater proportion of the species with large home ranges would probably come into contact with the edge of the reserve. This could, in theory, result in higher mortality and losses to that population. Because males of many species are more wide-ranging than females, this same concept could apply within species, with males affected more by fragmentation and resultant edge effects.

Human activities

The sexes also may differ in their susceptibility to human activities, whether direct or indirect. In bighorn sheep, for example, human presence may cause increased stress, as indicated by telemetered heart-rate and behavioural studies (MacArthur et al., 1982). The resulting effect may be influenced by patterns of sexual segregation, if human activities are focused in certain habitat types. In desert populations, for example, female bighorn sheep tend to occur closer to water (Bleich et al., 1997) and may be affected more severely by human activities centred near water sources. Cattle grazing occurs throughout much of the North American Rocky Mountains and southwest deserts, and often overlaps with bighorn sheep habitat. In some populations, male bighorn sheep may be more impacted by cattle grazing, because grazing allotments are typically located in gentle or flat terrain, which may be used more by males than females during periods of sexual segregation (Bleich et al., 1997). Heavy grazing by cattle can result in diminished availability or quality of forage in these areas. Because sexual segregation among native ungulates can result in differential effects on males and females, grazing strategies should include an analysis of that consequence, as suggested by Bleich et al. (1997).

Habitat modification or loss of movement corridors may also affect the degree of segregation observed (Bowyer et al., 2001). Social organization is influenced by the abundance and spatial distribution of resources (Komdeur & Deerenberg, 1997). As a result, changes in resource abundance or distribution could alter patterns of sexual segregation: one sex may be excluded from habitat traditionally used during periods of sexual segregation. Similarly, spatial configuration of resources differs among geographic areas, and the same degree of sexual segregation will not necessarily be evident across all populations within a species, even in the absence of habitat changes. Comparisons of such populations may provide productive research opportunities and insight into the mechanisms underlying sexual segregation by habitat. The effect of scale needs to be carefully considered when quantifying sexual segregation in these situations (Bowyer et al., 1996).

POPULATION MONITORING AND ECOLOGICAL STUDIES

Monitoring demographic parameters

Basic demographic data collected frequently include abundance, distribution, and age and sex structures. This information can be derived in a variety of ways including aerial (e.g. helicopter), ground (e.g. foot, vehicle) and trap surveys. The goal of the study should be considered when determining the proper timing of surveys. If the primary objective is to obtain data on abundance and population composition (e.g. sex ratios), then the most efficient time to survey may be during aggregation (Bleich *et al.*, 1997). In contrast, a survey intended to collect information on distribution might better be conducted during segregation, to document the full extent of habitat use. For example, bighorn sheep surveys conducted during the autumn rut (sexual aggregation) will only reflect habitat use in a subset of habitat actually used by the population during the remainder of the year. In some populations, males and females move between seasonal ranges at different times (Festa-Bianchet, 1988), further emphasizing the importance of choosing the appropriate survey schedule and scale.

Abundance estimates often are derived from mark–recapture data, where marked animals are 'recaptured' either literally in traps or visually during surveys (Krebs, 1989). If males and females have different chances of being 'captured', either because of differences in habitat use or behavioural differences that bias capture probabilities toward one sex, such a difference could cause errors in interpreting mark–recapture data. Bleich *et al.* (1997) reported that male bighorn sheep used areas with denser vegetation than did females and, as a result, males could have a lower chance of being 'captured' (seen) during helicopter surveys. In trapping surveys, males and females may have different capture probabilities if traps are placed in habitat used disproportionately by one sex (Arcese *et al.*, 1997). Social segregation of males and females may also cause biases in survey results. For example, male raccoons, *Procyon lotor*, had a higher rate of capture in live traps than did females, possibly because females travelled independently while males often travelled in all-male groups, resulting in multiple catches when a male group encountered traps (Gehrt & Fritzell, 1996).

Monitoring genetics and health

Population monitoring also includes health and genetic screening. As mentioned earlier, the design and timing of a survey can influence the observed sex ratio, and this may have important implications in assessing the genetic status of a population. In most populations, the total population size differs from the effective population size (N_e), which is the size of an ideal population that would undergo the same amount of genetic drift as the population under consideration (Crow & Kimura, 1970). The effective population size, which is an indicator of the ability of a population to maintain genetic variation, is influenced by a number of factors including sex ratio and mating system. One common method

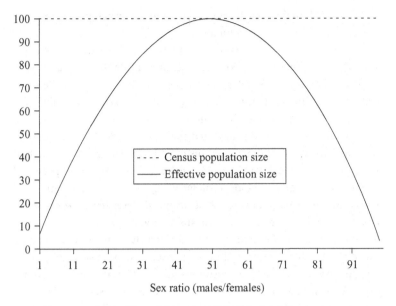

Figure 19.2   The effect of sex ratio on the relationship between effective population size (N_e) and census population size. The correct determination of sex ratio has implications for assessing the genetic status of a censused population. $N_e = 4[(1/N_m) + (1/N_f)]^{-1}$, where N_m and N_f are the numbers of males and females, respectively (Lande & Barrowclough, 1987).

of calculating N_e requires knowledge of the numbers of breeding males and females in the population (Lande & Barrowclough, 1987). Clearly, estimates of N_e will be influenced by the accuracy of the estimated sex ratio (Fig. 19.2).

Screening for genetic profiles can help evaluate taxonomic relationships, elucidate the substructure of a population or assess whether a loss of genetic variation has occurred. If a subset of a population is sampled, sex-based differences may be observed in genetic profiles, which may be due to sex-biased patterns of dispersal. Sampling only one sex may, therefore, not provide an accurate representation of the population. Timing of sampling is, therefore, again an important consideration. In bighorn sheep, females are philopatric and sampling from one locale during the mating season will probably reveal that females are closely related, but that older males (who may have moved in from neighbouring areas) are less related. In contrast, samples collected during segregation may represent females and young (two- to three-year-old) males that have not left their maternal group, and may not represent large breeding males.

Numerous studies have reported sex differences in disease exposure (Zuk & McKean, 1996). For example, male nyala, *Tragelaphus angasii*, were more heavily infested by arthropod parasites than were females (Horak et al., 1995). A number of explanations exist to explain such sex-based differences, and the underlying causes may differ among situations. Horak et al. (1995) suggested that differences observed between male and female nyala may be a result of body size, grooming or hormonal influences. Zuk and McKean (1996) provided a review of proposed explanations, which include ecological (e.g. feeding behaviours) and physiological (usually hormonal in origin) differences between sexes, and proposed that differences in susceptibility to disease may have evolved as a result of selection acting differently on males and females. Although differences must, in some instances, be based on physiological differences such as hormone-influenced immuno-competence (e.g. when differences are observed in a laboratory setting), Zuk and McKean (1996) suggested that further studies, including those on mating systems, may help elucidate the underlying causes. We suggest that sex-based differences in disease prevalence may be caused, in some instances, by sexual segregation if differences in habitat use cause sex-based differences in exposure to other species, disease vectors or environmental conditions. The researcher or veterinarian should explore this possibility, because it may help explain the epidemiology of the disease and, thereby, identify prevention or treatment protocols.

Investigators should be cognizant that disease levels in one sex may not be representative of the entire population. Dobson and Poole (1998) suggested that epidemiological models that divide a population into susceptible, infectious and recovered individuals, and are used to model and predict the transmission of infectious diseases, could be improved if they incorporated information on how individuals aggregate temporally and spatially. This suggestion would apply to aggregations of males and females if these groups differed in percentage of infected animals and rates of transmission, and could also pertain to species in which the sexes segregate socially rather than spatially, because transmission rates may still differ between the sexes. For example, females may shed parasites at higher rates during parturition, thereby increasing transmission rates within female groups; similarly, differences between densities of males and females during segregation could result in different rates of disease transmission. The level of environmental contaminants may also differ between the sexes (Merchant *et al.*, 1991) because exposure to environmental contaminants may be influenced by differences in habitat use by males and females. This possibility should receive consideration in management programmes (see also Chapter 6).

Behavioural studies

Sexual segregation also should be considered in the design of behavioural studies. Detailed behaviours may differ among animals in different age-classes, and the composition of groups may be influenced by sexual segregation. For example, in bighorn sheep, younger males associate with females for a longer time-period during sexual aggregation than do older males (Bleich *et al.*, 1997; Rubin *et al.*, 2000). Age may influence male breeding behaviour (Hogg, 1984) or activity budget (Ruckstuhl, 1999), or behaviour of individuals within the group may be influenced by group composition. The timing of the study relative to periods of segregation may, therefore, influence the conclusions of a behavioural study and could be manifested as differences in rates of feeding, vigilance or aggressive behaviours.

POPULATION MANAGEMENT

The use of models

Conservationists often rely on models to guide decision-making. Such models may be used to explore expected demographic changes under alternative management scenarios, effects of habitat loss, population viability, dispersal or genetic structure. Models may depict members of a population in various levels of detail, which will depend on the goals of the modelling exercise. Some applications use population-based demographic models, where demographic parameters may be general, as obtained from a life table (Burgman *et al.*, 1993). Other models consider the survival, reproduction, dispersal and, possibly, the genetics of each individual (e.g. Heppel *et al.*, 1994). In some situations, it is possible to model the dynamics of either males or females, if it is believed that the outcome of interest is driven by one sex (e.g. Rubin *et al.*, 2002). Behavioural data also can be incorporated explicitly into some models to make them more realistic (Beissinger, 1997). It may be necessary to assign different parameters to the sexes in such models, because males and females may differ in dispersal, seasonal movements, habitat use, survival and cause-specific mortality as a result of sexual segregation.

When modelling effects of habitat loss, the needs of both sexes should also be considered. For example, there may be a tendency to grant the most protection to those areas viewed as 'typical' habitat for a species, with less protection deemed necessary for other areas. Indeed, models of habitat loss in bighorn sheep would be incomplete if differential use by each sex of this habitat was not incorporated. In another example, male Harlequin ducks, *Histrionicus histrionicus*, used habitat further offshore than did females during winter and, thus, the sexes could be affected differently by habitat changes (Rodway *et al.*, 2003). Durell (2000) stressed that if habitat losses or changes influence one age- or sex-category more than another, those differences must be included in predictive models.

In some contexts, genetic models can be improved by incorporating patterns of sexual segregation. For example, it is important to know which age-classes of males segregate from females during periods of segregation, whether segregating males consistently disperse to breed with distantly related females and whether younger, non-segregating males breed in their natal group. Van Noordwijk (1994) used models to examine the interactions between inbreeding depression, environmental stochasticity and extinction risk in small populations with different social systems, and reported that social structure influenced genetic-related extinction risks.

Understanding predation risk

The understanding of interactions among species can benefit from increased knowledge of sexual segregation. Males and females may differ in their vulnerability to predation (Pierce *et al.*, 2000) or they can have different survival rates or sources of mortality (Schaefer *et al.*, 2000), which may be a result of differences in habitat use. These parameters should, therefore, be monitored in both sexes. At the very least, potential sex differences should be explored prior to pooling data for males and females. Interactions such as those in predator–prey systems may also differ by population. For example, in a study by Bleich *et al.* (1997), male and female bighorn sheep used separate areas during periods of sexual segregation and predation by mountain lions, *Puma concolor*, was documented only in areas used by males. Because predation by lions was limited to males, the authors concluded it did not significantly affect the reproductive base of their study population (Bleich *et al.*, 1997). In contrast, male and female bighorn sheep in another California mountain range were not separated by great distances during sexual segregation and both sexes were killed by mountain lions (E. Rubin, unpublished data; Hayes *et al.*, 2000). The same pattern was found in a Canadian population of bighorn sheep (Ross *et al.*, 1997). Differences in sexual segregation and lion predation between these study areas may be related to differences in habitat type and configuration as well as juxtaposition of resources, all of which may influence the overlap of bighorn sheep and mountain lions. The demographic consequences of lion predation on those populations may be very different and, as a result, management needs and protocols could differ. This example also illustrates the need for a better understanding of how resource availability and distribution influence sexual segregation.

Harvest programmes

Individuals may be removed from a population as part of legal harvest and population-control programmes or through illegal hunting, and understanding the effects on the population requires that the timing and extent of sexual segregation be considered. Harvesting can have direct consequences, not only by reducing overall numbers, but also by changing the sex ratio. Most legal harvest programmes have quotas for males and females but sexual segregation may cause individuals of one sex to be selected disproportionately even when the harvest is assumed to be non-selective. Shoaling behaviour of fishes is one example in which one sex may be selected non-intentionally during harvest (Reynolds & Jennings, 2000). If same-sex shoals use different habitat, and the habitat used by one sex allows fish to be caught more efficiently, harvests will be biased towards that sex (Vincent & Sadovy, 1998; see also Chapter 8 on sharks). In another example, males of some African ungulates may be captured in illegal snares disproportionately because of sex differences in habitat use (Arcese *et al.*, 1997).

The immediate consequence of sex-biased harvest is a change in the effective population size (Crow & Kimura, 1970), possibly followed by additional changes in behaviour and reproductive success of the remaining animals. To understand fully the effects of such removals, knowledge of behavioural ecology and sexual segregation should be used to determine which animals can be removed (i.e. selected by the removal techniques), and what influence the loss of those individuals (or various harvest strategies) will have on the demographics and viability of the altered population (Burgman *et al.*, 1993). Predictive harvest models, which typically assume an equal sex ratio in the catch, may not be accurate (Vincent & Sadovy, 1998). Beissinger (1997) noted that harvest models are often not successful in predicting population trends because detailed behavioural information has not been included.

Sex-biased removals may also alter patterns or intensity of sexual selection or change the breeding system of the population. For example, female coral trout, *Plectropomus areolatus*, are selectively removed during harvest, leading to increased harassment of remaining females by territorial males (Johannes *et al.*, 1994). Managers should also be mindful that harvest programmes could influence populations by changing population densities, which may alter habitat selection and, possibly, the degree of sexual segregation. Indeed, in red deer, *Cervus elaphus*, and white-tailed deer, *Odocoileus virginianus*, populations, changes in forage selection and degree of sexual segregation were observed when densities changed (Clutton-Brock *et al.*, 1987a; Kie & Bowyer, 1999).

Population augmentation and captive breeding programmes

Population augmentation is a tool increasingly used in conservation programmes, and such programmes should consider the patterns and extent of sexual segregation in the target species or population. Timing (season) of augmentation, types of habitat individuals are to be released into and sex composition of individuals to be released are all important considerations. In release programmes for rare butterflies, the timing of release and sex-composition of release groups is critically important, because male and female butterflies do not emerge at the same time; that is, there is some degree of temporal segregation (Thomas *et al.*, 2000).

Captive breeding can be a useful conservation tool and has been important for the recovery of some species (e.g. California condors, *Gymnogyps californianus*, and Arabian oryx, *Oryx leucoryx*). Maintaining a normal social structure for a species is requisite for the successful reproduction and rearing of offspring (Kleiman, 1980). As knowledge about social behaviour has increased, managers have improved husbandry by adjusting practices to mimic more closely natural social structure. Although most zoos historically exhibited animals in pairs or family groups, this approach may conflict with normal patterns of sexual segregation in some species. For example, wild talapoin monkeys, *Miopithecus talapoin*, segregate into same-sex groups outside of breeding periods; when housed in mixed-sex groups in captivity, however, females exhibited high levels of aggression toward males (Rowell, 1973). Survival of young tree shrews, *Tupaia belangeri*, was reduced when males and females were housed together without separate nest boxes

(Kleiman, 1980). Weckerly *et al.* (2001) reported increased aggression among female elk when groups included high numbers of males. In other species the absence of a dominant male (or absence of male aggression) may actually decrease female reproduction (Kleiman, 1980) or the presence of a male may be important for successful rearing of young, as with bush dogs, *Speothos venaticus*, (Jantschke, 1973) and lion tamarins, *Leontopithecus rosalia* (Kleiman, 1980).

Clearly, the temporal and spatial groupings of animals are important considerations for successful breeding programmes, and maintaining animals in inappropriate social groupings or forcing them to maintain certain social groupings for extended periods, could greatly affect behaviour and reproduction. One of the challenges facing managers of captive breeding programmes is to determine the correct social composition when the natural system is not well understood. This emphasizes the need for additional field studies on a variety of taxa to gain a better understanding of the patterns and extent of sexual segregation. Examination of captive or otherwise artificial settings, where the social grouping has been altered with resultant negative consequences, may provide opportunities to learn more about mechanisms underlying sexual segregation.

CONCLUSION

There remains much to be learned about the causes of sexual segregation and factors that affect the degree of sexual segregation. Regardless of the mechanisms that cause sexual segregation, however, the fact that males and females often live parts of their lives separated in time or space has implications for wildlife management and conservation strategies. Knowledge of patterns of sexual segregation, along with other behavioural attributes of a species, will make conservation programmes more effective. Ultimately, increasing our understanding of the mechanisms and significance of sexual segregation will help us better to understand the implications of this behavioural phenomenon for conservation.

ACKNOWLEDGEMENTS

We would like to thank Benjamin R. Miller for the use of his photographs of bighorn sheep.

Part VIII Outlook

PETER NEUHAUS, KATHREEN E. RUCKSTUHL
AND LARISSA CONRADT

20

Conclusions and future directions

OVERVIEW

Sexual segregation is predominant in sexually dimorphic ungulates and is generally a common feature in vertebrates and also occurs in some invertebrates as described recently in squids (Arkhipkin, 2002). When we first had the idea of this book we did not know if, and to what extent, non-ungulate species showed this phenomenon. We think that the contributors to this volume have done a great job in describing what is known about sexual segregation, not only in ungulates but also in a wide variety of other vertebrate species. However, this book is not only an update of what is known about this topic in vertebrates, but within the different chapters it also discusses what is not known and where future research could and should progress in clarifying the causes and consequences of sexual segregation. The novelty of this book is the collection and discussion of the many different consequences and probable causes of sexual segregation on many different levels. While different taxa have been looked at and described in this book, notably ungulates (Chapters 2, 3, 9, 10, 11 and 19), marsupials (Chapter 14), primates, including humans (Chapters 12 and 17), odontocetes (Chapter 16), bats (Chapter 15), birds (Chapters 5, 6 and 18), seals (Chapter 4), sharks (Chapter 8), reptiles (Chapter 13) and to a certain extent other fish (Chapter 7), other important groups such as rodents or carnivores (other than seals) have not been dealt with. Sexual segregation does, however, occur in many other species, as for example in African ground squirrels (Waterman, 1995) – the main reason why other such species have not been included in this book is primarily the lack of topic related studies

Sexual Segregation in Vertebrates: Ecology of the Two Sexes, eds. K. E. Ruckstuhl and P. Neuhaus. Published by Cambridge University Press. © Cambridge University Press 2005.

and data. This book illustrates that segregation by sex is much more common than previously thought and we think that the importance of this is not yet fully appreciated, something this book will hopefully help to achieve.

While working on this book, we became more and more aware of how widespread this phenomenon is and how little is known about this important topic. Reading this book, it should become clear that in order to understand life-history strategies, sexual and natural selection and even speciation, one has to look at males and females separately. The potential consequences for conservation are also straightforward and well discussed in Part VII of this book.

WHAT HAVE WE FOUND?

It is not surprising that a trait which is as widely distributed as sexual segregation has been described as having many different proximate and ultimate causes in the different species in which it has been observed. It is not always easy to categorize different hypotheses describing sexual segregation. One of the main messages before even starting to work on this topic has to be that one needs to make a clear distinction between social and solitary animals and, hence, between social segregation and segregation by other factors such as habitat and/or food (Chapter 2). In all vertebrate groups dealt with in this book there are some species that do segregate by sex, but also some species that do not. While the vast majority of sexually segregating species show a large sexual dimorphism in body size, there are some species, notably bats and some birds (Chapters 15 and 18), where the sexes do segregate despite a lack of pronounced sexual dimorphism in body size. In some bat species, the different needs of males and females in thermoregulation has been argued to be an important factor. In birds, different niche use of females and males either during or outside the nesting period have been discussed as main factors leading to the observed spatial or food segregation. In other social species, mainly primates (Chapter 17) and orcas (Chapter 16), no segregation by sex is found despite a large sexual dimorphism in body size.

Historically, most work to explain sexual segregation has been done on ungulates and it is, therefore, no surprise that it has been in this animal class where most attempts to explain this enigma have been carried out. An important goal of this book is to gain a better understanding of the different hypotheses because they are often misinterpreted in the scientific literature. Once the hypotheses are clearly understood one can then start evaluating to which extent each of them may explain sexual segregation in the different taxa.

THE DIFFERENT HYPOTHESES

First of all, this book should help to put an end to the variety of different names used to describe the same hypothesis. The ones we have used in this volume might not be, in all cases, the best fit but it is a huge advantage if everybody uses the same terms to explain the same hypothesis describing sexual segregation. Larissa Conradt (Chapter 2) makes it clear that sexual segregation has to be understood as a much more general term, meaning that females and males do things differently – sometimes in different habitats but also sometimes in the same habitat. While in this book we mostly talk of social sexual segregation (meaning gregarious animals living in social but uni-sex groups), segregation can also occur in solitary animals such as described in black rhinos, where males and females are mostly solitary, use the same habitat but segregate in their feeding habits (Chapter 3).

First we will discuss the forage selection hypothesis, which is also sometimes termed the sexual dimorphism body size hypothesis. Since there are other hypotheses also referring to sexual dimorphism being a major factor leading to segregation, the term forage selection hypothesis is preferred since it directly describes the proximate cause that leads to the observed segregation. The scramble competition hypothesis can be looked at as a special case of the forage selection hypothesis. The study by Clutton Brock et al. (1982, 1987a) on red deer that proposed this hypothesis was not supported by an experiment carried out on the same population with the aim to test it (Conradt et al., 1999b). Nonetheless, there are possibilities that in some species scramble competition might be an important factor leading to segregation (see Chapter 18). The originally termed reproductive strategy hypothesis is now termed the predation risk hypothesis. The main reason for this name is, again, the focus on the direct proximate cause leading to segregation. The term 'reproductive strategy' is much too broad – after all, the evolutionary cause for most behaviours is to optimize reproductive output. In his chapter, Main (Chapter 9) refers to the reproductive strategy hypothesis, using its broad meaning. Main rightly states that males and females do things differently because they have different ways of maximizing their reproductive success. This is in no doubt a valid way to describe the phenomenon of sexual segregation. However, it is very important and interesting to look at the more specific reasons leading to sexual segregation in some species as well as reasons why this phenomenon is not observed in others. The activity budget hypothesis is a more recent hypothesis put forward to explain sexual segregation. It also had another name (the body size predation hypothesis). The activity budget hypothesis aims to explain social sexual segregation. As in the forage selection hypothesis, body-size difference is an important factor, but the most direct proximate causes for segregation are differences in activity budgets and movement rates caused by this dimorphism. It is crucial to keep in mind that the activity budget hypothesis is dealing with social segregation, meaning that habitat segregation is facultative and its presence must have additional reasons. Therefore, it deals primarily with socially aggregating species. The same is true for the social affinity hypothesis, dealt with in Chapter 11. While it is definitely a possible explanation for segregation between the sexes, the problem here is that this hypothesis is unlikely to be falsifiable and is therefore un-testable.

By looking at other taxa outside the Artiodactyla, more important factors leading to sexual segregation do arise. Notable is the temperature sensitivity hypothesis described in Chapter 2. This hypothesis fits some birds, bats, sharks and snakes. Another special case is described in Chapter 18 – where in some birds, males and females rearing chicks together are morphologically different, using different niches in search for food. Differences in diet between females and males are not enough studied and should be considered an important topic. Another important but so far mainly neglected field, which should be a focus of future research in sexual segregation, is migration. Many birds, but also bats and some ungulates, sexually segregate during migration and in the wintering grounds but mix in the breeding grounds.

In addition to these potentially valuable hypotheses there are others, put forward to describe sexual segregation, such as the so called competition avoidance hypothesis (Geist & Petocz, 1977). This hypothesis predicts that males leave good foraging areas to the females and their offspring in an attempt to optimize their own reproductive success. This hypothesis can be discarded as extremely unlikely, not only because nobody has ever found any evidence of this happening, but because cheating would also be beneficial and it would therefore not be an evolutionary stable strategy. Another hypothesis, put forward particularly for cervids, states that males are more vulnerable to predation than females after the rut, and that they should therefore segregate from females if they have not shed their antlers because they are easily distinguished and picked out by predators (Geist & Bromely, 1978). Geist and Bromley argued that the shedding of the antlers by cervids is a form of female mimicry, to disguise that they are males. While this hypothesis has also not found any support in empirical data, similar mechanisms to avoid oddity are described for fish in Chapter 7. For ungulates however, it is unlikely to apply. It was further suggested that males would segregate from females outside the mating season to avoid energetically costly intra-sexual interactions (Morgantini & Hudson, 1981). Although Weckerly et al. (2001) found that female elk were more aggressive when males were present in female groups than when they were not, there is little other support for this hypothesis. Another hypothesis explaining sexual segregation in ungulates was that males seek out areas with fewer parasites than females do, because the latter have to disregard the costs due to the requirements of gestation and lactation (Clutton-Brock et al., 1987a). There are more detailed descriptions of these hypotheses, all primarily put forward for ungulate species, to be found in the literature (Main & Coblentz, 1990; Miquelle et al., 1992; Main et al., 1996; Ruckstuhl & Neuhaus, 2002).

THE DEBATE

There has been some discussion about which hypothesis describes sexual segregation best; in particular, the forage selection, predator avoidance, activity budget and social affinity hypotheses have been proposed to play a major role. First of all we have to be aware that these hypotheses are not trying to explain the same thing. While the forage selection and the predator avoidance hypotheses try to explain habitat segregation, the social affinity and the activity budget hypotheses are trying to explain social segregation. This means that while the first two cannot provide an adequate description of social segregation within the same habitat, the latter two are not able to explain habitat segregation. Further, the social affinity and the activity budget hypotheses relate to social animals, while segregation by habitat or diet can also be explained in solitary species. It is therefore important to know if social segregation is a by-product of habitat segregation or if it can occur independently. While Conradt showed in the second chapter of this book that theoretically both are possible, empirical data point in the direction that social segregation is an independent phenomenon, as it has been observed in many species without habitat segregation. We do, however, not know if species segregating by habitat would socially segregate if there were only one habitat type available. For now, at least, we cannot answer this question with certainty. It could well be, as some authors in this book have suggested, that there are different reasons why the sexes segregate socially or by habitat. We need to keep the two different types of segregation in mind when developing models on how sexual segregation could have evolved in social animals (see Chapter 10 for a model).

SIZE DIMORPHISM

Most hypotheses describing sexual segregation predict a positive corre- lation between body size dimorphism and the degree of segregation. This prediction seems to hold in most cases, especially in ungulates (Ruckstuhl & Neuhaus, 2002). There are, however, some other ani- mal groups such as many primate species (Chapter 17) and orcas (Chapter 16) where very sexually dimorphic species build year-round stable, mixed-sex groups. In primates we often see a potential trade-off for the animals, where males have to be large to optimize reproductive success but also have to stay with the females to protect the young from infanticide by other males. Also constrained by a very special social system is the orca. Here it would be important to know what selection pressure actually led to the observed body size differences between males and females. In bats, on the other hand, sexual segrega- tion has been described in species without marked sexual dimorphism. Different microclimatic needs might explain some of the observed seg- regation between the sexes (Chapter 15).

MATING SYSTEMS

Monogamous relationships and harems do lead to groups of mixed sex. In polygynous mating systems an important factor, which makes it either possible or not for the animals to segregate, is the duration of the mating season. In species with mating occuring year-round (aseasonal) it is less likely for males to segregate for longer time periods, because they have to stay around females to optimize their reproductive success. In oryx, where year-round mating can occur, mixed groups are the rule and there is no sexual body size dimorphism (Chapter 10). Differences in climate and food availability can lead to different mating systems, different body size dimorphism and different degrees of sexual segregation within a species as well as across taxa. High seasonality in climate and food availability is often associated with high seasonality in breeding and, therefore, high seasonality in mating, which in turn favours body size dimorphism and offers temporal opportunities for sexual segregation outside the mating season. However, size dimorphism is not restricted to species with seasonal mating. In the size dimorphic African buffalo, year-round mating occurs, with males often seen in mixed sex groups. The interesting observation in this species is that bachelor groups are quite common, and Prins (1996) argued that these groups form because males have to recuperate after spending time within the mixed groups. A very interesting approach is to look at this not only across, but also within, a species as done by MacFarlane and Coulson in marsupials (Chapter 14) or by Shine and Wall in snakes (Chapter 13). The mating system might therefore affect body size dimorphism and in combination with the seasonality adds pressure on each sex, determining whether they should aggregate or segregate.

POPULATION DENSITY

With the exception of extremely low or extremely high density, we would not expect a density dependent impact on the degree of segregation. (The scramble competition hypothesis predicts density dependence of sexual segregation). However, studies looking into this possibility are only few (McCullough, 1979; Bowyer, 1984) and further investigation into the impact of population density and sex ratio on the degree of sexual segregation would be very valuable. In his study on pronghorns (*Antilocapra america*), Byers (1997) showed that the mating system changed from territorial males to a harem system following a population crash. After the recovery of the population, males did not become territorial again, the harem system remained intact even at high density.

CONCLUSIONS

This book is a reminder of the importance of the different evolutionary pressures on males and females. As modern evolutionary theory predicts, all individuals try to optimize fitness, mostly measured in lifetime reproductive success. It shows clearly that decisions do not only differ between sexes, but also between age classes and often between different conditions. Hence, individuals that resemble each other are more likely to do things similarly than individuals that differ in sex, age, size, body condition or reproductive condition. Further, environmental situations, such as food availability, temperatures, topography etc., are important vectors, either increasing or decreasing behavioural differences between different individuals. An animal's optimal life-history strategy is therefore driven by factors within as well as from the outside. Sexual segregation is a phenomenon very central in understanding the ecology of the two sexes.

Here are some major summarizing points of this book:

(1)   Sexual segregation occurs in a wider variety of species and ecological conditions than have previously found general attention.

(2)   Social and habitat segregation have to be treated as independent phenomena.

(3)   It is unlikely that there is one general explanation that fits all species.

(4)   Sexual segregation is dependent on fundamental factors like body size dimorphism, mating systems, seasonality of climate and forage availability, and predation pressures.

(5)   Sexual segregation requires that for many ecological and evolutionary questions, the sexes have to be treated as different entities.

(6)   All these points and their implications can have a big impact, and have to be considered as important for conservation and management.

References

Abernethy, K.A., White, L.J.T. and Wickings, E.J. (2002). Hordes of mandrills (*Mandrillus sphinx*): extreme group size and seasonal male presence. *Journal of Zoology*, London, **258**, 131–7.

Acuna, H.O. and Francis, J.M. (1995). Spring and summer prey of the Juan-Fernandez fur seal, *Arctocephalus philippii*. *Canadian Journal of Zoology*, **73**, 1444–52.

Adamopoulou, C. and Legakis, A. (2002). Diet of a lacertid lizard (*Podarcis milensis*) in an insular dune ecosystem. *Israel Journal of Zoology*, **48**, 207–19.

Afanasyev, V. (2004). A miniature daylight level and activity data recorder for tracking animals over long periods. *Memoirs of the National Institute of Polar Research*, Special Issue, **58**, 227–233.

Agrimi, U. and Luiselli, L. (1992). Feeding strategies of the viper *V. u. ursinii* (Reptilia: Viperidae) in the Apennines. *Herpetological Journal*, **2**, 37–42.

Alberts, S.A. and Altmann, J. (1995). Balancing costs and benefits: dispersal in male baboons. *American Naturalist*, **145**, 279–306.

Albon, S.D., Clutton-Brock, T.H. and Langvatn, R. (1992). Cohort variation in reproduction and survival: implications for population demography. In *The Biology of Deer*, ed. R.D. Brown. Berlin: Springer Verlag.

Alexander, C., Lynch, J.J. and Mottershead, B.E. (1979). Use of shelter and selection of lambing sites by shorn and unshorn ewes in paddocks with closely or widely spaced shelters. *Applied Animal Ethology*, **5**, 51–69.

Allen, B.A., Burghardt, G.M. and York, D.S. (1984). Species and sex differences in substrate preference and tongue-flick rate in three sympatric species of water snakes (Nerodia). *Journal of Comparative Psychology*, **98**, 358–67.

Altmann, J. (1980). *Baboon Mothers and Infants*. Cambridge, MA: Harvard University Press.

  (1990). Primate males go where the females are. *Animal Behaviour*, **39**, 193–5.

Altringham, J.D. (1996). *Bats: Biology and Behaviour*. Oxford: Oxford University Press.

Altringham, J.D. and Fenton, M.B. (2003). Sensory ecology and communication in the Chiroptera. In *Bat Ecology*, eds. T.H. Kunz and M.B. Fenton. Chicago: University of Chicago Press, pp. 90–127.

Anderson, D.J., Reeve, J., Gomez, J.E.M. *et al.* (1993). Sexual size dimorphism and food requirements of nestling birds. *Canadian Journal of Zoology*, **71**, 2541–5.

Anderson, D.J., Schwandt, A.J. and Douglas, H.D. (1998). Foraging ranges of waved albatrosses in the Eastern tropical Pacific. In *Albatross Biology and*

Conservation, eds. G. Robertson and R. Gales. Chipping Norton: Surrey Beatty and Sons, pp. 180–5.

Andersson, M. (1994). *Sexual Selection*, Princeton: Princeton University Press.

Andrews, R.M. (1971). Structural habitat and time budget of a tropical Anolis lizard. *Ecology*, **52**, 262–70.

Andrews, R.M., de la Cruz, F.R.M. and Santa Cruz, M.V. (1997). Body temperatures of female *Sceloporus grammicus*: Thermal stress or impaired mobility? *Copeia*, 1997, 108–15.

Angelici, F.M., Effah, C., Inyang, M.A. and Luiselli, L. (2000). A preliminary radio-tracking study of movements, activity patterns and habitat use of free-ranging Gaboon vipers, *Bitis gabonica*. *Terre et la Vie*, **55**, 45–55.

Anibaldi, C., Luiselli, L. and Angelici, F.M. (1998). Notes on the ecology of a suburban population of rainbow lizards in coastal Kenya. *African Journal of Ecology*, **36**, 199–206.

Antonelis, G.A., Lowry, M.S., Fiscus, C.H., Stewart, B.S. and Delong, R.L. (1994). Diet of the northern elephant seal. In *Elephant Seals: Population Ecology, Behaviour, and Physiology*, ed. B.J. Le Boeuf and R.M. Laws. Berkley: University of California Press, pp. 211–23.

Appleby, M.C. (1982). The consequences and causes of high social rank in red deer stags. *Behaviour*, **80**, 259–73.

(1983). Competition in a red deer stag social group – rank, age and relatedness of opponents. *Animal Behaviour*, **31**, 913–18.

Arcese, P., Keller, L.F. and Cary, J.R. (1997). Why hire a behaviorist into a conservation or management team? In *Behavioral Approaches to Conservation in the Wild*, eds. J.R. Clemmons and R. Buchholz. Cambridge, United Kingdom: Cambridge University Press, pp. 48–71.

Archer, J. and Lloyd, B. (2002). *Sex and Gender*. Cambridge, UK: Cambridge University Press.

Archibald, G.W. and Meine, C.D. (1996). Family Gruidae (Cranes). In *Handbook of the Birds of the World*, Vol. 3., eds. J. del Hoyo, A. Elliot and J. Sargatal. Barcelona: Lynx Edicions.

Ardia, D.R. and Bildstein, K.L. (1997). Sex-related differences in habitat selection in wintering American kestrels, *Falco sparverius*. *Animal Behaviour*, **53**, 1305–11.

(2001). Sex-related differences in habitat use in wintering American kestrels. *Auk*, **118**, 746–50.

Arkhipkin, A.I. and Middleton, D.A.J. (2002). Sexual segregation in ontogenetic migrations by the squid *Loligo gahi* around the Falkland Islands. *Bulletin of Marine Science*, **71**, 109–27.

Arlettaz, R. (1999). Habitat selection as a major resource partitioning mechanism between the two sympatric sibling bat species *Myotis myotis* and *Myotis blythii*. *Journal of Animal Ecology*, **68**, 460–71.

Arnbom, T. and Lundberg, S. (1995). Notes on *Lepas australis* (Cirripedia, Lepadidae) recorded on the skin of southern elephant seal (*Mirounga leonina*). *Crustaceana*, **68**, 655–8.

Arnbom, T., Papastavrou, V., Weilgart, L.S. and Whitehead, H. (1987). Sperm whales react to an attack by killer whale. *Journal of Mammalogy*, **68**, 450–3.

Arnold, G.W., Boundy, C.A.P., Morgan, P.D. and Bartle, G. (1975). The roles of sight and hearing in the lamb in the location and discrimination between ewes. *Applied Animal Ethology*, **5**, 43–50.

Arnold, G.W., Steven, D.E. and Grassia, A. (1990). Associations between individuals and classes in groups of different size in a population of western grey kangaroos, *Macropus fuliginosus*. *Australian Wildlife Research*, **17**, 551–62.

Arnold, G.W., Steven, D.E. and Weeldenberg, J.R. (1994). Comparative ecology of western grey kangaroos (*Macropus fuliginosus*) and Euros (*M. robustus erubescens*) in Durokoppin Nature Reserve, isolated in the central wheatbelt of Western Australia. *Wildlife Research*, **21**, 307–22.

Arnold, P.W. (1972). Predation on harbour porpoise, *Phocoena phocoena*, by a white shark, *Carcharodon carcharias*. *Journal of Fisheries Research Board of Canada*, **29**, 1213–14.

Arnold, S.J. (1993). Foraging theory and prey-size-predator-size relations in snakes. In *Snakes: Ecology and Behavior*, eds. R.A. Seigel and J.T. Collins. New York: McGraw-Hill, pp. 87–116.

Arnold, S.J. and Peterson, C.R. (2002). A model for optimal reaction norms: the case of the pregnant garter snake and her temperature-sensitive embryos. *American Naturalist*, **160**, 306–16.

Arnould, J.P.Y. and Duck, C.D. (1997). The cost and benefits of territorial tenure, and factors affecting mating success in male Antarctic fur seals. *Journal of Zoology*, **241**, 649–64.

Arnould, J.P.Y., Green, J.A. and Rawlins, D.R. (2001). Fasting metabolism in Antarctic fur seal (*Arctocephalus gazella*) pups. *Comparative Biochemistry and Physiology. Part A, Molecular and Integrative Physiology*, **129**, 829–41.

Atsalis, K. (1999). The diet of the brown mouse lemur, *Microcebus rufus*, in Ranomafana National Park, Madagascar. *International Journal of Primatology*, **20**, 193–229.

Backus, R.H., Springer, S. and Arnold, E.L. (1956). A contribution to the natural history of the white-tip shark, *Pterolamiops longimanus* (Poey). *Deep-Sea Research*, **3**, 178–88.

Badyaev, A.V. (2002). Growing apart: an ontogenetic perspective on the evolution of sexual size dimorphism. *Trends in Ecology and Evolution*, **17**, 369–78.

Baird, R.W. (2000). The killer whale: Foraging specializations and group hunting. In *Cetacean Societies: Field Studies of Dolphins and Whales*, eds. J. Mann, R.C. Connor, P.L. Tyack and H. Whitehead. Chicago: Academic Press, pp. 127–53.

Baird, R.W. and Whitehead, H. (2000). Social organization of mammal-eating killer whales: group stability and dispersal patterns. *Canadian Journal of Zoology*, **78**, 2096–105.

Baker, J.R. and McCann, T.S. (1989). Pathology and bacteriology of adult male Antarctic fur seals, *Arctocephalus gazella*, dying at Bird Island, South Georgia. *British Veterinary Journal*, **145**, 263–75.

Baldassarre, G.A. and Bolen, E.G. (1994). *Waterfowl Ecology and Management*. New York: John Wiley and Sons.

Baldellou, M. and Aden, A. (1997). Time, gender, and seasonality in vervet activity: a chronobiological approach. *Primates*, **38**, 33–43.

Barboza, P.S. and Bowyer, R.T. (2000). Sexual segregation in dimorphic deer: a new gastrocentric hypothesis. *Journal of Mammalogy*, **81**, 473–89.

Barclay, R.M.R. (1991). Population structure of temperate zone insectivorous bats in relation to foraging behaviour and energy demand. *Journal of Animal Ecology*, **60**, 165–78.

Barclay, R.M.R. and Harder, L.D. (2003). Life histories of bats: life in the slow lane. In *Bat Ecology*, eds. T.H. Kunz and M.B. Fenton. Chicago: University of Chicago Press, pp. 209–53.

Barrette, C. (1991). The size of Axis deer fluid groups in Wilpattu National Park, Sri Lanka. *Mammalia*, **2**, 207–20.

Barros, N.B. and Odell, D.K. (1990). Food habits of bottlenose dolphins in Southeastern United States. In *The Bottlenose Dolphin*, eds. S. Leatherwood and R.R. Reeves. San Diego: Academic Press, pp. 309–28.

Barry, R.E., Botje, M.A. and Grantham, L.B. (1984). Vertical stratification of *Peromyscus leucopus* and *P. maniculatus* in southwestern Virginia. *Journal of Mammalogy*, **65**, 145–8.

Bartle, J.A. (1990). Sexual segregation of foraging zones in procellariiform birds: implications of accidental capture on commercial fishery longlines of grey petrels (*Procellaria cinerea*). *Notornis*, **37**, 146–50.

Bass, A.J., D'Aubrey, J. and Kistnasamy, N. (1973). Sharks of the east coast of southern Africa. I. The genus *Carcharhinus* (Carcharhinidae). *Investigation Report of the Oceanographic Research Institute*, **33**, 1–168.

Bateson, P.P.G. and Martin, P. (1999). *Design for a Life: How Behaviour Develops.* London: Jonathan Cape.

Bauwens, D. and Thoen, C. (1981). Escape tactics and vulnerability to predation associated with reproduction in the lizard *Lacerta vivipara*. *Journal of Animal Ecology*, **50**, 733–43.

Beamish, F.W.H. (1978). Swimming capacity of fish. In *Fish Physiology*, Vol. 7, ed. D.J. Randall. New York: Academic Press, pp. 101–87.

Bearder, S.K. (1987). Lorises, bushbabies, and tarsiers: diverse societies in solitary foragers. In *Primate Societies*, eds. B.B. Smuts, D.L. Cheney, R.M. Seyfarth, R.W. Wrangham, and T.T. Struhsaker. Chicago: University of Chicago Press, pp. 11–24.

Beck, C.A., Bowen, W.D. and Iverson, S.J. (2003a). Sex differences in the seasonal patterns of energy storage and expenditure in a phocid seal. *Journal of Animal Ecology*, **72**, 280–91.

Beck, C.A., Bowen, W.D., McMillan, J.I. and Iverson, S.J. (2003b). Sex differences in the diving behaviour of a size-dimorphic capital breeder: the grey seal. *Animal Behaviour*, **66**, 777–89.

Becker, C.D. and Ginsberg, J.R. (1990). Mother-infant behavior of wild Grevy's zebra: adaptation for survival in semi-desert East Africa. *Animal Behaviour*, **40**, 1111–18.

Becker, D.S. (1988). Relationship between sediment character and sex segregation in English sole, *Parophrys vetulus*. *U.S. Fishery Bulletin*, **86**, 517–24.

Becker, P.H., González-Solís, J., Behrends, B. and Croxall, J.P. (2002). Feather mercury levels in seabirds at South Georgia: Influence of trophic position, sex and age. *Marine Ecology Progress Series*, **243**, 261–9.

Beer, J.R. and Richards, A.G. (1956). Hibernation of the big brown bat. *Journal of Mammalogy*, **37**, 31–41.

Beier, P. (1987). Sex differences in quality of white-tailed deer diets. *Journal of Mammalogy*, **68**, 323–9.

Beier, P. and McCullough, D.R. (1990). Factors influencing white-tailed deer activity patterns and habitat use. *Wildlife Monographs*, **109**, 1–51.

Beissinger, S.R. (1997). Integrating behavior into conservation biology: potential and limitations. In *Behavioral Approaches to Conservation in the Wild*, eds. J.R. Clemmons and R. Buchholz. Cambridge, United Kingdom: Cambridge University Press, pp. 23–47.

Béland, P., Faucher, A. and Corbeil, P. (1990). Observations on the birth of a beluga whale (*Delphinapterus leucas*) in the St. Lawrence Estuary, Quebec, Canada. *Canadian Journal of Zoology*, **68**, 1327–9.

Belcher, C.A. (2003). Demographics of tiger quoll (*Dasyurus maculatus maculatus*) populations in south-eastern Australia. *Australian Journal of Zoology*, **51**, 611–26.

Bell, R.H.V. (1969). The use of the herbaceous layer by grazing ungulates in the Serengeti National Park, Tanzania. Ph.D. thesis, University of Manchester. (1971). A grazing ecosystem in the Serengeti. *Scientific American*, **225**, 86–93.

Benham, P.F.J. (1982). Synchronization of behaviour in grazing cattle. *Applied Animal Ethology*, **8**, 403–4.

Bennett, P.M. and Owens, P.F. (2002). *Evolutionary Ecology of Birds*. Oxford: Oxford University Press.

Benshemesh, J. and Johnson, K. (2003). Biology and conservation of marsupial moles (Notoryctes). In *Predators with Pouches: The Biology of Carnivorous Marsupials*, eds. M.E. Jones, C. Dickman and M. Archer. Melbourne: CSIRO Publishing, pp. 464–74.

Bercovitch, F.B. (1983). Time budgets and consorts in olive baboons (*Papio anubis*). *Folia Primatologica*, **41**, 180–90.

(1997). Reproductive strategies of rhesus macaques. *Primates*, **38**, 247–63.

Bergan, J.F. and Smith, L.M. (1989). Differential habitat use by diving ducks wintering in South Carolina. *Journal of Wildlife Management*, **53**, 1117–26.

Berger, J. (1986). *Wild Horses of the Great Basin*. Chicago: University of Chicago Press.

(1991). Pregnancy incentives, predation constraints, and habitat shifts: experimental and field evidence for wild bighorn sheep. *Animal Behaviour*, **41**, 66–77.

Berger, J. and Cunningham, C. (1988). Size-related effects on search times in North American grassland female ungulates. *Ecology*, **69**, 177–83.

(1995). Predation, sensitivity, and sex: why female black rhinoceroses outlive males. *Behavioral Ecology*, **6**, 57–64.

Berger, J. and Gompper, M.E. (1999). Sex ratios in extant ungulates: products of contemporary predation or past life histories? *Journal of Mammalogy*, **80**, 1084–113.

Berger, J., Swenson, J.E. and Persson, I.-L. (2001). Recolonizing carnivores and naïve prey: conservation lessons from Pleistocene extinctions. *Science*, **291**, 1036–9.

Bergerud, A.T. (1974). Rutting behaviour of Newfoundland caribou. In *The Behaviour of Ungulates and its Relationship to Management*, eds. V. Giest and F. Walther, pp. 395–435. International Union for Conservation of Nature and Natural Resources: New Series, **24**, 1–940.

Bergerud, A.T., Butler, H.E. and Miller, D.R. (1984). Antipredator tactics of calving caribou: dispersion in mountains. *Canadian Journal of Zoology*, **62**, 1566–75.

Bernard, H.J. and Hohn, A.A. (1989). Differences in feeding habits between pregnant and lactating spotted dolphins (*Stenella attenuata*). *Journal of Mammalogy*, **70**, 211–15.

Bernstein, N.P. and Maxson, S.J. (1984). Sexually distinct activity patterns of blue-eyed shags in Antarctica. *Condor*, **86**, 151–6.

Berrow, S.D. and Croxall, J.P. (2001). Provisioning rate and attendance patterns of wandering albatrosses at Bird Island, South Georgia. *Condor*, **103**, 230–9.

Berrow, S.D., Humpidge, R. and Croxall, J.P. (2000). Influence of adult breeding experience on growth and provisioning of wandering albatross *Diomedea exulans* chicks at South Georgia. *Ibis*, **142**, 199–207.

Berry, J.F. and Shine, R. (1980). Sexual size dimorphism and sexual selection in turtles (Order Chelonia). *Oecologia*, **44**, 185–91.

Berta, A. (2002). Pinnipedia, overview. In *Encyclopedia of Marine Mammals*, eds. W.F. Perrin, B. Wursig and J.G.M. Thewissen. San Diego: Academic Press, pp. 931–9.

Berteaux, D. (1993). Female-biased mortality in a sexually dimorphic ungulate: feral cattle of Amsterdam Island. *Journal of Mammalogy*, **74**, 732–7.

Best, P. B. (1979). Social organization in sperm whales, *Physeter macrocephalus*. In *Behavior of Marine Mammal: Current Perspectives in Research*, vol. 3, *Cetaceans*, eds. H. E. Winn and B. L. Olla. New York: Plenum Press, pp. 227–89.

Best, R. C. and da Silva, V. M. F. (1993). *Inia geoffrensis*. *Mammalian Species*, **423**, 1–8.

Best, T. L. and Gennaro, A. L. (1984). Feeding ecology of the lizard, *Uta stansburiana*, in southwestern Mexico. *Journal of Herpetology*, **18**, 291–301.

Best, T. L. and Pfaffenberger, G. S. (1987). Age and sexual variation in the diet of collared lizards (*Crotaphytus collaris*). *Southwestern Naturalist*, **32**, 415–26.

Beuchat, C. A. (1986). Reproductive influences on the thermoregulatory behavior of a live-bearing lizard. *Copeia*, **1986**, 971–9.

Bibby, C. J. and Thomas, D. K. (1984). Sexual dimorphism in size, moult and movements of Cetti's warbler *Cettia cetti*. *Bird Study*, **31**, 28–34.

Bigg, M. A. (1973). Census of California sea lions on southern Vancouver Island, British Columbia. *Journal of Mammal Research*, **54**, 285–7.

(1990). Migration of northern for seals (*Callorhinus ursinus*) off western North America. *Canadian Technical Report of Fisheries and Aquatic Science*, **1764**, 1–64.

Bigg, M. A., Olesiuk, P. F., Ellis, G. M., Ford, J. K. B. and Balcomb III, K. C. (1990). Social organization and genealogy of resident killer whales (*Orcinus orca*) in the coastal waters of British Columbia and Washington State. *Report of the International Whaling Commission*, Special issue, **12**, 383–405.

BirdLife International (2000). *Threatened Birds of the World*. Barcelona and Cambridge, UK: Lynx edicions and BirdLife International.

Blackburn, D. G. (1985). Evolutionary origins of viviparity in the Reptilia. II. Serpentes, Amphisbaenia, and Icthyosauria. *Amphibia-Reptilia*, **6**, 259–91.

Blatchford, P., Baines, and Pellegrini, A. D. (2003). The social context of school playground games: sex and ethnic differences and changes over time after entry into junior school. *British Journal of Developmental Psychology*, **21**, 459–71.

Blaxter, J. H. S. and Holliday, F. G. T. (1969). The behaviour and physiology of herring and other clupeids. *Advances in Marine Biology*, **1**, 261–393.

Bleich, V. C., Bowyer, R. T. and Wehausen, J. D. (1997). Sexual segregation in mountain sheep: resources or predation? *Wildlife Monographs*, **134**, 1–50.

Bleich, V. C., Wehausen, J. D. and Holl, S. A. (1990). Desert-dwelling mountain sheep: conservation implications of a naturally fragmented distribution. *Conservation Biology*, **4**, 383–90.

Bleich, V. C., Wehausen, J. D., Ramey, R. R. and Rechel, J. L. (1996). Metapopulation theory and mountain sheep: implications for conservation. In *Metapopulations and Wildlife Conservation*, ed. D. R. McCullough. Covelo, California: Island Press, pp. 353–73.

Bleske, A. L. and Buss, D. M. (2000). Can men and women just be friends? *Personal Relationships*, **7**, 131–51.

Bleske-Rechek, A. L. and Buss, D. M. (2001). Opposite-sex friendship: sex differences and similarities in initiation, selection, and interpretation. *Personality and Social Psychology Bulletin*, **27**, 1310–23.

Blondel, J., Perret, P., Anstett, M.-C. and Thébaud, C. (2002). Evolution of sexual size dimorphism in birds: test of hypotheses using blue tits in contrasted Mediterranean habitats. *Journal of Evolutionary Biology*, **15**, 440–50.

Blouin-Demers, G. and Weatherhead, P. J. (2001). Thermal ecology of black rat snakes (*Elaphe obsoleta*) in a thermally challenging environment. *Ecology*, **82**, 3025–43.

Boase, H. (1926). Proportion of male and female duck on Tay estuary. *British Birds*, **20**, 169–72.

Boesch, C. and Boesch-Achermann, H. (2000). *The Chimpanzees of the Tai Forest.* Oxford: Oxford University Press.

Boinski, S. (1987). Birth synchrony in squirrel monkeys: a strategy to reduce neonatal predation. *Behavioral Ecology and Sociobiology,* **21,** 393–400.

(1988). Sex differences in the foraging behavior of squirrel monkeys. *Behavioral Ecology and Sociobiology,* **23,** 177–86.

(1994). Affiliation patterns among Costa Rican squirrel monkeys. *Behaviour,* **130,** 191–209.

(1999). The social organizations of squirrel monkeys: implications for ecological models of social evolution. *Evolutionary Anthropology,* **8,** 101–12.

Boinski, S., Sughrue, K., Selvaggi, L. *et al.* (2002). An expanded test of the ecological model of primate social evolution: competitive regimes and female bonding in three species of squirrel monkeys *(Saimiri oerstedi, Saimiri boliviensis,* and *Saimiri sciureus). Behaviour,* **139,** 227–61.

Bon, R. (1991). Social and spatial segregation of males and females in polygamous ungulates: proximate factors. In *Ongulés/Ungulates 1991,* eds. F. Spitz, G. Janeau, G. Gonzalez, and S. Aulagnier. Paris/Toulouse: SFEPM-IRGM, pp. 195–8.

(1998). Perspective éthologique de l'étude de la ségrégation sexuelle chez les ongulés. *Habilitation à Diriger les Recherches.* Université Paul Sabatier, Toulouse.

Bon, R. and Campan, R. (1989). Social tendencies of the Corsican mouflon *Ovis ammon musimon* in the Caroux-Espinouse (South of France). *Behavioural Processes,* **19,** 57–78.

(1996). Unexplained sexual segregation in polygamous ungulates: a defense of an ontogenetic approach. *Behavioural Processes,* **38,** 131–54.

Bon, R., Dubois, M. and Maublanc, M.L. (1993). Does age influence between-rams companionship in mouflon *(Ovis gmelini)? Revue d'Ecologie (Terre vie),* **47,** 57–64.

Bon, R., Joachim, J. and Maublanc, M.L. (1995). Do lambs affect feeding habitat use of lactating female mouflons in spring within an area free of predators? *Journal of Zoology, London,* **235,** 43–51.

Bon R., Rideau, C., Villaret, J.-C. and Joachim, J. (2001). Segregation is not only a matter of sex in Alpine ibex *(Capra ibex ibex). Animal Behaviour,* **62,** 495–504.

Boness, D.J., Clapham, P.J. and Mesnick, S.L. (2002). Life history and reproductive strategies. In *Marine Mammal Biology: An Evolutionary Approach,* ed. A.R. Hoelzel. Oxford: Blackwell Science, pp. 278–324.

Bonner, W.N. (1968). The fur seal of South Georgia. *British Antarctic Survey Scientific Report,* **56,** 81.

Bonnet, X., Guy, N. and Shine, R. (1999a). The dangers of leaving home: dispersal and mortality in snakes. *Biological Conservation,* **89,** 39–50.

Bonnet, X., Naulleau, G., Shine, R. and Lourdais, O. (1999b). What is the appropriate timescale for measuring costs of reproduction in a 'capital breeder' such as the aspic viper? *Evolutionary Ecology,* **13,** 485–97.

Bonnet, X., Shine, R., Naulleau, G. and Thiburce, C. (2001). Plastic vipers: genetic and environmental influences on the size and shape of Gaboon vipers, *Bitis gabonica. Journal of Zoology, London,* **255,** 341–51.

Booth, J. and Peters, J.A. (1972). Behavioural studies on the green turtle *(Chelonia mydas)* in the sea. *Animal Behaviour,* **20,** 808–12.

Bourne, W.R.P. and Warham, J. (1966). Geographical variation in the giant petrels of the genus *Macronectes. Ardea,* **54,** 45–67.

Bowen, W.D., Oftedal, O.T. and Boness, D.J. (1985). Birth to weaning in 4 days: remarkable growth in the hooded seal, *Cystophora cristata. Canadian Journal of Zoology,* **63,** 2841–6.

Bowen, W. D., Read, A. J. and Estes, J. A. (2002). Feeding ecology. In *Marine Mammal Biology: An Evolutionary Approach*, ed. A. R. Hoelzel. Oxford: Blackwell Science, pp. 217–46.

Bowers, M. A. (1994). Dynamics of age- and habitat-structured populations. *Oikos*, 69, 327–33.

Bowman, J., Jaeger, J. A. G. and Fahrig, L. (2002). Dispersal distance of mammals is proportional to home range size. *Ecology*, 83, 2049–55.

Bowyer, R. T. (1984). Sexual segregation in southern mule deer. *Journal of Mammalogy*, 65, 410–17.

Bowyer R. T., Kie, J. G. and Van Ballengerghe, V. (1996). Sexual segregation in black-tailed deer: effects of scale. *Journal of Wildlife Management*, 60, 10–17.

Bowyer, R. T., Pierce, B. M., Duffy, L. K. and Haggstrom, D. A. (2001). Sexual segregation in moose: effects of habitat manipulation. *Alces*, 37, 109–22.

Boyd, I. L. (1993a). Pup production and distribution of breeding Antarctic fur seals *Arctocephalus gazella* at South Georgia. *Antarctic Science*, 5, 17–24.

(1993b). Trends in marine mammal science. In *Marine Mammals: Advances in Behavioural and Population Biology*, ed. I. L. Boyd. Oxford: Clarendon Press, pp. 1–12.

Boyd, I. L. and Croxall, J. P. (1996). Dive durations in pinnipeds and seabirds. *Canadian Journal of Zoology*, 74, 1696–1705.

Boyd, I. L., Arnbom, T. A. and Fedak, M. A. (1994). Biomass and energy consumption of the South Georgia population of southern elephant seals. In *Elephant Seals: Population Ecology, Behaviour, and Physiology*, eds. B. J. Le Boeuf and R. M. Laws. Berkley: University of California Press.

Boyd, I. L., Lockyer, C. and Marsh, H. D. (1996). Reproduction in marine mammals. In *Biology of Marine Mammals*, eds. J. E. Reynolds and S. A. Rommel. Washington: Smithsonian Institution Press.

Boyd, I. L., McCafferty, D. J., Reid, K., Taylor, R. and Walker, T. R. (1998). Dispersal of male and female Antarctic fur seals *Arctocephalus gazella*. *Canadian Journal of Fisheries and Aquatic Sciences*, 55, 845–52.

Boyd, I. L., Staniland, I. J. and Martin, A. R. (2002). Distribution of foraging by female Antarctic fur seals. *Marine Ecology, Progress Series*, 242, 285–94.

Bradbury, J. W. (1977). Lek mating behaviour in the hammer-headed bat. *Zeitschrift für Tierpsychologie*, 45, 225–55.

Bradbury, J. W. and Vehrencamp, S. L. (1976a). Social organisation and foraging in emballonurid bats. I. Field studies. *Behavioral Ecology and Sociobiology*, 1, 337–81.

(1976b). Social organisation and foraging in emballonurid bats. II. A model for the determination of group size. *Behavioral Ecology and Sociobiology*, 1, 383–404.

(1977). Social organisation and foraging in emballonurid bats. III. Mating systems. *Behavioral Ecology and Sociobiology*, 2, 1–17.

Bradshaw, S. D. and Bradshaw, F. J. (2002). Short-term movements and habitat use of the marsupial honey possum (*Tarsipes rostratus*). *Journal of Zoology*, 258, 343–8.

Brana, F. (1993). Shifts in body temperature and escape behaviour of female *Podarcis muralis* during pregnancy. *Oikos*, 66, 216–22.

(1996). Sexual dimorphism in lacertid lizards: male head increase vs. female abdomen increase? *Oikos*, 75, 511–23.

Breitwisch, R. (1989). Mortality patterns, sex ratios and parental investment in monogamous birds. In *Current Ornithology 6*, ed. D. Power. New York and London: Plenum Press.

Bried, J. and Jouventin, P. (2002). Site and mate choice in seabirds: an evolutionary approach. In *Biology of Marine Birds*, eds. E. A. Schreiber and J. Burger. Boca Raton, Florida: CRC Press, pp. 263–305.

Brodie, E. D. I. (1989). Behavioural modification as a means of reducing the cost of reproduction. *American Naturalist*, **134**, 225–38.

Brodie, P. F. (1971). A reconsideration of aspects of growth, reproduction, and behavior of the white whale (*Delphinapterus leucas*), with reference to the Cumberland Sound, Baffin Island, population. *Journal of the Fisheries Research Board of Canada*, **28**, 1309–18.

Brooke, A. P. (1990). Tent selection, roosting ecology and social organization of the tent-making bat, *Ectophylla alba*, in Costa Rica. *Journal of Zoology, London*, **221**, 11–19.

Brooks, G. R. and Mitchell, J. C. (1989). Predator-prey size relations in three species of lizard from Sonora, Mexico. *Southwestern Naturalist*, **34**, 541–6.

Broome, L. (2001). Density, home range, seasonal movements and habitat use of the mountain pygmy-possum *Burramys parvus* (Marsupialia: Burramyidae) at Mount Blue Cow, Kosciuszko National Park. *Austral Ecology*, **26**, 275–92.

Brothers, N. (1990). Albatross mortality and associated bait loss in the Japanese longline fishery in the Southern Ocean. *Biological Conservation*, **55**, 255–68.

Brothers, N., Gales, R. and Reid, T. (1999). The influence of environmental variables and mitigation measures on seabird catch rates in the Japanese tuna longline fishery within the Australian Fishing Zone, 1991–1995. *Biological Conservation*, **88**, 85–101.

Brown, D. R., Strong, C. M. and Stouffer, P. C. (2002). Demographic effects of habitat selection by hermit thrushes wintering in a pine plantation landscape. *Journal of Wildlife Management*, **66**, 407–16.

Brown, G. P. and Brooks, R. J. (1993). Sexual and seasonal differences in activity in a northern population of snapping turtles, *Chelydra serpentina*. *Herpetologica*, **49**, 311–18.

Brown, G. P. and Shine, R. (2004). Effects of reproduction on the antipredator tactics of snakes (*Tropidonophis mairii*, Colubridae). *Behavioral Ecology and Sociobiology*, **56**, 257–62.

Brown, G. P. and Weatherhead, P. J. (2000). Thermal ecology and sexual size dimorphism in northern water snakes, *Nerodia sipedon*. *Ecological Monographs*, **70**, 311–30.

Brownell Jr, R. L. and Cipriano, F. (1989). Dusky dolphin – *Lagenorhynchus obscurus* (Gray, 1828). In *Handbook of Marine Mammals: The Second Book of Dolphins and the Porpoises*, vol. 6, eds. S. H. Ridgway and R. Harrison. London: Academic Press, pp. 85–104.

Bubenik, A. B. (1984). *Ernaehrung, Verhalten und Umwelt des Schalenwildes*. München: BLV Verlagsgesellschaft.

Bull, J. J. (1983). *Evolution of Sex Determining Mechanisms*. Menlo Park, California: Benjamin/Cummings Publ. Co.

Bull, J. J. and Charnov, E. L. (1989). Enigmatic reptilian sex ratios. *Evolution*, **43**, 1561–6.

Bullis, H. R. (1967). Depth segregations and distribution of sex-maturity groups in the marbled catshark, *Galeus arae*. In *Sharks, Skates and Rays*, eds. P. W. Gilbert, R. F. Mathewson and D. P. Rall. Baltimore, MD: Johns Hopkins University Press, pp. 141–8.

Bunnell, F. L. and Gillingham, M. P. (1985). Foraging behavior: dynamics of dining out. In *Bioenergetics of Wild Herbivores*, eds. R. J. Hudson and R. G. White. Boca Raton: CRC Press, Inc., pp. 53–75.

Bunnell, F.L. and Harestad, A.S. (1983). Dispersal and dispersion of black-tailed deer: models and observations. *Journal of Mammalogy*, **64**, 201–9.

Burger, A.E. (1991). Maximum diving depths and underwater foraging in alcids and penguins. In Studies of High Latitude Seabirds I. Behavioural, Energetic and Oceanographic Aspects of Seabird Feeding Ecology, eds. W.A. Montevecchi A.J. Gaston. *Canadian Wildlife Service Occasional Papers*, pp. 9–15.

Burgman, M.A., Ferson, S. and Akcakaya, H.R. (1993). *Risk Assessment in Conservation Biology*. London: Chapman and Hall.

Burke, R.L. and Mercurio, R.J. (2002). Food habits of a New York population of Italian wall lizards, *Podarcis sicula* (Reptilia, Lacertidae). *American Midland Naturalist*, **147**, 368–75.

Burland, T.M., Barratt, E.M., Beaumont, M.A. and Racey, P.A. (1999). Population genetic structure and gene flow in a gleaning bat, *Plecotus auritus*. *Proceedings of the Royal Society, London B*, **266**, 975–80.

Burland, T.M., Barratt, E.M., Nichols, R.A. and Racey, P.A. (2001). Mating patterns, relatedness and the basis of natal philopatry in the brown long-eared bat, *Plecotus auritus*. *Molecular Ecology*, **10**, 1309–21.

Burns, J.M., Trumble, S.J., Castellini, M.A. and Testa, J.W. (1998). The diet of Weddell seals in McMurdo Sound, Antarctica as determined from scat collections and stable isotope analysis. *Polar Biology*, **19**, 272–82.

Bury, R.B. (1972). Habits and home range of the Pacific pond turtle, *Clemmys marmorata*, in a stream community. Ph.D. Thesis, Berkeley, CA: University of California.

    (1986). Feeding ecology of the turtle, *Clemmys marmorata*. *Journal of Herpetology*, **20**, 515–21.

Busack, S.D. and Jaksic, F.M. (1982). Autecological observations of *Acanthodactylus erythrurus* (Sauria: Lacertidae) in southern Spain. *Amphibia-Reptilia*, **3**, 237–55.

Busse, C.D. (1984). Spatial structure of chacma baboon groups. *International Journal of Primatology*, **5**, 247–61.

Butler, P.J. and Taylor, E.W. (1975). The effect of progressive hypoxia on respiration in the dogfish (*Scyliorhinus canicula*) at different seasonal temperatures. *Journal of Experimental Biology*, **63**, 117–30.

Buttemer, W.A. and Dawson, W.R. (1993). Temporal pattern of foraging and microhabitat use by Galapagos marine iguanas, *Amblyrhynchus cristatus*. *Oecologia*, **96**, 56–64.

Byers, J.A. (1997a). *American Pronghorn: Social Adaptations and the Ghosts of Predators Past*. Chicago, University of Chicago Press.

    (1997b). Mortality risk to young pronghorns from handling. *Journal of Mammalogy*, **78**, 894–9.

Byers, J.A. and Walker, C. (1995). Refining the motor training hypothesis for the evolution of play. *American Naturalist*, **146**, 25–40.

Caister, L.E., Shields, W.M. and Gosser, A. (2003). Female tannin avoidance: a possible explanation for habitat and dietary segregation of giraffes (*Giraffa camelopardalis peralta*) in Niger. *African Journal of Ecology*, **41**, 201–10.

Cameron, E., Linklater, W.L., Stafford, K.J. and Minot, E.O. (2003). Social grouping and maternal behaviour in feral horses (*Equus caballus*): the influence of males on maternal protectiveness. *Behavioral Ecology and Sociobiology*, **53**, 92–101.

Cameron, E.Z. and du Toit, J.T. (2005). Social influences on vigilance behaviour in giraffes (*Giraffa camelopardalis*). *Animal Behaviour* **69**, 1337–44.

Camilleri, C. and Shine, R. (1990). Sexual dimorphism and dietary divergence: differences in trophic morphology between male and female snakes. *Copeia*, **1990**, 649–58.

Campagna, C., Werner, R., Karesh, W. *et al.* (2001). Movements and location at sea of South American sea lions (*Otaria flavescens*). *Journal of Zoology*, 2, 205–20.

Campbell, A. (1999). Staying alive: evolution, culture, and women's intrasexual aggression. *Behavioral and Brain Sciences*, 22, 203–52.

Campbell, A., Shirley, L., Heywood, C. and Crook, C. (2000). Infants' visual preference for sex-congruent babies, children, toys, and activities: a longitudinal study. *British Journal of Developmental Psychology*, 18, 479–98.

Campbell, D.W. and Eaton, W.O. (1999). Sex differences in the activity level of infants. *Infant and Child Development*, 8, 1–17.

Campredon, P. (1983). Sexe et age ratios chez le canard siffleur *Anas penelope* en periode hivernale en Europe de l'ouest. *Revue d'Ecologie (Terre et Vie)*, 37, 117–28.

Carbone, C. and Owen, M. (1995). Differential migration of the sexes of pochard *Aythya ferina*: results from a European survey. *Wildfowl*, 46, 99–108.

Caro, T. (1998). *Behavioral Ecology and Conservation Biology*. New York: Oxford University Press.

Carpenter, C.C. (1956). Body temperatures of three species of *Thamnophis*. *Ecology*, 37, 732–5.

Carrier, J.C., Pratt, H.L. and Martin, L.K. (1994). Group reproductive behaviours in free-living nurse sharks, *Ginglymostoma cirratum*. *Copeia*, 1994, 646–56.

Carson, J., Burks, V. and Parke, R. (1993). Parent-child physical play: determination and consequences. In *Parent-child Play*, ed. K. MacDonald. Albany: State University of New York Press, pp. 197–220.

Carter, S.L., Haas, C.A. and Mitchell, J.C. (1999). Home range and habitat selection of bog turtles in southwestern Virginia. *Journal of Wildlife Management*, 63, 853–60.

Catry, P., Campos, A., Almada, V. and Cresswell, W. (2004). Winter segregation of migrant European robins *Erithacus rubecula* in relation to sex, age and size. *Journal of Avian Biology*, 35, 204–9.

Catry, P., Phillips, R.A. and Furness, R.W. (1999). Evolution of reversed sexual size dimorphism in skuas and jaegers. *Auk*, 116, 158–68.

Caughley, G. and Sinclair, A.R.E. (1994). *Wildlife Ecology and Management*. Cambridge, MA: Blackwell Science.

Cederlund, B.-M. (1987). Parturition and early development of moose (*Alces alces* L.) calves. *Swedish Wildlife Research Supplement*, 1, 399–422.

Cederlund, G. and Sand, H.K.G. (1992). Dispersal of subadult moose (*Alces alces*) in nonmigratory population. *Canadian Journal of Zoology*, 70, 1309–14.

Cederlund, G., Sandegren, F. and Larsson, K. (1987). Summer movements of female moose and dispersal of their offspring. *Journal of Wildlife Management*, 51, 342–52.

Censky, E.J. (1995). The evolution of sexual size dimorphism in the teiid lizard *Ameiva plei*. *Behaviour*, 132, 529–57.

Cerling, T.E., Harris, J.M. and Passey, B.H. (2003). Diets of East African Bovidae based on stable isotope analysis. *Journal of Mammalogy*, 84, 456–70.

Chambers (The) Encyclopedic English Dictionary. (1994). Edinburgh: Chambers.

Chapman, C.A. (1990). Association patterns of spider monkeys: the influence of ecology and sex on social organization. *Behavioral Ecology and Sociobiology*, 26, 409–14.

Chapman, C.A. and Wrangham, R.W. (1993). Range use of the forest chimpanzees of Kibale: implications for the understanding of chimpanzee social organization. *American Journal of Primatology*, 31, 263–73.

Chapman, C.A., Wrangham, R.W. and Chapman, L.J. (1995). Ecological constraints on group size: an analysis of spider monkey and chimpanzee subgroups. *Behavioral Ecology and Sociobiology*, **32**, 199–209.

Charland, M.B. and Gregory, P.T. (1990). The influence of female reproductive status on thermoregulation in a viviparous snake, *Crotalus viridis. Copeia*, **1990**, 1089–98.

Chase, J.D., Dixon, K.R., Gates, J.E., Jacobs, D. and Taylor, G.J. (1989). Habitat characteristics, population size, and home range of the bog turtle, *Clemmys muhlenbergii*, in Maryland. *Journal of Herpetology*, **23**, 356–62.

Cheney, D.L. (1978). The play partners of immature baboons. *Animal Behaviour*, **26**, 1038–50.

Chism, J. and Rowell, T. (1986). Mating and residence patterns of male patas monkeys (*Erythrocebus patas*). *Ethology*, **103**, 109–26.

Chivers, D.J. (1974). *The Siamang in Malaysia*. Basel: S. Karger.

Chivers, D.P., Brown, G.E. and Smith, R.J.F. (1995). Familiarity and shoal cohesion in fathead minnows (*Pimephales promelas*) – implications for antipredator behaviour. *Canadian Journal of Zoology – Revue Canadienne De Zoologie*, **73**, 955–60.

Choudhury, S. and Black, J.M. (1991). Testing the behavioural dominance and dispersal hypothesis in pochard. *Ornis Scandinavica*, **22**, 155–9.

Christal, J. and Whitehead, H. (1997). Aggregations of mature male sperm whales on the Galápagos Islands breeding ground. *Marine Mammal Science*, **13**, 59–69.

Christal, J., Whitehead, H. and Lettevall, E. (1998). Sperm whale social units: variation and change. *Canadian Journal of Zoology*, **76**, 1431–40.

Ciofi, C. and Chelazzi, G. (1994). Analysis of homing pattern in the colubrid snake *Coluber viridiflavus*. *Journal of Herpetology*, **28**, 477–84.

Clapham, P.J. (1996). The social and reproductive biology of humpback whales: an ecological perspective. *Mammal Review*, **26**, 27–49.

Clarke, J., Manly, B., Kerry, K.R. *et al.* (1998). Sex differences in Adélie penguin foraging strategies. *Polar Biology*, **20**, 248–58.

Clauss, M., Frey, R., Kiefer, B. *et al.* (2003). The maximum attainable body size of herbivorous mammals: morphophysiological constraints on foregut, and adaptations of hindgut fermenters. *Oecologia*, **136**, 14–27.

Clemmons, J.R. and Buchholz, R. (1997). *Behavioural Approaches to Conservation in the Wild*. Cambridge: Cambridge University Press.

Clinton, W.L. (1994). Sexual selection and growth in male northern elephant seals. In *Elephant Seals: Population Ecology, Behaviour and Physiology*, eds. B.J. Le Boeuf and R.M. Laws. Berkley: University of California Press, pp. 154–68.

Clutton-Brock, J., Dennis-Bryan, K., Armitage, P.L. and Jewell, P.A. (1990). Osteology of the Soay sheep. *Bulletin of the British Museum of Natural History*, **56**, 1–56.

Clutton-Brock, T.H. (1977). Some aspects of intra-specific variation in feeding and ranging behaviour in primates. In *Primate Ecology*, ed. T.H. Clutton-Brock. Cambridge: Cambridge University Press, pp. 539–56.

(1983). Selection in relation to sex. In *Evolution: From Molecules to Men*, ed. D.S. Bendall: Cambridge University Press, pp. 457–81.

(1988). Reproductive success. In *Reproductive Success*, ed. T.H. Clutton-Brock. Chicago: University of Chicago Press, pp. 472–85.

(1989). Mammalian mating systems. *Proceedings of the Royal Society of London, Series B*, **236**, 339–72.

Clutton-Brock, T.H., Albon, S.D. and Guinness F.E. (1985). Parental investment and sex differences in juvenile mortality in birds and mammals. *Nature*, **313**, 131–3.

(1987b). Interactions between population density and maternal characteristics affecting fecundity and juvenile survival in red deer (Cervus elaphus). Journal of Animal Ecology, 56, 857–71.

(1988). Reproductive success in male and female red deer. In Reproductive Success, ed. T. H. Clutton-Brock. Chicago: University of Chicago Press, pp. 325–43.

Clutton-Brock, T. H., Guinness, F. E. and Albon, S. D. (1982). Red Deer: behavior and ecology of two sexes. Chicago: University of Chicago Press.

Clutton-Brock, T. H., Iason, G. R. and Guinness, F. E. (1987a). Sexual segregation and density-related changes in habitat use in male and female red deer (Cervus elaphus). Journal of Zoology, London, 211, 275–89.

Clutton-Brock, T. H., Major, M. and Guinness, F. E. (1985). Population regulation in male and female red deer. Journal of Animal Ecology, 54, 831–46.

Cochran, P. A. and McConville, D. R. (1983). Feeding by Trionyx spiniferus in backwaters of the upper Mississippi River. Journal of Herpetology, 17, 82–6.

Cockburn, A. and Lazenby-Cohen, K. A. (1992). Use of nest trees by Antechinus stuartii a semelparous lekking marsupial. Journal of Zoology, London, 226, 657–80.

Cockcroft, V. G. and Ross, G. J. B. (1990). Food and feeding of the Indian Ocean bottlenose dolphin off Southern Natal, South Africa. In The Bottlenose Dolphin, eds. S. Leatherwood and R. R. Reeves. San Diego: Academic Press, pp. 295–308.

Cogger, H. G. (2000). Reptiles and Amphibians of Australia. Sydney: Reed New Holland.

Collar, N. J. (1996). Family Otididae (Bustards). In Handbook of the Birds of the World, vol. 3., eds. J. del Hoyo, A. Elliot and J. Sargatal. Barcelona: Lynx Edicions.

Compagno, L. J. V. (1984). FAO Species Catalogue. Volume 4 Sharks of the World, Parts 1 and 2. Rome: Food and Agriculture Organization of the United Nations.

(1999). Checklist of living elasmobranchs. In Sharks, Skates and Rays: The Biology of Elasmobranch Fishes, ed. W. C. Hamlett. Baltimore, MD: Johns Hopkins University Press, pp. 471–98.

Connor, R. C., Mann, J., Tyack, P. L. and Whitehead, H. (1998). Social evolution in toothed whales. Trends in Ecology and Evolution, 13, 228–32.

Connor, R. C., Read, A. J. and Wrangham, R. (2000). Male reproductive strategies and social bonds. In Cetacean Societies: Field Studies of Dolphins and Whales, eds. J. Mann, R. C. Connor, P. L. Tyack and H. Whitehead. Chicago: University of Chicago Press, pp. 247–70.

Connor, R. C., Richards, A. F., Smolker, R. A. and Mann, J. (1996). Patterns of female attractiveness in Indian Ocean bottlenose dolphins. Behaviour, 133, 37–69.

Connor, R. C., Smolker, R. A. and Richards, A. F. (1992). Two levels of alliance formation among male bottlenose dolphins (Tursiops sp.). Proceedings of the Academy of Science of United States, 89, 987–90.

Conradt, L. (1997). Causes of sex differences in habitat use in red deer (Cervus elaphus, L.). Ph.D. thesis, University of Cambridge.

(1998a). Could asynchrony in activity between the sexes cause intersexual social segregation in ruminants? Proceedings of the Royal Society of London, Series B, Biological Sciences, 265, 1359–63.

(1998b). Measuring the degree of sexual segregation in group-living animals. Journal of Animal Ecology, 67, 217–26.

(1999). Social segregation is not a consequence of habitat segregation in red deer and feral Soay sheep. Animal Behaviour, 57, 1151–7.

(2000). Use of a seaweed habitat by red deer (*Cervus elaphus* L.). *Journal of Zoology*, **250**, 541–9.

Conradt, L. and Roper, T. J. (2000). Activity synchrony and social cohesion: a fisson-fusion model. *Proceedings of the Royal Society of London, Series B*, **267**, 2213–18.

(2003). Group decision-making in animals. *Nature*, **421**, 155–8.

Conradt, L., Clutton-Brock, T. H. and Guinness, F. E. (1999a). The relationship between habitat choice and lifetime reproductive success in female red deer. *Oecologia*, **120**, 218–24.

(2000). Sex differences in weather sensitivity can cause habitat segregation: red deer as an example. *Animal Behaviour*, **59**, 1049–60.

Conradt, L., Clutton-Brock, T. H. and Thomson, D. (1999b). Habitat segregation in ungulates: are males forced into suboptimal foraging habitats through indirect competition by females? *Oecologia*, **119**, 367–77.

Conradt, L., Gordon, I. J., Clutton-Brock, T. H., Thomson, D. and Guinness, F. E. (2001). Could the indirect competition hypothesis explain inter-sexual site segregation in red deer (*Cervus elaphus* L.)? *Journal of Zoology, London*, **254**, 185–93.

Conroy, J. W. H. (1972). Ecological aspects of the biology of the giant petrel *Macronectes giganteus* (Gmelin) in the maritime Antarctic. *Scientific Report of the British Antarctic Survey*, **75**, 1–74.

Constable, A. J. and Nicol, S. (2002). Defining smaller-scale management units to further develop the ecosystem approach in managing large-scale pelagic krill fisheries in Antarctica. *CCAMLR Science*, **9**, 117–31.

Conway, C. J., Powell, G. V. N. and Nichols, J. D. (1995). Overwinter survival of neotropical migratory birds in early-successional and mature tropical forests. *Conservation Biology*, **9**, 855–64.

Cooper, J., Brooke, M. L., Burger, A. E., Crawford, R. J. M., Hunter, S. and Williams, T. (2001). Aspects of the breeding biology of the Northern Giant Petrel (*Macronectes halli*) and the Southern Giant Petrel (*M. giganteus*) at sub-Antarctic Marion Island. *International Journal of Ornithology*, **4**, 53–68.

Cooper, J., Henley, S. and Klages, N. (1992). The diet of the wandering albatross *Diomedea exulans* at sub-Antarctic Marion Island. *Polar Biology*, **12**, 477–84.

Cooper, J., Wilson, R. P. and Adams, N. J. (1993). Timing of foraging by the wandering albatross *Diomedea exulans*. *Proceedings National Institute Polar Research Symposium on Polar Biology*, **6**, 55–61.

Cooper, W. E. (2003). Sexual dimorphism in distance from cover but not escape behavior by the keeled earless lizard *Holbrookia propinqua*. *Journal of Herpetology*, **37**, 374–8.

Cords, M. (1984). Mating patterns and social structure in redtail monkeys (*Cercopithecus ascanius*). *Zietschrift für Tierpsycholgie*, **64**, 213–39.

Corfield, T. (1973). Elephant mortality in Tsavo National Park, Kenya. *East African Wildlife Journal*, **11**, 339–68.

Cork, S. J. (1991). Meeting the energy-requirements for lactation in a macropodid marsupial – current nutrition versus stored body reserves. *Journal of Zoology*, **225**, 567–76.

Corkeron, P. J. and Connor, R. C. (1999). Why do baleen whales migrate? *Marine Mammal Science*, **15**, 1228–45.

Côté, S. D., Schaefer, J. A. and Messier, F. (1997). Time budgets and synchrony of activities in muskoxen: the influence of sex, age, and season. *Canadian Journal of Zoology*, **75**, 1628–35.

Coulson, G. (1993). The influence of population density and habitat on grouping in the western grey kangaroo, *Macropus fuliginosus*. *Wildlife Research*, **20**, 151–62.

Coulson, J.C. (2002). Colonial breeding in seabirds. In *Biology of Marine Birds*, eds. E.A. Schreiber and J. Burger. Boca Raton, Florida: CRC Press.

Couzin, I.D. and Krause, J. (2003). Self-organization and collective behaviour in vertebrates. *Advances in the Study of Behaviour*, **32**, 1–75.

Craig, M.J. (1992). Radio-telemetry and tagging study of movement patterns, activity cycles, and habitat utilization in Cagle's map turtle, *Graptemys caglei*. M.Sc. thesis, Canyon, TX: West Texas State University.

Cransac, N., Gerard, J.-F., Maublanc, M.L. and Pépin, D. (1998). An example of segregation between age and sex classes only weakly related to habitat use in mouflon sheep (*Ovis gmelini*). *Journal of Zoology, London*, **244**, 371–8.

Crawley, M.C. (1973). A live-trapping study of Australian brush-tailed possums, *Trichosurus vulpecula* (Kerr), in the Orongorongo Valley, Wellington, New Zealand. *Australian Journal of Zoology*, **21**, 75–90.

Creer, S., Chou, W.H., Malhotra, A. and Thorpe, R.S. (2002). Offshore insular variation in the diet of the Taiwanese bamboo viper *Trimeresurus stejnegeri* (Schmidt). *Zoological Science*, **19**, 907–13.

Cristol, D.A., Baker, M.B. and Carbone, C. (1999). Differential migration revisited. Latitudinal segregation by age and sex class. In *Current Ornithology*, vol. 15, eds. V. Nolan Jr, E.D. Ketterson and C.F. Thompson. New York: Kluwer Academic/Plenum Publishers.

Croft, D.B. (1981). Social behaviour of the euro, *Macropus robustus* (Gould), in the Australian Arid Zone. *Australian Wildlife Research*, **8**, 13–49.

(1987). Socio-ecology of the antilopine wallaroo, *Macropus antilopinus*, in the Northern Territory, with observations on sympatric M. *robustus woodwardii* and M. *agilis*. *Australian Wildlife Research*, **14**, 243–55.

(1989). Social organisation of the Macropodoidea. In *Kangaroos, Wallabies and Rat-kangaroos*, eds. G. Grigg, P. Jarman and I. Hume. New South Wales, Australia: Surrey Beatty and Sons, pp. 505–25.

(1991a). Home range of the euro, *Macropus robustus erubescens*. *Journal of Arid Environments*, **20**, 99–111.

(1991b). Home range of the red kangaroo, *Macropus rufus*. *Journal of Arid Environments*, **20**, 83–98.

Croft, D.P., Arrowsmith, B.J., Bielby, J. et al. (2003). Mechanisms underlying shoal composition in the Trinidadian guppy (*Poecilia reticulata*). *Oikos*, **100**, 429–38.

Croft, D.P., Botham, M.S. and Krause, J. (2004). Is sexual segregation in the guppy, *Poecilia reticulata*, consistent with the predation risk hypothesis? *Environmental Biology of Fishes*, **71**(2), 127–33.

Crow, J.F. and Kimura, M. (1970). *An Introduction to Population Genetics Theory*. New York: Harper and Row.

Croxall, J.P. (1991). Constraints on reproduction in albatrosses. *Proceedings of the 20th International Ornithological Congress*, 281–302.

(1995). Sexual size dimorphism in seabirds. *Oikos*, **73**, 399–403.

(1998). Research and conservation: a future for albatrosses? In *Albatross Biology and Conservation*, eds. G. Robertson and R. Gales. Chipping Norton, Australia: Surrey Beatty and Sons, pp. 267–88.

Croxall, J.P. and Gales, R. (1998). An assessment of the conservation status of albatrosses. In *Albatross Biology and Conservation*, eds. G. Robertson and R. Gales. Chipping Norton, Australia: Surrey Beatty and Sons, pp. 46–65.

Croxall, J.P. and Prince, P.A. (1990). Recoveries of wandering albatrosses *Diomedea exulans* ringed at South Georgia (1958–1986). *Ringing and Migration*, **11**, 43–51.

418   References

(1996). Potential interactions between wandering albatrosses and longline fisheries for Patagonian toothfish at South Georgia. *CCAMLR Science*, **3**, 101–10.

Croxall, J.P., Black, A.D. and Wood, A.G. (1999). Age, sex and status of wandering albatrosses *Diomedea exulans* L. in Falkland Islands waters. *Antarctic Science*, **11**, 150–6.

Croxall, J.P., Prince, P.A. and Reid, K. (1997). Dietary segregation of krill-eating South Georgia seabirds. *Journal of Zoology, London*, **242**, 531–56.

Croxall, J.P., Prince, P.A., Rothery, P. and Wood, A.D. (1998). Population changes in albatrosses at South Georgia. In *Albatross Biology and Conservation*, eds. G. Robertson and R. Gales. Chipping Norton: Surrey Beatty and Sons, pp. 69–83.

Croxall, J.P., Rothery, P., Pickering, S.P.C. and Prince, P.A. (1990). Reproductive performance, recruitment and survival of wandering albatrosses *Diomedea exulans* at Bird Island, South Georgia. *Journal of Animal Ecology*, **59**, 775–96.

Croxall, J.P., Silk, J.R.D., Phillips, R.A., Afanasyev, V. and Briggs, D.R. (2005). Global circumnavigations: tracking year-round ranges of nonbreeding albatrosses. *Science*, **307**, 249–50.

Cryan, P.M., Bogan, M.A. and Altenbach, J.S. (2000). Effect of elevation on distribution of female bats in the Black Hills, South Dakota. *Journal of Mammalogy*, **81**, 719–25.

Cuadrado, M. (1997). Why are migrant Robins *Erithacus rubecula* territorial in winter?: the importance of the anti-predation behaviour. *Ethology, Ecology and Evolution*, **9**, 77–88.

Curlewis, J.D. (1989). The breeding season of Bennett's wallaby (*Macropus rufogriseus rufogriseus*) in Tasmania. *Journal of Zoology, London*, **218**, 337–9.

Cutter, J. (2002). Dietary Segregation in the Western Grey Kangaroo, *Macropus fuliginosus*, at Hattah-Kulkyne National Park. Unpublished honours thesis, University of Melbourne.

D'Onghia, G., Matarrese, A., Tursi, A. and Sion, L. (1995). Observations on the depth distribution pattern of the small-spotted catshark in the north Aegean Sea. *Journal of Fish Biology*, **47**, 421–6.

Dahlheim, M.E. and Heyning, J.E. (1999). Killer whale – *Orcinus orca* (Linnaeus, 1758). In *Handbook of Marine Mammals: The Second Book of Dolphins and the Porpoises*, vol. 6, eds. S.H. Ridgway and R. Harrison. London: Academic Press, pp. 281–322.

Dalrymple, G.H. (1977). Intraspecific variation in the cranial feeding mechanism of the turtles of the genus *Trionyx* (Reptilia, Testudines, Trionychidae). *Journal of Herpetology*, **11**, 255–85.

Daltry, J.C., Wuster, W. and Thorpe, R.S. (1998). Intraspecific variation in the feeding ecology of the crotaline snake *Calloselasma rhodostoma* in Southeast Asia. *Journal of Herpetology*, **32**, 198–205.

Dalziell, J. and de Poorter, M. (1993). Seabird mortality in longline fisheries around South Georgia. *Polar Record*, **29**, 143–5.

Daneri, G.A. and Coria, N.R. (1992). The diet of Antarctic fur seals, *Arctocephalus gazella*, during the summer-autumn period at Mossman Peninsula, Laurie Island (South Orkneys). *Polar Biology*, **11**, 565–6.

Darwin, C. (1871). *The Decent of Man and Selection in Relation to Sex*, London: Murray.

Daunt, F., Monaghan, P., Wanless, S., Harris, M.P. and Griffiths, R. (2001). Sons and daughters: age-specific differences in parental rearing capacities. *Functional Ecology*, **15**, 211–16.

Daut, E.F. and Andrews, R.M. (1993). The effect of pregnancy on thermoregulatory behavior of the viviparous lizard *Chalcides ocellatus*. *Journal of Herpetology*, **27**, 6–13.

Davis, L.S. and Speirs, E.A.H. (1990). Mate choice in penguins. In *Penguin Biology*, eds. S. Davis and J.T. Darby. San Diego, California: Academic Press, pp. 377–97.

Davis, R.B., Herreid, C.F. and Short, H.L. (1962). Mexican free-tailed bats in Texas. *Ecological Monographs*, **32**, 311–46.

Dawbin, W.H. (1982). The tuatara *Sphenodon punctatus* (Reptilia: Rhynchocephalia): a review. In *New Zealand Herpetology*, ed. D. G. Newman. Wellington, New Zealand: New Zealand Wildlife Service, pp. 149–81.

Day, T., O'Connor, C. and Matthews, L. (2000). Possum social behaviour. In *The Brushtail Possum: Biology, Impact and Management of an Introduced Marsupial*, ed. T.L. Montague. Lincoln, NZ: Manaaki Whenua Press, pp. 35–46.

Deere, J.A. (2001). The use of black rhinoceros (*Diceros bicornis*) dung sampling to investigate sexual differences in diet quality and midden site selection in Pilanesberg National Park. Unpublished dissertation, University of Pretoria.

del Hoyo, J. (1994). Family Cracidae (Chachalacas, Guans and Curassows). In *Handbook of the Birds of the World*, vol. 2., eds. J. del Hoyo, A. Elliot and J. Sargatal. Barcelona: Lynx Edicions.

Delgado, R.A. and van Schaik, C.P. (2000). The behavioral ecology and conservation of the orangutan (*Pongo pygmaeus*): a tale of two islands. *Evolutionary Anthropology*, **9**, 201–18.

Demment, M.W. and van Soest, P.J. (1985). A nutritional explanation for body-size patterns of ruminant and nonruminant herbivores. *American Naturalist*, **125**, 641–72.

Demski, L.S. and Northcutt, R.G. (1996). The brain and cranial nerves of the white shark: an evolutionary perspective. In *Great White Sharks: The Biology of Carcharodon carcharias*, eds. A.P. Klimley and D.G. Ainley. San Diego, CA: Academic Press, pp. 121–30.

Deneubourg, J.L. and Goss, S. (1989). Collective patterns and decision making. *Ethology, Ecology and Evolution*, **1**, 295–311.

Deneubourg, J.L., Goss, S., Franks, N. *et al.* (1991). The dynamics of collective sorting robot-like ants and ant-like robots. In *Simulation of Adaptive Behavior: From Animals to Animats*, eds. J.A. Meyer and S.W. Wilson. Cambridge, MA: MIT Press/Bradford Books, pp. 356–65.

Dennis, A.J. and Marsh, H. (1997). Seasonal reproduction in musky rat-kangaroos, *Hypsiprymnodon moschatus*: a response to changes in resource availability. *Wildlife Research*, **24**, 561–78.

Desrochers, A. (1989). Sex, dominance, and microhabitat use in wintering black-capped chickadees: a field experiment. *Ecology*, **70**, 636–45.

Deutsch, C.J., Crocker, D.E., Costa, D.P. and Le Boeuf, B.J. (1994). Sex- and age-related variation in reproductive effort of northern elephant seals. In *Elephant Seals: Population Ecology, Behaviour and Physiology*, eds. B.J. Le Boeuf and R.M. Laws. Berkeley: University of California Press, pp. 169–210.

Deutsch, C.J., Haley, M.P. and Leboeuf, B.J. (1990). Reproductive effort of male northern elephant seals – estimates from mass-loss. *Canadian Journal of Zoology*, **68**, 2580–93.

Di Bitetti, M.S. and Janson, C.H. (2000). When will the stork arrive? Patterns of birth seasonality in neotropic primates. *American Journal of Primatology*, **50**, 109–30.

Dickman, C.R. (1995). Agile Antechinus. In *The Mammals of Australia*, ed. R. Strahan. Sydney: Reed New Holland, pp. 99–101.

Diego-Rasilla, F.J. and Perez-Mellado, V. (2003). Home range and habitat selection by *Podarcis hispanica* (Squamata, Lacertidae) in western Spain. *Folia Zoologica*, **52**, 87–98.

DiFiore, A. and Rodman, P. S. (2001). Time allocation patterns of lowland woolly monkeys (*Lagothrix lagotricha poepigii*) in a neotropical terra firma forest. *International Journal of Primatology*, **22**, 449–80.

Dobson, A. and Poole, J. (1998). Conspecific aggregation and conservation biology. In *Behavioral Ecology and Conservation Biology*, ed. T. Caro. New York: Oxford University Press, pp. 193–208.

Dodd C. K., Jr, Enge, K. M. and Stuart, J. N. (1988). Aspects of the biology of the flattened musk turtle, *Sternotherus depressus*, in northern Alabama. *Bulletin of the Florida State Museum, Biological Sciences Series*, **34**, 1–64.

Dodd, J. M. (1983). Reproduction in cartilaginous fishes (Chondrichthyes). In *Fish Physiology, Vol. 9 – Reproduction: Endocrine Tissues and Hormones*, eds. W. S. Hoar, D. J. Randall and D. M. Donaldson. New York: Academic Press, pp. 31–87.

Doidge, D. W. (1990). Integumentary heat loss and blubber distribution in the beluga, *Delphinapterus leucas*, with comparisons to narwhal, *Monodon monoceros*. *Canadian Bulletin of Fisheries and Aquatic Sciences*, **224**, 129–40.

Doidge, D. W., McCann, T. S. and Croxall, J. P. (1986). Attendance behavior of Antarctic fur seals. In *Fur Seals: Maternal Strategies on Land and at Sea*, eds. R. L. Gentry and G. L. Kooyman. Princeton: Princeton University Press, pp. 102–14.

Doody, J. S., Young, J. E. and Georges, A. (2002). Sex differences in activity and movements in the pig-nosed turtle, *Carettochelys insculpta*, in the wet-dry tropics of Australia. *Copeia*, **2002**, 93–103.

Dressen, W. (1993). On the behaviour and social organisation of agile wallabies, *Macropus agilis* (Gould, 1842) in two habitats of northern Australia. *Zeitschrift für Säugetierkunde*, **58**, 201–11.

Du, W. G., Yan, S. J. and Ji, X. (2000). Selected body temperature, thermal tolerance and thermal dependence of food assimilation and locomotor performance in adult blue-tailed skinks, *Eumeces elegans*. *Journal of Thermal Biology*, **25**, 197–202.

du Toit, J. T. (1988). Patterns of resource use within the browsing ruminant guild in the central Kruger National Park. Unpublished Ph.D. thesis, Johannesburg: University of the Witwatersrand.

(1990). Feeding height stratification among African browsing ruminants. *African Journal of Ecology*, **28**, 55–61.

(1995). Sexual segregation in kudu: sex differences in competitive ability, predation risk or nutritional needs? *South African Journal of Wildlife Research*, **25**, 127–32.

(2003). Large herbivores and savanna heterogeneity. In *The Kruger Experience: Ecology and Management of Savanna Heterogeneity*, eds. J. T. du Toit, K. H. Rogers and H. C. Biggs. Washington, DC: Island Press, pp. 292–309.

Dubois, M., Bon, R., Cransac, N. and Maublanc, M. L. (1994). Dispersion patterns among ewes of Corsican mouflon: importance and proximate influences. *Applied Animal Behaviour Science*, **42**, 29–40.

Dubois, M., Quenette, P. Y., Bideau, E. and Magnac, M. P. (1993). Seasonal range use by European mouflon rams in medium altitude mountains. *Acta Theriology*, **38**, 185–98.

Duellman, W. E. and Trueb, L. (1986). *Biology of Amphibians*. New York: McGraw-Hill.

Dufault, S. and Whitehead, H. (1995a). An encounter with recently wounded sperm whales (*Physeter macrocephalus*). *Marine Mammal Science*, **11**, 560–3.

(1995b). The geographic stock structure of female and immature sperm whales in the South Pacific. *Report of the International Whaling Commission*, **45**, 401–5.

Dulvy, N. K. and Reynolds, J. D. (1997). Evolutionary transitions among egg-laying, live-bearing and maternal inputs in sharks and rays. *Proceedings of the Royal Society of London B*, **264**, 1309–15.

Dunbar, R. I. M. (1977). Feeding ecology of gelada baboons: a preliminary report. In *Primate Ecology*, ed. T. H. Clutton-Brock. Cambridge: Cambridge University Press, pp. 251–73.

(1984). *Reproductive Decisions: An Economic Analysis of Gelada Baboon Social Organization*. Princeton: Princeton University Press.

Dunbar, R. I. M. and Dunbar, P. (1988). Maternal time budgets of gelada baboons. *Animal Behaviour*, **76**, 970–80.

Dunn, D. G., Barco, S. G., Pabst, D.-A. and McLellan, W. A. (2002). Evidence of infanticide in bottlenose dolphins of the Western North Atlantic. *Journal of Wildlife Diseases*, **38**, 505–10.

Durell, S. E. A. Le V. dit (2000). Individuals feeding specialisation in shorebirds: population consequences and conservation implications. *Biological Reviews of the Cambridge Philosophical Society*, **75**, 503–18.

Durell, S. E. A. Le V. dit and Goss-Custard, J. D. (1996). Oystercatcher *Haematopus ostralegus* sex ratios on the wintering grounds: the case of the Exe estuary. *Ardea*, **84A**, 373–81.

Durell, S. E. A. Le V. dit, Goss-Custard, J. D. and Caldow, R. W. G. (1993). Sex-related differences in diet and feeding method in the oystercatcher *Haematopus ostralegus*. *Journal of Animal Ecology*, **62**, 205–15.

Durell, S. E. A. Le V. dit, Goss-Custard, J. D., Caldow, R. W. G., Malcolm, H. M. and Osborn, D. (2001a). Sex, diet and feeding method-related differences in body condition in the oystercatcher *Haematopus ostralegus*. *Ibis*, **143**, 107–19.

Durell, S. E. A. Le V. dit, Goss-Custard, J. D. and Clarke, R. T. (2001b). Modelling the population consequences of age- and sex-related differences in winter mortality in the oystercatcher, *Haematopus ostralegus*. *Oikos*, **95**, 69–77.

Durell, S. E. A. Le V. dit, Ormerod, S. J. and Dare, P. J. (1996). Differences in population structure between two oystercatcher *Haematopus ostralegus* roosts on the Burry Inlet, South Wales. *Ardea*, **84A**, 383–8.

Durner, G. M. and Gates, J. E. (1993). Spatial ecology of black rat snakes on Remington Farms, Maryland. *Journal of Wildlife Management*, **57**, 812–26.

Durtsche, R. D. (1992). Feeding time strategies of the fringe-toed lizard, *Uma inornata*, during breeding and non-breeding seasons. *Oecologia*, **89**, 85–9.

Duvall, D., King, M. B. and Gutzwiller, M. J. (1985). Behavioral ecology and ethology of the prairie rattlesnake. *National Geographic Research*, **1**, 80–111.

Dwyer, D. P. (1966). The population pattern of *Miniopterus schreibersii* (Chiroptera) in north-eastern New South Wales. *Australian Journal of Zoology*, **14**, 1073–137.

Eaton, W. C. and Enns, L. R. (1986). Sex differences in human motor activity level. *Psychology Bulletin*, **100**, 19–28.

Eaton, W. C. and Yu, A. (1989). Are sex function of sex differences in maturational status? *Child Development*, **60**, 1005–11.

Eaton, W. C., Enns, L. R. and Presse, M. (1987). Scheme for observing activity. *Journal of Psychoeducational Assessment*, **3**, 273–80.

Eberle, M. and Kappeler, P. (2002). Mouse lemurs in space and time: a test of the socioecological model. *Behavioral Ecology and Sociobiology*, **51**, 131–9.

Ebert, D. A. (2002). Ontogenetic changes in the diet of the sevengill shark (*Notorynchus cepedianus*). *Marine and Freshwater Research*, **53**, 517–23.

Economakis, A. E. and Lobel, P. S. (1998). Aggregation behaviour of the grey reef shark, *Carcharhinus amblyrhynchos*, at Johnston Atoll, Central Pacific Ocean. *Environmental Biology of Fishes*, **51**, 129–39.

Edwards, J. (1983). Diet shifts in moose due to predator avoidance. *Oecologia*, **60**, 185–9.

Eifler, D. A. and Eifler, M. A. (1999a). Foraging behavior and spacing patterns of the lizard *Oligosoma grande*. *Journal of Herpetology*, **33**, 632–9.

(1999b). The influence of prey distribution on the foraging strategy of the lizard *Oligosoma grande* (Reptilia: Scincidae). *Behavioral Ecology and Sociobiology*, **45**, 397–402.

Eisenberg, J. F. (1981). *The Mammalian Radiations: An Analysis of Trends in Evolution, Adaptation, and Behavior*. Chicago: The University of Chicago Press.

Ellis, J. R., Pawson, M. G. and Shackley, S. E. (1996). The comparative feeding ecology of six species of shark and four species of ray (*Elasmobranchii*) in the north-east Atlantic. *Journal of the Marine Biological Association of the United Kingdom*, **76**, 89–106.

Emlen, S. T. and Oring, L. W. (1977). Ecology, sexual selection, and the evolution of mating systems. *Science*, **197**, 215–23.

Ena, V., Lucio, A. and Purroy, F. J. (1985). The great bustard in Leon, Spain. *Bustard Studies*, **2**, 35–52.

Endler, J. A. (1980). Natural selection on colour patterns in *Poecillia reticulata*. *Evolution*, **34**, 76–91.

Entwistle, A. C., Racey, P. A. and Speakman, J. R. (1996). Habitat exploitation by a gleaning bat, *Plecotus auritus*. *Philosophical Transactions of the Royal Society London, B*, **351**, 921–31.

(2000). Social and population structure of a gleaning bat, *Plecotus auritus*. *Journal of Zoology, London*, **252**, 11–17.

Ernst, C. H., Lovich, J. E. and Barbour, R. W. (1994). *Turtles of the United States and Canada*. Washington, D.C.: Smithsonian Institution Press.

Estep, D. Q., Crowell-Davis, S. L., Earl-Costello, S.-A. and Beatey, S. A. (1993). Changes in the social behaviour of drafthorse (*Equus caballus*) mares coincident with foaling. *Applied Animal Behaviour Science*, **35**, 199–213.

Estes, R. D. (1991a). *The Behavior Guide to African Mammals: including hoofed mammals, carnivores, primates*. Berkley: University of California Press.

(1991b). The significance of horns and other male secondary sexual characters in female bovids. *Applied Animal Behaviour Science*, **29**, 403–51.

Fabes, R. A. (1994). Physiological, emotional, and behavioral correlates of gender segregation. In *Childhood Gender Segregation: Causes and Consequences*, ed. C. Leaper. San Francisco: Jossey-Bass, pp. 33–50.

Fagot, B. I. (1994). Peer relations and the development of competence in boys and girls. In *Childhood Gender Segregation: Causes and Consequences*, ed. C. Leaper. San Francisco: Jossey-Bass, pp. 53–66.

Fairbanks, L. A. (1993). Juvenile vervet monkeys: establishing relationships and practicing skills for the future. In *Juvenile Primates*, eds. M. E. Pereira and L. A. Fairbanks. Oxford: Oxford University Press, pp. 211–27.

Fashing, P. (2001). Activity and ranging patterns of guerezas in the Kakamega Forest: inter-group variation and implications for intra-group feeding competition. *International Journal of Primatology*, **22**, 549–78.

Feare, C. J., Gill, E. L., McKay, H. V. and Bishop, J. D. (1995). Is the distribution of starlings *Sturnus vulgaris* within roosts determined by competition? *Ibis*, **137**, 379–82.

Fein, G. (1981). Pretend play in childhood: an integrative review. *Child Development*, **52**, 1095–118.

Feldheim, K. A., Gruber, S. H. and Ashley, M. V. (2002). The breeding biology of lemon sharks at a tropical nursery lagoon. *Proceedings of the Royal Society of London B*, **269**, 1655–61.

Fenton, M.B. (1990). The foraging behaviour and ecology of animal-eating bats. *Canadian Journal of Zoology*, **68**, 411–22.

Fenton, M.B., Rautenbach, I.L., Smith, S.E., Swanepoel, C.M., Grosell, J. and van Jaarsveld, J. (1994). Raptors and bats: threats and opportunities. *Animal Behaviour*, **48**, 9–18.

Festa-Bianchet, M. (1986). Site fidelity and seasonal range use by bighorn rams. *Canadian Journal of Zoology*, **64**, 2126–32.

(1988). Seasonal range selection in bighorn sheep: conflicts between forage quality, forage quantity, and predator avoidance. *Oecologia*, **75**, 580–6.

(1991). The social system of bighorn sheep: grouping patterns, kinship and female dominance rank. *Animal Behaviour*, **42**, 71–82.

Festa-Bianchet, M. and Apollonio, M. (2003). *Animal Behavior and Wildlife Conservation*. Washington, D.C.: Island Press.

Fietz, C. (1999). Mating systems of *Microcebus murinus*. *American Journal of Primatology*, **48**, 127–33.

Fisher, D.O. and Owens, I.P.F. (2000). Female home range size and the evolution of social organization in macropod marsupials. *Journal of Animal Ecology*, **69**, 1083–98.

Fisher, D.O., Blomberg, S.P. and Owens, I.P.F. (2002). Convergent maternal care strategies in ungulates and macropods. *Evolution*, **56**, 167–76.

Fitch, H.S. (1960). Autecology of the copperhead. *University of Kansas Publications, Museum of Natural History*, **13**, 85–288.

(1982). Resources of a snake community in prairie-woodland habitat of northeastern Kansas. In *Herpetological Communities*, ed. N.J. Scott Jr. Washington, D.C.: U.S. Department of the Interior, Fish and Wildlife Service, pp. 83–98.

(1999). *A Kansas Snake Community: Composition and Changes Over 50 Years*. Malabar, Florida: Krieger.

Fitch, H.S. and Shirer, H.W. (1971). A radiotelemetric study of spatial relationships in some common snakes. *Copeia*, **1971**, 118–28.

Fleming, M.R. and Frey, H. (1984). Aspects of the natural history of Feathertail Gliders *Acrobates pygmaeus* (Marsupialia: Burramyidae) in Victoria. In *Possums and Gliders*, eds. A.P. Smith and I.D. Hume. Chipping Norton, NSW: Surrey Beatty and Sons, pp. 403–8.

Fleming, T.H. and Eby, P. (2003). Ecology of bat migration. In *Bat Ecology*, eds. T.H. Kunz and M.B. Fenton. Chicago: University of Chicago Press, pp. 156–208.

Fleming, T.H. and Hooker, R.S. (1975). *Anolis cupreus*: the response of a lizard to tropical seasonality. *Ecology*, **56**, 1243–61.

Fletcher, T. and Selwood, L. (2000). Possum reproduction and development. In *The Brushtail Possum: Biology, Impact and Management of an Introduced Marsupial*, ed. T.L. Montague. Lincoln, NZ: Manaaki Whenua Press, pp. 62–81.

Ford, D. (1999). Foraging ecology and demography of *Sternotherus odoratus* in a southwestern Missouri population. M.Sc. thesis, Springfield: Southwest Missouri State University.

Ford, E. (1921). A contribution to our knowledge of the life-histories of the dogfishes landed at Plymouth. *Journal of the Marine Biological Association of the United Kingdom*, **12**, 468–505.

Forero, M.G. and Hobson, K.A. (2003). Using stable isotopes of Nitrogen and Carbon to study seabird ecology: applications in the Mediterranean seabird community. *Scientia Marina*, **67**, 23–32.

Forero, M.G., Hobson, K.A., Bortolotti, G.R. *et al.* (2002). Food resource utilisation by the Magellanic penguin evaluated through stable-isotope analysis: segregation by sex and age and influence on offspring quality. *Marine Ecology Progress Series*, **234**, 289–99.

Fox, S.F., Conder, J.M. and Smith, A.E. 1998. Sexual dimorphism in the case of tail autotomy: *Uta stansburiana* with and without previous tail loss. *Copeia*, **1998**, 376–82.

Fragaszy, D.M. and Boinski, S. (1995). Patterns of individual diet choice and efficiency of foraging of wedge-capped capuchin monkeys (*Cebus olivaceous*). *Journal of Comparative Psychology*, **109**, 339–48.

Fraker, M.A., Gordon, C.D., McDonald, J.W., Ford, J.K.B. and Cambers, G. (1979). White whale (*Delphinapterus leucas*) distribution and abundance and the relationship to physical and chemical characteristics of the Mackenzie Estuary. *Canadian Fisheries and Marine Service Technical Report*, **863**, v–56.

Francisci, F., Focardi, S. and Boitani, L. (1985). Male and female Alpine ibex: phenology of space use and herd size. In *The Biology and Management of Mountain Ungulates*, ed. S. Lovari. London: Croom Helm, pp. 124–33.

Freake, M.J. (1998). Variation in homeward orientation performance in the sleepy lizard (*Tiliqua rugosa*): effects of sex and reproductive period. *Behavioral Ecology and Sociobiology*, **43**, 339–44.

Frid, A. (1999). Huemul (*Hippocamelus bisulcus*) sociality at a periglacial site: sexual segregation and habitat effects on group size. *Canadian Journal of Zoology*, **77**, 1083–91.

Friend, J. (1989). Myrmecobiidae: In *Fauna of Australia*, eds. D. Walton and B. Richardson. Canberra: Australian Government Publishing Service, pp. 583–90.

(1995). Numbat: In *The Mammals of Australia*, ed. R. Strahan. Sydney: Reed New Holland, pp. 160–2.

(1996). Numbats on a junk food diet. *Nature Australia*, **29**, 40–9.

Friend, J. and Whitford, D. (1993). Maintenance and breeding of the numbat (*Myrmecobius fasciatus*) in captivity. In *Biology and Management of Australasian Carnivorous Marsupials*, eds. M. Roberts, J. Carnio, G. Crawshaw and M. Hutchins. Toronto: Metropolitan Toronto Zoo and Monotreme and Marsupial Advisory Group of AAZPA, pp. 103–24.

Frischknecht, M. (1993). The breeding coloration of male 3-spined sticklebacks (*Gasterosteus aculeatus*) as an indicator of energy investment in vigor. *Evolutionary Ecology*, **7**, 439–50.

Frost, K.J., Russel, R.B. and Lowry, L.F. (1992). Killer whales, *Orcinus orca*, in the Southeastern Bearing Sea: recent sightings and predation on other marine mammals. *Marine Mammal Science*, **8**, 110–19.

Furness, R.W. (1993). Birds as monitors of pollutants. In *Birds as Monitors of Environment Change*, eds. R.W. Furness and J.J.D. Greenwood. London: Chapman and Hall, pp. 86–143.

Gadsen, H. and Palacios-Orona, L.E. (1997). Seasonal dietary patterns of the Mexican fringe-toed lizard (*Uma paraphygas*). *Journal of Herpetology*, **31**, 1–9.

Galan, P. (1999). Demography and population dynamics of the lacertid lizard *Podarcis bocagei* in north-west Spain. *Journal of Zoology, London*, **249**, 203–18.

Galdikas, B. (1988). Orangutan diet, range, and activity at Tanjung Putting, Central Borneo. *International Journal of Primatology*, **9**, 1–35.

Gales, N. (2002). New Zealand sea lion. In *Encyclopedia of Marine Mammals*, eds. W.F. Perrin, B. Wursig and J.G.M. Thewissen. San Diego: Academic Press.

Gales, R. (1998). Albatross populations: status and threats. In *Albatross Biology and Conservation*, eds. G. Robertson and R. Gales. Chipping Norton, Australia: Surrey Beatty and Sons, pp. 20–45.

(1993). *Co-operative Mechanisms for the Conservation of Albatrosses*. Hobart, Australia: Australian Nature Conservation Agency and Australian Antarctic Foundation.

Gales, R., Brothers, N. and Reid, T. (1998). Seabird mortality in the Japanese tuna longline fishery around Australia. *Biological Conservation*, **86**, 37–56.

Galois, P., Leveille, M., Bouthillier, L., Daigle, C. and Parren, S. (2002). Movement patterns, activity and home range of the eastern spiny softshell turtle (*Apalone spinifera*) in northern Lake Champlain, Quebec, Vermont. *Journal of Herpetology*, **36**, 402–11.

Gannon, V.P.J. and Secoy, D.M. (1985). Seasonal and daily activity patterns in a Canadian population of the prairie rattlesnake, *Crotalus viridis viridis*. *Canadian Journal of Zoology*, **63**, 86–91.

Garrick, L.D. (1974). Reproductive influences on behavioral thermoregulation in the lizard, *Sceloporus cyanogenys*. *Physiology and Behaviour*, **12**, 85–91.

Garstka, W.R., Camazine, B. and Crews, D. (1982). Interactions of behavior and physiology during the annual reproductive cycle of the red-sided garter snake (*Thamnophis sirtalis parietalis*). *Herpetologica*, **38**, 104–23.

Garton, J.S. and Dimmick, R.W. (1969). Food habits of the copperhead in middle Tennessee. *Journal of the Tennessee Academy of Science*, **44**, 113–17.

Gates, S. (2002). Review of methodology of quantitative reviews using meta-analysis in ecology. *Journal of Animal Ecology*, **71**, 547–57.

Gauthreaux, S.A. Jr (1978). The ecological significance of behavioural dominance. *Perspectives in Ethology*, **3**, 17–54.

(1982). The ecology and evolution of avian migration systems. In *Avian Biology* Vol. 6, eds. D.S. Farner, J.R. King and K.C. Parkes. New York: Academic Press.

Gautier-Hion, A. (1980). Seasonal variations of diet related to species and sex in a community of Cercopithecus monkeys. *Journal of Animal Ecology*, **49**, 237–69.

Gehrt, S.D. and Fritzell, E.K. (1996). Sex-biased response of raccoons (*Procyon lotor*) to live traps. *American Midland Naturalist*, **135**, 23–32.

Geisler, J.H. and Uhen, M.D. (2003). Morphological support for a close relationship between hippo and whales. *Journal of Vertebrate Paleontology*, **23**, 991–6.

Geist, V. (1968). On delayed social and physical maturation in mountain sheep. *Canadian Journal of Zoology*, **46**, 899–904.

(1971). *Mountain Sheep: a Study in Behavior and Evolution*. Chicago: University Chicago Press.

(1974). On the relationship of social evolution and ecology in ungulates. *American Zoologist*, **14**, 205–20.

Geist, V. and Bromley, P.T. (1978). Why deer shed antlers. *Zeitschrift für Säugetierkunde – International Journal of Mammalian Biology*, **43**, 223–31.

Geist, V. and Petocz, R.G. (1977). Bighorn sheep in winter: do rams maximize reproductive fitness by spatial and habitat segregation from ewes? *Canadian Journal of Zoology*, **55**, 1802–10.

Gerard, J.-F. and Loisel, P. (1995). Spontaneous emergence of a relationship between habitat openness and mean group size and its possible evolutionary consequences in large herbivores. *Journal of Theoretical Biology*, **176**, 511–22.

Gerard, J.-F., Bideau, E., Maublanc, M.-L., Loisel, P. and Marchal, C. (2002). Herd size in large herbivores: encoded in the individual or emergent? *Biological Bulletin*, **202**, 275–82.

Gerard, J.-F., Le Pendu, Y., Maublanc, M.-L., Vincent, J.-P., Poulle, M.-L. and Cibien, C. (1995). Large group formation in European roe deer, an adaptive feature? *Revue d'Ecologie (Terre Vie)*, **50**, 391–401.

Gerpe, M.S., de Moreno, J.E., Moreno, V.J. and Patat, M.L. (2000). Cadmium, zinc and copper accumulation in the squid *Illex argentinus* from the Southwest Atlantic Ocean. *Marine Biology*, **136**, 1039–44.

Gibbons, J.W. (1986). Movement patterns among turtle populations: applicability to management of the desert tortoise. *Herpetologica*, **42**, 104–13.

Gibson, R. and Falls, J.B. (1979). Thermal biology of the common garter snake *Thamnophis sirtalis* (L.). 1. Temporal variation, environmental effects and sex differences. *Oecologia*, **43**, 79–97.

Gilardi, J.D. (1992). Sex-specific foraging distributions of brown boobies in the eastern tropical Pacific. *Colonial Waterbirds*, **15**, 148–51.

Gillis, R. (1991). Thermal biology of two populations of red-chinned lizards (*Sceloporus undulatus erythrocheilus*) living in different habitats in south-central Colorado. *Journal of Herpetology*, **25**, 18–23.

Gillooly, J.F., Brown, J.H., West, G.B., Savage, V.M. and Charnov, E.L. (2001). Effects of size and temperature on metabolic rate. *Science*, **293**, 2248–51.

Gilmore, R.G., Dodrill, J.W. and Linley, P.A. (1983). Reproduction and embryonic development of the sand tiger shark, *Odontaspis taurus. U.S. Fishery Bulletin*, **81**, 201–25.

Ginnett, T.F. and Demment, M.W. (1997). Sex differences in giraffe foraging behavior at two spatial scales. *Oecologia*, **110**, 291–300.

(1999). Sexual segregation by Masai giraffes at two spatial scales. *African Journal of Ecology*, **37**, 93–106.

Gjerde, I. (1991). Cues in winter habitat selection by capercaillie. I. Habitat characteristics. *Ornis Scandinavica*, **22**, 197–204.

Goldfoot, D.A., Wallen, K., Neff, K., McBrair, D.A. and Goy, M.C. (1984). Social influences on the display of sexually dimorphic behavior in rhesus monkeys: isosexual rearing. *Archives of Sexual Behaviour*, **13**, 395–412.

Goldingay, R.L. and Kavanagh, R.P. (1990). Socioecology of the yellow-bellied glider (*Petaurus australis*) at Waratah Creek, NSW. *Australian Journal of Zoology*, **38**, 327–41.

Goldsworthy, S. (1995). Differential expenditure of maternal resources in Antarctic fur seals, *Arctocephalus gazella*, at Heard Island, southern Indian Ocean. *Behavioral Ecology*, **6**, 218–28.

Gompper, M.E. (1996). Sociality and asociality in white-nosed coatis (*Nasua narica*): foraging costs and benefits. *Behavioral Ecology*, **7**, 254–63.

González-Solís, J. (2004a). Sexual size dimorphism in northern giant petrels: ecological correlates and scaling. *Oikos*, **105**(2), 247–54.

(2004b). The regulation of incubation shifts near hatching: a timed mechanism, embryonic signalling or food availability? *Animal Behaviour*, **67**, 663–71.

González-Solís, J., Croxall, J.P. and Briggs, D.R. (2002a). Activity patterns of giant petrels *Macronectes* spp. using different foraging strategies. *Marine Biology*, **140**, 197–204.

González-Solís, J., Croxall, J.P. and Wood, A.G. (2000a). Sexual dimorphism and sexual segregation in foraging strategies of northern giant petrels, *Macronectes halli*, during incubation. *Oikos*, **90**, 390–8.

(2000b). Foraging partitioning between giant petrels *Macronectes* spp. and its relationship with breeding population changes at Bird Island, South Georgia. *Marine Ecology Progress Series*, **204**, 279–288.

González-Solís, J., Sanpera, C. and Ruiz, X. (2002b). Metals and selenium as bioindicators of geographic and trophic segregation in giant petrels *Macronectes* spp. *Marine Ecology Progress Series*, **244**, 257–64.

Goodall, J. (1986). *The Chimpanzees of Gombe*. Cambridge, MA: Harvard University Press.

Goodman-Lowe, G.D. (1998). Diet of the Hawaiian monk seal (*Monachus schauinslandi*) from the Northwestern Hawaiian islands during 1991 to 1994. *Marine Biology*, **132**, 535–46.

Gordon, D.M. and MacCulloch, R.D. (1980). An investigation of the ecology of the map turtle, *Graptemys geographica* (Le Seur), in the northern part of its range. *Canadian Journal of Zoology*, **58**, 2210–19.

Gordon, I.J. and Illius, A.W. (1988). Incisor arcade structure and diet selection in ruminants. *Functional Ecology*, **2**, 15–22.

Gosler, A.G. (1987). Pattern and process in the bill morphology of the great tit *Parus major*. *Ibis*, **129**, 451–76.

(1990). The variable niche hypothesis revisited; an analysis of intra- and inter-specific differences in bill variation in *Parus*. In *Population Biology of Passerine Birds*, eds. J. Blondel, A.G. Gosler, J. Lebreton and R. McCleery. Berlin: Springer-Verlag, pp. 167–74.

(1991). On the use of greater covert moult and pectoral muscle as measures of condition in passerines with data for the great tit *Parus major*. *Bird Study*, **31**, 1–9.

(1996). Environmental and social determinants of winter fat storage in the great tit *Parus major*. *Journal of Animal Ecology*, **65**, 1–17.

Gosler, A.G. and Carruthers, T.D. (1994). Bill size and niche breadth in the Irish coal tit *Parus ater hibernicus*. *Journal of Avian Biology*, **25**, 171–7.

(1999). Body reserves and social dominance in the great tit *Parus major* in relation to winter weather in Southwest Ireland. *Journal of Avian Biology*, **30**, 447–59.

Gosling, L.M. (1969). Parturition and related behaviour in Coke's Hartebeest, *Alcelaphus buselaphus cokei* Günther. *Journal of Reproduction and Fertility, Supplement*, **6**, 265–86.

Gosling L.M. and Sutherland, W.J. (2000). *Behaviour and Conservation*. Cambridge, United Kingdom: Cambridge University Press.

Gottman, J.M. (1983). How children become friends. *Monographs of the Society for Research in Child Development*, **48** (3) Serial No. 201.

Gowans, S., Whitehead, H. and Hooker, S.K. (2001). Social organization in northern bottlenose whales, *Hyperoodon ampullatus*: not driven by deep-water foraging? *Animal Behaviour*, **62**, 369–77.

Graham, T.E. and Graham, A.A. (1992). Metabolism and behavior of wintering common map turtles, *Graptemys geographica*, in Vermont. *Canadian Field-Naturalist*, **104**, 517–19.

Grand Usuel Larousse (1997). Paris: Larousse-Bordas.

Grant, G.S. and Banak, S.A. (1995). Harem structure and reproductive behaviour of *Pteropus tonganus* in American Samoa. Department of Marine and Wildlife Research, American Samoan Government. *Biological Reports*, **69**, 214–44.

Grassi, C. (2002). Sex differences in feeding, height, and space use in *Hapalemur griseus*. *International Journal of Primatology*, **23**, 677–93.

Graves, B.M. and Duvall, D. (1987). An experimental study of aggregation and thermoregulation in prairie rattlesnakes (*Crotalus viridis viridis*). *Herpetologica*, **43**, 259–64.

(1995). Aggregation of squamate reptiles associated with gestation, oviposition, and parturition. *Herpetological Monographs*, **9**, 102–19.

Greenberg, R. (1986). Competition in migrant birds in the nonbreeding season. In *Current Ornithology 3*, ed. R.F. Johnston. New York and London: Plenum Press, pp. 281–307.

Greene, H.W. (1997). *Snakes: The Evolution of Mystery in Nature*. Berkeley, CA: University of California Press.

Gregory, P.T. (1974). Patterns of spring emergence of the red-sided garter snake (*Thamnophis sirtalis parietalis*) in the Interlake region of Manitoba. *Canadian Journal of Zoology*, **52**, 1063–9.

Gregory, P.T. and Isaac, L.A. (2004). Food habits of the grass snake in southeastern England: is *Natrix natrix* a generalist predator? *Journal of Herpetology*, **38**, 88–95.

Gregory, P.T., Crampton, L.H. and Skebo, K.M. (1999). Conflicts and interactions among reproduction, thermoregulation and feeding in viviparous reptiles: are gravid snakes anorexic? *Journal of Zoology, London*, **248**, 231–41.

Gregory, P. T., Macartney, J. M. and Larsen, K. W. (1987). Spatial patterns and movements. In *Snakes: Ecology and Evolutionary Biology*, eds. R.A. Seigel, J.T. Collins and S.S. Novak. New York: Macmillan, pp. 366–95.

Griffiths, R. (1992). Sex-biased mortality in the Lesser Black-backed Gull *Larus fuscus* during the nestling stage. *Ibis*, **134**, 237–44.

Griffiths, S. W. (2003). Learned recognition of conspecifics by fishes. In *Fish are Smarter than You Think: Learning in Fishes*, eds. C. Brown, K.N. Laland and J. Krause. *Fish and Fisheries, Special edn.*, **4**, 256–68.

Griffiths, S.W. and Magurran, A.E. (1998). Sex and schooling behaviour in the Trinidadian guppy. *Animal Behaviour*, **56**, 698–93.

Grindal, S.D., Morissette, J.L. and Brigham, R.M. (1999). Concentration of bat activity in riparian habitats over an elevational gradient. *Canadian Journal of Zoology*, **77**, 972–7.

Gross, J.E. (1998). Sexual segregation in ungulates: a comment. *Journal of Mammalogy*, **79**, 1404–9.

Gross, J.E., Alkon, P.U. and Demment, M.W. (1996). Nutritional ecology of dimorphic herbivores: digestion and passage rates in Nubian ibex. *Oecologia*, **107**, 170–8.

(1995a). Grouping patterns and spatial segregation by Nubian ibex. *Journal of Arid Environments*, **30**, 423–39.

Gross, J.E., Demment, M.W., Alkon, P.U. and Kotzman, M. (1995b). Feeding and chewing behaviours of Nubian ibex: compensation for sex-related differences in body size. *Functional Ecology*, **9**, 385–93.

Grubb, P. (1974). Mating activity and the social significance of rams in a feral sheep community. In *The Behaviour of Ungulates and its Relation to Management*, vol. 2, eds. V. Geist and F. Walther. Morges, Switzerland: IUCN No 50, pp. 457–76.

Grubb, P. and Jewell, P.A. (1966). Social grouping and home range in feral Soay sheep. *Symposium of Zoology, Society of London*, **18**, 179–210.

Grubb, T.C. and Woodrey, M.S. (1990). Sex, age, intraspecific dominance status, and the use of food by birds wintering in temperate-deciduous and cold-coniferous woodlands: a review. *Studies in Avian Biology*, **13**, 270–9.

Gruber, S.H., Nelson, D.R. and Morrissey, J.F. (1988). Patterns of activity and space utilization of lemon sharks, *Negaprion brevirostris*, in a shallow Bahamian lagoon. *Bulletin of Marine Science*, **43**, 61–76.

Gruys, R.C. (1993). Autumn and winter movements and sexual segregation of willow ptarmigan. *Arctic*, **46**, 228–39.

Gueron, S., Levin, S.A. and Rubenstein, D.I. (1996). The dynamics of herds: from individuals to aggregations. *Journal of Theoretical Biology*, **182**, 85–98.

Gustin, K. and McCracken, G.F. (1987). Scent recognition in the Mexican free-tailed bat, *Tadarida brasiliensis mexicana*. *Animal Behaviour*, **35**, 13–19.

Hamilton, I.A. and Barclay, R.M.R. (1994). Patterns of daily torpor and day-roost selection by male and female big brown bats (*Eptesicus fuscus*). *Canadian Journal of Zoology*, **72**, 744–9.

Hamilton, W.D. (1971). Geometry of the selfish herd. *Journal of Theoretical Biology*, **31**, 295–311.

Hammerson, G.A. (1978). Observations on the reproduction, courtship, and aggressive behavior of the striped racer, *Masticophis lateralis euryxanthus* (Reptilia, Serpentes, Colubridae). *Journal of Herpetology*, **12**, 253–5.

Hammond, K.A., Spotila, J.R. and Standora, E.A. (1988). Basking behavior of the turtle *Pseudemys scripta*: effects of digestive state, acclimation temperature, sex, and season. *Physiological Zoology*, **61**, 69–77.

Handasyde, K.A. and Martin, R.W. (1996). Field observations on the common striped possum (*Dactylopsila trivirgata*) in North Queensland. *Wildlife Research*, **23**, 755–66.

Hanley, T.A. (1982). The nutritional basis for food selection by ungulates. *Journal of Range Management*, **35**, 146–51.

Hanski, I. and Gilpin, M. (1991). Metapopulation dynamics: brief history and conceptual domain. *Biological Journal of the Linnean Society*, **42**, 3–16.

Haramis, G.M., Nichols, J.D., Pollock, K.H. and Hines, J.E. (1986). The relationship between body mass and survival of wintering canvasbacks. *Auk*, **103**, 506–14.

Harestad, A.S. and Bunnell, F.L. (1979). Home range and body weight – a reevaluation. *Ecology*, **60**, 389–402.

Harper, P.C. (1987). Feeding behaviour and other notes on 20 species of Procellariiformes at sea. *Notornis*, **34**, 169–92.

Harrel, J.B., Allen, C.M. and Hebert, S.J. (1996). Movements and habitat use of subadult alligator snapping turtles (*Macroclemys temminckii*) in Louisiana. *American Midland Naturalist*, **135**, 60–7.

Harris, J.E. (1952). A note on the breeding season, sex ratio and embryonic development of the dogfish *Scyliorhinus canicula* (L.). *Journal of the Marine Biological Association of the United Kingdom*, **31**, 269–70.

Hart, D.R. (1983). Dietary and habitat shift with size of red-eared turtles (*Pseudemys scripta*) in a southern Louisiana population. *Herpetologica*, **39**, 285–90.

Hashimoto, C., Furuichi, T., Tashiro, Y. (2001). What factors influence the size of chimpanzee parties in the Kalinzu Forest, Uganda? Examination of fruit abundance and number of estrous females. *International Journal of Primatology*, **22**, 947–59.

Hasegawa, T. (1990). Sex differences in ranging patterns. In *The Chimpanzees of the Mahale Mountains*, ed. T. Nishida. Tokyo: University of Tokyo Press, pp. 99–114.

Hass, C.C. and Jenni, D.A. (1993). Social play among juvenile bighorn sheep: structure, development, and relationship to adult behavior. *Ethology*, **93**, 105–16.

Hauksson, E. and Bogason, V. (1997). Comparative feeding of grey (*Halichoerus grypus*) and common seals (*Phoca vitulina*) in coastal waters of iceland, with a note on the diet of hooded (*Cystophora cristata*) and harp seals (*Phoca groenlandica*). *Journal of Northwest Atlantic Fishery Science*, **22**, 125–35.

Hayes, C.L., Rubin, E.S., Jorgensen, M.C., Botta, R.A. and Boyce, W.M. (2000). Mountain lion predation of bighorn sheep in the Peninsular Ranges, California. *Journal of Wildlife Management*, **64**, 954–9.

Heatwole, H. (1968). Relationship of escape behavior and camouflage in anoline lizards. *Copeia*, **1968**, 109–13.

Hebrard, J.J. and Madsen, T. (1984). Dry season intersexual habitat partitioning by flap-necked chameleons (*Chamaeleo dilepis*) in Kenya. *Biotropica*, **16**, 69–72.

Hedd, A., Gales, R. and Brothers, N. (2001). Foraging strategies of shy albatross *Thalassarche cauta* breeding at Albatross Island, Tasmania, Australia. *Marine Ecology Progress Series*, **224**, 267–82.

Hedrick, A.V. and Temeles, E.J. (1989). The evolution of sexual dimorphism in animals: hypotheses and tests. *Trends in Ecology and Evolution*, **4**, 136–8.

Heinsohn, G. E. (1966). Ecology and reproduction of the Tasmanian bandicoots (*Perameles gunni* and *Isoodon obesulus*). *University of California Publications in Zoology*, **80**, 1–96.

Helfman, G. S., Collette, B. B. and Facey, D. E. (1997). *The Diversity of Fishes*. Malden, MA: Blackwell Science.

Henderson, R. W. (1993). Foraging and diet in West Indian *Corallus enydris* (Serpentes: Boidae). *Journal of Herpetology*, **27**, 24–8.

Henderson, R. W. and Binder, M. H. (1980). The ecology and behavior of vine snakes (*Ahaetulla, Oxybelis, Thelotornis, Uromacer*): a review. *Milwaukee Public Museum Contributions in Biology and Geology*, **37**, 1–38.

Henry, S. R. (1984). Social organization of the greater glider (*Petauroides volans*) in Victoria. In *Possums and Gliders*, eds. A. P. Smith and I. D. Hume. Chipping Norton, NSW: Surrey Beatty and Sons, pp. 221–8.

Hepp, G. R. and Hair, J. D. (1984). Dominance in wintering waterfowl (*Anatini*): effects on distribution of sexes. *Condor*, **86**, 251–7.

Heppel, S. S., Walters, J. R. and Crowder, L. B. (1994). Evaluating management alternatives for red-cockaded woodpeckers: a management approach. *Journal of Wildlife Management*, **58**, 479–87.

Herbinger, I., Boesch, C. and Rothe, H. (2001). Territory characteristics among three neighboring chimpanzee communities in the Taï National Park, Côte d'Ivoire. *International Journal of Primatology*, **22**, 143–67.

Herremans, M. (1997). Habitat segregation of male and female red-backed shrikes *Lanius collurio* and lesser grey shrikes *Lanius minor* in the Kalahari basin, Botswana. *Journal of Avian Biology*, **28**, 240–8.

Hertz, P. E., Huey, R. B. and Stevenson, R. D. (1993). Evaluating temperature regulation by field-active ectotherms: the fallacy of the inappropriate question. *American Naturalist*, **142**, 796–818.

Heulin, B., Surget-Groba, Y., Guiller, A., Guillaume, C. P. and Deunff, J. (1999). Comparisons of mitochondrial DNA (mtDNA) sequences (16S rRNA gene) between oviparous and viviparous strains of *Lacerta vivipara*: a preliminary study. *Molecular Ecology*, **8**, 1627–31.

Hickey, M. B. C. and Fenton, M. B. (1990). Foraging by red bats (*Lasiurus borealis*) – do intraspecific chases mean territoriality? *Canadian Journal of Zoology*, **68**, 2477–82.

Hickling, C. F. (1930). A contribution towards the life-history of the spur-dog. *Journal of the Marine Biological Association of the United Kingdom*, **16**, 529–76.

Hillman, J. C. (1987). Group size and association patterns of the common eland (*Tragelaphus oryx*). *Journal of Zoology, London*, **213**, 641–63.

Hinch, G. N., Lynch, J. J., Elwin, R. L. and Green, G. C. (1990). Long-term associations between Merino ewes and their offspring. *Applied Animal Behaviour Science*, **27**, 93–103.

Hinde, R. A. (1974). Biological bases of human social behaviour. New York: McGraw-Hill.

Hirth, D. H. (1977). Social behavior of white-tailed deer in relation to habitat. *Wildlife Monographs*, **53**, 1–55.

Hjelm, J. and Persson, L. (2001). Size-dependent attack rate and handling capacity: inter-cohort competition in zooplantivorous fish. *Oikos*, **95**, 520–32.

Hjelset, A. M., Andersen, M., Gjertz, I., Lydersen, C. and Gulliksen, B. (1999). Feeding habits of bearded seals (*Erignathus barbatus*) from the Svalbard area, Norway. *Polar Biology*, **21**, 186–93.

Hobson, E. S. (1968). Predatory behaviour of some shore fishes in the Gulf of California. *United States Bureau of Sport Fisheries and Wildlife Research Report*, **73**, 1–91.

Hobson, K.A., Piatt, J.F. and Pitocchelli, J. (1994). Using stable isotopes to determine seabird trophic relationships. *Journal of Animal Ecology*, **63**, 786–98.

Hobson, K.A., Sease, J.L., Merrick, R.L. and Piatt, J.F. (1997). Investigating trophic
relationships of pinnipeds in Alaska and Washington using stable isotope
ratios of nitrogen and carbon. *Marine Mammal Science*, **13**, 114–32.

Hocking, G.J. (1981). The population ecology of the brush-tail possum, *Trichosurus
vulpecula* (Kerr) in Tasmania. Unpublished Masters thesis, University of
Tasmania.

Hodum, P.J. and Hobson, K.A. (2000). Trophic relationships among Antarctic ful-
marine petrels: insights into dietary overlap and chick provisioning strategies inferred from stable-isotope (d15N and d13C) analyses. *Marine Ecology
Progress Series*, **198**, 273–81.

Hoek, W. (1992). An unusual aggregation of harbor porpoises (*Phocoena phocoena*).
Marine Mammal Science, **8**, 152–4.

Hoelzel, A.R., Le Boeuf, B.J., Reiter, J. and Campagna, C. (1999). Alpha-male paternity in elephant seals. *Behavioral Ecology and Sociobiology*, **46**, 298–306.

Hofmann, R.R. (1989). Evolutionary steps of ecophysiological adaptation and
diversification of ruminants: a comparative view of their digestive system.
Oecologia, **78**, 443–57.

Hoge, A.R. and Federsoni, P.A.J. (1977). Observations on a brood of *Bothrops
atrox* (Linnaeus, 1758): (Serpentes: Viperidae: Crotalinae). *Memorias do Instituto
Butantan (Sao Paulo)*, **40/41**, 19–36.

Hogg, J.T. (1984). Mating in bighorn sheep: multiple creative male strategies.
Science, **225**, 526–9.

Hohn, A.A. and Brownell, R.L. (1990). Harbor porpoise in central Californian
waters: life history and incidental catches. Paper SC/42/SM47 presented at
42nd meeting of the scientific committee, International Whaling Commission, Nordwijk, Holland.

Holekamp, K.E. and Sherman, P.W. (1989). Why do male ground squirrels
disperse? *American Scientist*, **77**, 232–9.

Holmes, R.T. (1986). Foraging patterns of forest birds: male-female differences.
Wilson Bulletin, **98**, 196–213.

Hölzenbein, S. and Marchinton, L. (1992). Spatial integration of maturing-male
white-tailed deer into the adult population. *Journal of Mammalogy*, **73**, 326–34.

Honda, K., Yamamoto, Y. and Tatsukawa, R. (1987). Distribution of heavy metals
in Antarctic marine ecosystems. *Proceedings of the NIPR Symposium of Polar
Biology*, **1**, 184–97.

Hooge, P.N. and Eichenlaub, B. (1997). Animal movement extension to Arcview
ver. 1.1. Alaska Science Center – Biological Science Center, U.S. Geological
Survey, Anchorage, AK, USA.

Horak, I.G., Boomker, J. and Flamand, J.R.B. (1995). Parasites of domestic and
wild animals in South Africa. XXXIV. Arthropod parasites of nyalas in north-
eastern KwaZulu-Natal. *Onderstepoort Journal of Veterinary Research*, **62**, 171–9.

Houde, A.E. (1997). *Sex, Color, and Mate Choice in Guppies*. Princeton: Princeton
University Press.

Houston, D.L. and Shine, R. (1993). Sexual dimorphism and niche divergence:
feeding habits of the Arafura filesnake. *Journal of Animal Ecology*, **62**, 737–49.

(1994). Population demography of Arafura filesnakes (Serpentes, Acrochordi-
dae) in tropical Australia. *Journal of Herpetology*, **28**, 273–80.

How, R.A. (1972). The ecology and management of *Trichosurus* species (*Marsupialia*)
in New South Wales. Unpublished Ph.D. thesis, University of New England.

How, R.A., Barnett, J.L., Bradley, A.J., Humphreys, W.F. and Martin, R. (1984). The
population biology of *Pseudocheirus peregrinus* in a *Leptospermum laevigatum*

thicket. In *Possums and Gliders*, eds. A.P. Smith and I.D. Hume. Chipping Norton, NSW: Surrey Beatty and Sons, pp. 261–8.

Howell, D.J. (1979). Flock-foraging in nectar-feeding bats: advantages to the bats and to the host plants. *American Naturalist*, **114**, 23–49.

Hoyer, R.E. and Stewart, G.R. (2000). Biology of the rubber boa (*Charina bottae*) with emphasis on *C. b. umbratica*. Part I: Capture, size, sexual dimorphism, and reproduction. *Journal of Herpetology*, **34**, 348–54.

Hrabar, H. and du Toit, J. T. (in press). Dynamics of an introduced population of black rhinoceros (*Diceros bicornis*): Pilanesberg National Park, South Africa. *Animal Conservation*.

Hudson, R. J. (1985). Body size, energetics, and adaptive radiation. In *Bioenergetics of Wild Herbivores*, eds. R.J. Hudson and R.G. White. Boca Raton: CRC Press, Inc., pp. 1–24.

Hudson, R.J. and White, R.G. (1985). *Bioenergetics of Wild Herbivores*. Boca Raton: CRC Press Inc.

Huey, R. and Slatkin, M. (1976). Costs and benefits of lizard thermoregulation. *Quarterly Review of Biology*, **51**, 363–84.

Hulscher, J.B. (1985). Growth and abrasion of the oystercatcher bill in relation to dietary switches. *Netherlands Journal of Zoology*, **35**, 124–54.

Humple, D.L., Nur, N., Geupel, G.R. and Lynes, M.P. (2001). Female-biased sex ratio in a wintering population of ruby-crowned kinglets. *Wilson Bulletin*, **113**, 419–24.

Hungate, R.E. (1975). The rumen microbial ecosystem. *Annual Review of Ecology and Systematics*, **6**, 39–66.

Hunter, S. (1983). The food and feeding of the giant petrels *Macronectes halli* and *M. giganteus* at South Georgia. *Journal of Zoology, London*, **200**, 521–38.

  (1984). Breeding biology and population dynamics of giant petrels *Macronectes* at South Georgia (Aves: Procellariiformes). *Journal of Zoology, London*, **203**, 441–60.

  (1985). The role of giant petrels in the Southern Ocean ecosystem. In *Antarctic Nutrient Cycles and Food Webs*, eds. W.R. Siegfried, R.M. Laws and P.R. Condy. Berlin: Springer-Verlag, pp. 534–42.

  (1987). Species and sexual isolation mechanisms in sibling species of giant petrels *Macronectes*. *Polar Biology*, **7**, 295–301.

Hunter, S. and Brooke, M.L. (1992). The diet of giant petrels *Macronectes* spp. at Marion Island, Southern Indian Ocean. *Colonial Waterbirds*, **15**, 56–65.

Huntingford, F.A. and Turner, A.K. (1987). *Animal Conflict*. London: Chapman and Hall.

Hyrenbach, K.D., Fernández, P. and Anderson, D.J. (2002). Oceanographic habitats of two sympatric North Pacific albatrosses during the breeding season. *Marine Ecology Progress Series*, **233**, 283–301.

Illius, A.W. and Gordon, I.J. (1987). The allometry of food intake in grazing ruminants. *Journal of Animal Ecology*, **56**, 989–99.

  (1992). Modelling the nutritional ecology of ungulate herbivores: evolution of body size and competitive interactions. *Oecologia*, **89**, 428–34.

Irwin, L.N. (1965). Diel activity and social interaction of the lizard *Uta stansburiana steinegeri*. *Copeia*, **1965**, 99–101.

Isaac, J. and Johnson, C. (2004). Sexual dimorphism and synchrony of breeding: variation in polygyny potential among populations in the common brushtail possum, *Trichosurus vulpecula*. *Behavioral Ecology*, **14**, 818–22.

Isbell, L. A., Cheney, D. L. and Seyfarth, R. M. (2002). Why vervet monkeys (*Cercopithecus aethiops*) live in multimale groups. In *The Guenons: Diversity and Adap-*

tation in African Monkeys, eds. M.E. Glenn and M. Cords. New York: Kluwer, pp. 173–88.

Jack, K.M. and Pavelka, M.S.M. (1997). The behavior of peripheral males during the mating season in *Macaca fuscata*. *Primates*, 38, 369–77.

Jackes, A.D. (1973). The use of wintering ground by red deer in Ross-shire, Scotland. M.Phil. thesis, University of Edinburgh.

Jacklin, C.N. and Maccoby, E.E. (1978). Social behaviour at thirty-three months in same-sex and mid-sex dyads. *Child Development*, 49, 557–69.

Jakimchuk, R.D., Ferguson, S.H. and Sopuck, L.G. (1987). Differential habitat use and sexual segregation in the Central Arctic caribou herd. *Canadian Journal of Zoology*, 65, 534–41.

Janson, C.H. (1990). Ecological consequences of individual spatial choice in foraging brown capuchin monkeys (*Cebus apella*). *Animal Behaviour*, 38, 922–34.

(1992). Evolutionary ecology of primate social structure. In *Evolutionary Ecology and Human Behavior*, eds. E.A. Smith and B. Winterhalder. Hawthorne, NY: Aldine.

(1998). Testing the predation hypothesis for vertebrate sociality: prospects and pitfalls. *Behaviour*, 135, 389–410.

Jantschke, F. (1973). On the breeding and rearing of bush dogs, *Speothos venaticus*, at the Frankfurt Zoo. *International Zoo Yearbook*, 13, 141–3.

Jaremovic, R.V. and Croft, D.B. (1991a). Social organisation of eastern grey kangaroos in southeastern New South Wales. II. Associations within mixed groups. *Mammalia*, 55, 543–54.

(1991b). Social organization of the eastern grey kangaroo (Macropodidae, Marsupialia) in southeastern New South Wales. I. Groups and group home ranges. *Mammalia*, 55, 169–85.

Jarman, P.J. (1968). The effect of the creation of Lake Kariba upon the terrestrial ecology of the middle Zambezi Valley, with particular references to the large mammals. Ph.D. thesis, University of Manchester.

(1974). The social organisation of antelope in relation to their ecology. *Behaviour*, 48, 215–67.

(1983). Mating system and sexual dimorphism in large, terrestrial, mammalian herbivores. *Biological Review*, 58, 485–520.

(1989). Sexual dimorphism in Macropodoidea. In *Kangaroos, Wallabies and Rat-kangaroos*, eds. G. Grigg, P. Jarman and I. Hume. New South Wales, Australia: Surrey Beatty and Sons, pp. 433–47.

(1991). Social behaviour and organization in the Macropodoidea. *Advances in the Study of Behavior*, 20, 1–50.

Jarman, P.J. and Coulson, G. (1989). Dynamics and adaptiveness of grouping in macropods. In *Kangaroos, Wallabies and Rat-kangaroos*, eds. G. Grigg, P. Jarman and I. Hume. New South Wales, Australia: Surrey Beatty and Sons, pp. 527–47.

Jarman, P.J. and Southwell, C.J. (1986). Grouping, associations, and reproductive strategies in eastern grey kangaroos. In *Ecological Aspects of Social Evolution*, eds. D.I. Rubenstein and R.W. Wrangham. Princeton: Princeton University Press, pp. 399–428.

Jefferson, T.A., Stacey, P.J. and Baird, R.W. (1991). A review of killer whale interactions with other marine mammals: predation to co-existence. *Mammal Review*, 21, 151–80.

Jenni, L. (1993). Structure of a brambling *Fringilla montifringilla* roost according to sex, age and body-mass. *Ibis*, 135, 85–90.

Jensen, K. H., Jakobsen, P. J. and Kleiven, O. T. (1998). Fish kairomone regulation of internal swarm structure in *Daphnia pulex* (Cladocera: Crustacea). *Hydrobiologia*, **368**, 123–7.

Jenssen, T. A. (1970). The ethoecology of *Anolis nebulosusi* (Sauria, Iguanidae). *Journal of Herpetology*, **4**, 1–38.

Jenssen, T. A. and Nunez, S. C. (1998). Spatial and breeding relationships of the lizard, *Anolis carolinensis*: evidence of intrasexual selection. *Behaviour*, **135**, 981–1003.

Jessopp, M. J., Forcada, J., Reid, K., Trathan, P. N., Murphy, E. J. (2004). Winter dispersal of Leopard seals (*Hydrurga leptonyx*): environmental factors influencing demographics and seasonal abundance. *Journal of Zoology*, **263**, 251–8.

Jewell, P. A. (1986). Survival in a feral population of primitive sheep on St. Kilda, Outer Hebrides, Scotland. *National Geographic Research*, **2**, 402–6.

(1997). Survival and behaviour of castrated Soay sheep (*Ovis aries*) in a feral island population on Hirta, St. Kilda, Scotland. *Journal of Zoology, London*, **243**, 623–36.

Jiang, Z., Liu, B., Zeng, Y., Han, G. and Hu, H. (2000). Attracted by the same sex, or repelled by the opposite sex? – Sexual segregation in Pere David's deer. *Chinese Science Bulletin*, **45**, 485–91.

Johannes, R. E., Squire, L. and Graham, T. (1994). Developing a protocol for monitoring spawning aggregations of *Palauan serranids* to facilitate the formulation and evaluation of strategies for their management. *South Pacific Forum Fisheries Agency Report 94/28*. Honiara, Solomon Islands.

Johnson, C. N. (1983). Variations in group size and composition in red and western grey kangaroos, *Macropus rufus* (Desmarest) and *M. fuliginosus* (Desmarest). *Australian Wildlife Research*, **10**, 25–31.

(1989). Grouping and the structure of association in the red-necked wallaby. *Journal of Mammalogy*, **70**, 18–26.

Johnson, C. N. and Bayliss, P. G. (1981). Habitat selection by sex, age and reproductive class in the red kangaroo, *Macropus rufus*, in western New South Wales. *Australian Wildlife Research*, **8**, 465–74.

Johnson F. A. and Moore, C. T. (1996). Harvesting multiple stocks of ducks. *Journal of Wildlife Management*, **60**, 551–9.

Johnson, P. M. (1980). Field observations on group compositions in the agile wallaby, *Macropus agilis* (Gould) (Marsupialia: Macropodidae). *Australian Wildlife Research*, **7**, 327–31.

Johnson, P. M. and Strahan, R. (1982). A further description of the Musky Rat-kangaroo, *Hypsiprymnodon moschatus* (Ramsay, 1876) (Marsupialia, Potoroidae), with notes on its biology. *Australian Zoologist*, **21**, 27–46.

Johnstone, G. W. (1977). Comparative feeding ecology of the giant petrels *Macronectes giganteus* (Gmelin) and *M. halli* (Mathews). In *Adaptations within Antarctic Ecosystems*, ed. G. Llano. Houston: Gulf Publishing Company, pp. 647–68.

Jones, G. and Rydell, J. (1994). Foraging strategy and predation risk as factors influencing emergence time in echolocating bats. *Philosophical Transactions of the Royal Society London, B*, **346**, 445–55.

Jones, M. (1995). Tasmanian devil. In *The Mammals of Australia*, ed. R. Strahan. Sydney: Reed New Holland, pp. 82–4.

Jones, M. and Barmuta, L. (1998). Diet overlap and relative abundance of sympatric dasyurid carnivores: a hypothesis of competition. *Journal of Animal Ecology*, **67**, 410–21.

(2000). Niche differentiation among sympatric Australian dasyurid carnivores. *Journal of Mammalogy*, **81**, 434–47.

Jones, M., Grigg, G. and Beard, L. (1997). Body temperatures and activity patterns of Tasmanian devils (*Sarcophilus harrisii*) and eastern quolls (*Dasyurus viverrinus*) through a subalpine winter. *Physiological Zoology*, **70**, 53–60.

Jones, R. L. (1996). Home range and seasonal movements of the turtle *Graptemys flavimaculata*. *Journal of Herpetology*, **30**, 376–85.

Jonsson, K. I. (1997). Capital and income breeding as alternative tactics of resource use in reproduction. *Oikos*, **78**, 57–66.

Joung, S.-J., Chen, S.-T., Clark, E., Uchida, S. and Huang, W.Y.P. (1996). The whale shark, *Rhincodon typus*, is a livebearer: 300 embryos found in one 'megamamma' supreme. *Environmental Biology of Fishes*, **46**, 219–23.

Jouventin, P. and Weimerskirch, H. (1990a). Long-term changes in seabird and seal populations in the Southern Ocean. In *Antarctic Ecosystems: Ecological Change and Conservation*, eds. K. R. Kerry and G. Hempel. Berlin: Springer-Verlag, pp. 208–13.

(1990b). Satellite tracking of wandering albatrosses. *Nature*, **343**, 746–8.

Jouventin, P., Lequette, B. and Dobson, F. S. (1999). Age-related mate choice in the wandering albatross. *Animal Behaviour*, **57**, 1099–106.

Kajimura, H. (1985). Opportunistic feeding by the northern fur seal (*Callorhinus ursinus*). In *Marine Mammals and Fisheries*, eds. J. R. Beddington, R. J. H. Beverton and D. M. Lavigne. London: George Allen and Unwin Ltd, pp. 300–18.

Kastelein, R. A. (2002). Walrus. In *Encyclopedia of Marine Mammals*, eds. W. F. Perrin, B. Wursig and J. G. M. Thewissen. San Diego: Academic Press, pp. 1294–300.

Kato, A., Watanuki, Y., Mishiumi, I., Kuroki, M., Shaughnessy, P. and Naito, Y. (2000). Variation in foraging and parental behaviour of king cormorants. *Auk*, **117**, 718–30.

Katsikaros, K. and Shine, R. (1997). Sexual dimorphism in the tusked frog, *Adelotus brevis* (Anura: Myobatrachidae): the roles of natural and sexual selection. *Biological Journal of the Linnean Society*, **60**, 39–51.

Kaufman, G. W., Siniff, D. B. and Reichle, R. (1975). Colony behaviour of Weddell seals, *Leptonychotes weddellii*, at Hutton Cliffs, Antarctica. *Rapport et Procès-verbeax des Réunions du Conseil international pour l'Exploration de la mer*, **11**, 228–46.

Kaufman, J. H. (1992). Habitat use by wood turtles in central Pennsylvania. *Journal of Herpetology*, **26**, 315–21.

Kaufmann, J. H. (1974). Social ethology of the whiptail wallaby, *Macropus parryi*, in northeastern New South Wales. *Animal Behaviour*, **22**, 281–369.

Keenlyne, K. D. (1972). Sexual differences in feeding habits of *Crotalus horridus horridus*. *Journal of Herpetology*, **6**, 234–7.

Kendrick, K. M., Atkins, K., Hinton, M. R. *et al.* (1995). Facial and vocal discrimination in sheep. *Animal Behaviour*, **49**, 1665–76.

Kerle, J. A. (1983). The population biology of the Northern brushtail possum. Unpublished Ph.D. thesis, Macquarie University.

(1984). Variation in the ecology of *Trichosurus*: its adaptive significance. In *Possums and Gliders*, eds. A. P. Smith and I. D. Hume. Chipping Norton, NSW: Surrey Beatty and Sons, pp. 115–28.

Kerle, J. A., McKay, G. M. and Sharman, G. B. (1991). A systematic analysis of the brushtail possum, *Trichosurus vulpecula* (Kerr, 1792) (Marsupialia: Phalangeridae). *Australian Journal of Zoology*, **39**, 313–31.

Kerth, G. and Reckardt, K. (2003). Information transfer about roosts in female Bechstein's bats: an experimental field study. *Proceedings of the Royal Society London, B*, **270**, 511–15.

Kerth, G., Mayer, F. and Konig, B. (2000). Mitochondrial DNA (mtDNA) reveals that female Bechstein's bats live in closed societies. *Molecular Ecology*, **9**, 793–800.

Kerth, G., Wagner, M. and Konig, B. (2001a). Roosting together, foraging apart: information transfer about food is unlikely to explain sociality in female Bechstein's bats (Myotis bechsteinii). Behavioural Ecology and Sociobiology, 50, 283–91.

Kerth, G., Weissman, K. and König, B. (2001b). Day roost selection in female Bechstein's bats (Myotis bechsteinii): a field experiment to determine the influence of roost temperature. Oecologia, 126, 1–9.

Ketterson, E. D. and Nolan, V. Jr (1983). The evolution of differential bird migration. In Current Ornithology 1, ed. R. F. Johnston. New York: Plenum Press.

Kie, J. G. and Bowyer, R. T. (1999) Sexual segregation in white-tailed deer: density-dependent changes in use of space, habitat selection, and dietary niche. Journal of Mammalogy, 80, 1004–20.

King, J. E. (1983). Seals of the World, 2nd edn. Oxford: Oxford University Press.

King, R. B. (1986). Population ecology of the Lake Erie water snake, Nerodia sipedon insularum. Copeia, 1986, 757–72.

(1989). Sexual dimorphism in snake tail length: sexual selection, natural selection, or morphological constraint? Biological Journal of the Linnean Society, 38, 133–54.

(1993). Microgeographic, historical, and size-correlated variation in water snake diet composition. Journal of Herpetology, 27, 90–4.

Kinzey, W. G. (1977). Diet and feeding behavior of Callicebus torquatus. In Primate Ecology, ed. T. H. Clutton-Brock. Cambridge: Cambridge University Press, pp. 127–51.

Kirsch, J. A. W., Lapointe, F. J. and Springer, M. S. (1997). DNA-hybridisation studies of marsupials and their implications for metatherian classification. Australian Journal of Zoology, 45, 211–80.

Kissner, K. J., Forbes, M. R. L. and Secoy, D. M. (1997). Rattling behavior of prairie rattlesnakes (Crotalus viridis viridis, Viperidae) in relation to sex, reproductive status, body size, and body temperature. Ethology, 103, 1042–50.

Kissner, K. J., Weatherhead, P. J. and Francis, C. M. (2003). Sexual size dimorphism and timing of spring migration in birds. Journal of Evolutionary Biology, 16, 154–62.

Kitchen, D. W. (1974). Social behavior and ecology of the pronghorn. Wildlife Monographs, 38, 1–96.

Kitching, J. A. and Ebling, F. J. (1967). Ecological studies in Lough Ine. Advances in Ecological Research, 4, 197–291.

Kleiber, M. (1975). An Introduction to Animal Energetics. Huntington, NY: R. E. Kreiger Publishing Co.

Kleiman, D. G. (1980). The sociobiology of captive propagation. In Conservation Biology: An Evolutionary-Ecological Perspective, eds. M. Soule and B. A. Wilcox. Sunderland, Massachusetts: Sinauer Associates, Inc., pp. 243–61.

Kleinenberg, S. E., Yablokov, A. V., Bel'kovich, B. M. and Tarasevich, M. N. (1964). Beluga (Delphinapterus leucas): investigation of the species. Translated by Israel Progr. Sci. Translat., Jerusalem 1969. Moscow: Academy Nauk USSR, p. 376.

Klimley, A. P. (1985). Schooling in Sphyrna lewini, a species with low risk of predation: a non-egalitarian state. Zeitschrift für Tierpsychologie, 70, 297–319.

(1987). The determinants of sexual segregation in the scalloped hammerhead shark, Sphyrna lewini. Environmental Biology of Fishes, 18, 27–40.

Kodric-Brown, A. (1990). Mechanisms of sexual selection – insights from fishes. Annales Zoologici Fennici, 27, 87–100.

Kohlmann, S. G., Müller, D. M. and Alkon, P. U. (1996). Antipredator constraints on lactating nubian ibexes. Journal of Mammalogy, 77, 1122–31.

Komdeur, J. and Deerenberg, C. (1997). The importance of social behavior studies for conservation. In Behavioral Approaches to Conservation in the Wild, eds.

J.R. Clemmons and R. Buchholz. Cambridge, United Kingdom: Cambridge University Press, pp. 262–76.

Komers, P.E., Messier, F. and Gates C. C. (1993). Group structure in wood bison: nutritional and reproductive determinants. *Canadian Journal of Zoology*, **71**, 1367–71.

Koopman, K. F. (1993). Bats. In *Mammal Species of the World: Taxonomic and Geographic Reference*, eds. D.E. Wilson and D.M. Reeder. Washington, DC: Smithsonian Institution Press, pp. 137–241.

Koskimies, J. (1957). Flocking behaviour in capercaillie *Tetrao urogallus* and blackgame *Lyrurus tetrix*. *Papers on Game Research Published by the Finish Game Foundation*, **18**, 1–31.

Kovacs, K.M., Lydersen, C., Hammill, M. and Lavigne, D.M. (1996). Reproductive effort of male hooded seals (*Cystophora cristata*): estimates from mass loss. *Canadian Journal of Zoology*, **74**, 1521–30.

Krause, J. (1994). The influence of food competition and predation risk on size-assortative shoaling in juvenile chub (*Leuciscus cephalus*). *Ethology*, **96**, 105–16.

Krause, J. and Godin, J.-G. J. (1994). Shoal choice in banded killifish (*Fundulus diaphanus*, Teleostei, Cyprinodontidae) – Effects of predation risk, fish size, species composition and size of shoals. *Ethology*, **98**, 128–36.

(1996a). Influence of parasitism on shoal choice in the banded killifish (*Fundulus diaphanus*, Teleostei, Cyprinodontidae). *Ethology*, **102**, 40–9.

(1996b). Phenotypic variability within and between fish shoals. *Ecology*, **77**, 1586–91.

Krause, J. and Ruxton, G.D. (2002). *Living in Groups*. Oxford: Oxford University Press.

Krause, J., Butlin, R., Peuhkuri, N. and Pritchard, V.L. (2000a). The social organisation of fish shoals: a test of the predictive power of laboratory experiments for the field. *Biological Reviews*, **75**, 477–501.

Krause, J., Godin, J.-G. J. and Brown, D. (1996). Size-assortativeness in multi-species fish shoals. *Journal of Fish Biology*, **49**, 221–5.

Krause, J., Hoare, D.J., Croft, D., Lawrence, J., Ward, A., Ruxton, G.D., Godin, J.G.J. and James, R. (2000b). Fish shoal composition: mechanisms and constraints. *Proceedings of the Royal Society of London Series B, Biological Sciences*, **267**, 2011–17.

Krause, J., Loader, S.P., McDermott, J. and Ruxton, G.D. (1998). Refuge use by fish as a function of body length-related metabolic expenditure and predation risks. *Proceedings of the Royal Society of London Series B, Biological Sciences*, **265**, 2373–9.

Krebs, C.J. (1989). *Ecological Methodology*. New York: Harper Collins Publishers.

Krohmer, R.W. and Aldridge, R.D. (1985). Female reproductive cycle of the lined snake (*Tropidoclonium lineatum*). *Herpetologica*, **41**, 39–44.

Krützen, M., Sherwin, W.B., Berggren, P. and Gales, N. (2004). Population structure in an inshore cetacean revealed by microsatellite and mtDNA analysis: bottlenose dolphins (*Tursiops* sp.) in Shark Bay Western Australia. *Marine Mammal Science*, **20**, 28–47.

Kruuk, H. (1972) *The Spotted Hyena*. Chicago: University of Chicago Press.

Kunz, T.H. (1974). Feeding ecology of a temperate insectivorous bat (*Myotis velifer*). *Ecology*, **55**, 693–711.

(1982). Roosting ecology of bats. In *Ecology of Bats*, ed. T.H. Kunz. New York: Plenum Press, pp. 1–55.

Kunz, T. H. and Lumsden, L. F. (2003). Ecology of cavity and foliage roosting bats. In *Bat Ecology*, eds. T.H. Kunz and M.B. Fenton. Chicago: University of Chicago Press, pp. 3–89.

Lack, D. (1968). *Ecological Adaptations for Breeding in Birds*. London: Methuen and Co.

Lagarde F., Bonnet, X., Corbin, J., Henen, B. and Nagy, K. (2002). A short spring before a long jump: the ecological challenge to the steppe tortoise (*Testudo horsfieldi*). *Canadian Journal of Zoology*, **80**, 493–502.

Lagarde, F., Bonnet, X., Henen, B. et al. (2003). Sex divergence in space utilisation in the steppe tortoise (*Testudo horsfieldi*). *Canadian Journal of Zoology*, **81**, 380–7.

LaGory, K.E., Bagshaw, C. III and Brisbin, I. L. Jr (1991). Niche differences between male and female white-tailed deer on Ossabaw Island, Georgia. *Applied Animal Science*, **29**, 205–14.

Lahanas, P.N. (1982). Aspects of the life history of the southern black-knobbed sawback, *Graptemys nigrinoda delticola* Folkerts and Mount. M.Sc. thesis, Auburn, AL: Auburn University.

Laidlaw, W.S., Hutchings, S. and Newell, G.R. (1996). Home range and movement patterns of *Sminthopsis leucopus* (*Marsupialia: Dasyuridae*) in coastal dry Heathland, Anglesea, Victoria. *Australian Mammalogy*, **19**, 1–9.

Lailvaux, S.P., Alexander, G.J. and Whiting, M.J. (2003). Sex-based differences and similarities in locomotor performance, thermal preferences, and escape behaviour in the lizard *Platysaurus intermedius wilhelmi*. *Physiological and Biochemical Zoology*, **76**, 511–21.

Lamb, T. (1984). The influence of sex and breeding condition on microhabitat and diet in the pig frog *Rana grylio*. *American Midland Naturalist*, **111**, 311–18.

Lande, R. and Barrowclough, G. F. (1987). Effective population size, genetic variation, and their use in population management. In *Viable Populations for Conservation*, ed. M.E. Soule. Cambridge, United Kingdom: Cambridge University Press, pp. 87–123.

Landeau, L. and Terborgh, J. (1986). Oddity and the confusion effect in predation. *Animal Behaviour*, **34**, 1372–80.

Landys-Ciannelli, M.M., Piersma, T. and Jukema, J. (2003). Strategic size changes of internal organs and muscle tissue in the bar-tailed godwit during fat storage on a spring stopover site. *Functional Ecology*, **17**, 151–9.

Langman, V.A. (1977). Cow-calf relationships in giraffe (*Giraffa camelopardalis giraffa*). *Zeitschrift für Tierpsychology*, **43**, 264–86.

Laska, A.L. (1970). The structural niche of *Anolis scripta* on Inagua. *Breviora*, **349**, 1–6.

Latta, S.C. and Faaborg, J. (2002). Demographic and population responses of Cape May warbler wintering in multiple habitats. *Ecology*, **83**, 2502–15.

Launchbaugh, K.L., Provenza, F.D. and Pfister, J.A. (2001). Herbivore response to anti-quality factors in forages. *Journal of Range Management*, **54**, 431–40.

Lausen, C.L. and Barclay, R.M.R. (2002). Roosting behaviour and roost selection of female big brown bats (*Eptesicus fuscus*) roosting in rock crevices in southeastern Alberta. *Canadian Journal of Zoology*, **80**, 1069–76.

(2003). Thermoregulation and roost selection by reproductive female big brown bats (*Eptesicus fuscus*) roosting in rock crevices. *Journal of Zoology, London*, **260**, 235–44.

Lawrence, A.L. (1990). Mother-daughter and peer relationships of Scottish hill sheep. *Animal Behaviour*, **39**, 481–6.

Laws, R.M. (1993). *Antarctic Seals: Research Methods and Techniques*, 1st edn., Cambridge: Cambridge University Press.

Lazenby-Cohen, K.A. and Cockburn, A. (1988). Lek promiscuity in a semelparous mammal, *Antechinus stuartii* (*Marsupialia: Dasyuridae*)? *Behavioral Ecology and Sociobiology*, **22**, 195–202.

(1991). Social and foraging components of the home range in *Antechinus stuartii* (Dasyuridae: Marsupialia). *Australian Journal of Ecology*, **16**, 301–8.

Le Boeuf, B. J. (1971). The aggression of the breeding bulls. *Natural History*, **70**, 83–94.

(1991). Pinniped mating systems on land, ice and in the water: emphasis on the *Phocidae*. In *Behaviour of Pinnipeds*, ed. D. Renouf. London: Chapman and Hall, pp. 45–65.

Le Boeuf, B. J. and Crocker, D. E. (1996). Diving behaviour of elephant seals: implications for predator avoidance. In *Great White Sharks: The Biology of Carcharodon carcharias*, eds. A. P. Klimley and D. G. Ainley. Berkley: University of California Press, pp. 193–206.

Le Boeuf, B. J. and Laws, R. M. (1994). Elephant seals: An introduction to the genus. In *Elephant Seals: Population Ecology, Behaviour, and Physiology*, eds. B. J. Le Boeuf and R. M. Laws. Berkley: University of California Press, pp. 1–26.

Le Boeuf, B. J. and Reiter, J. (1988). Lifetime reproductive success in northern elephant seals. In *Reproductive Success: Studies of Individual Variation in Contrasting Breeding Systems*, ed. T. H. Clutton-Brock. Chicago: University of Chicago Press, pp. 344–62.

Le Boeuf, B. J., Crocker, D. E., Blackwell, S. B., Morris, P. A. and Thorson, P. H. (1993). Sex differences in foraging in northern elephant seals. In *Marine Mammals: Advances in Behavioural and Population Biology*, ed. I. L. Boyd. London: Oxford University Press.

Le Boeuf, B. J., Crocker, D. E., Costa, D. P. *et al.* (2000). Foraging ecology of northern elephant seals. *Ecological Monographs*, **70**, 353–82.

Le Boeuf, B. J., Morris, P. A., Blackwell, S. B., Crocker, D. E. and Costa, D. P. (1996). Diving behavior of juvenile northern elephant seals. *Canadian Journal of Zoology*, **74**, 1632–44.

Le Pendu, Y., Guilhem, C., Briedermann, L., Maublanc, M.-L. and Gerard, J.-F. (2000). Interactions and associations between age and sex classes in mouflon sheep (*Ovis gmelini*) during winter. *Behavioural Processes*, **52**, 97–107.

Leaper, C. (1994). Exploring the consequences of gender segregation on social relationships. In *Childhood Gender Segregation: Causes and Consequences*, ed. C. Leaper. San Francisco: Jossey-Bass, pp. 67–86.

Lee, A. and Cockburn, A. (1985). *Evolutionary Ecology of Marsupials*. Cambridge: Cambridge University Press.

Lee, A., Woolley, P. and Braithwaite, R. (1982). Life history strategies of dasyurid marsupials. In *Carnivorous Marsupials*, ed. M. Archer. Sydney: Royal Zoological Society of New South Wales, pp. 1–11.

Lee, D. S., Franz, R. and Sanderson, R. A. (1975). A note on the feeding habits of male Barbour's map turtles. *Florida Field Naturalist*, **3**, 45–6.

Lee, Y. F. and McCracken, G. F. (2002). Foraging activity and food resource use of Brasilian free-tailed bats, *Tadarida brasiliensis* (Molossidae). *Ecoscience*, **9**, 306–13.

Legault, F. and Strayer, F. F. (1991). Genèse de la ségrégation sexuelle et différences comportementales chez des enfants d'âge préscolaire. *Behaviour*, **119**, 285–301.

Legler, J. M. L. (1985). Australian chelid turtles: reproductive patterns in wide-ranging taxa. In *The Biology of Australasian Frogs and Reptiles*, eds. G. C. Grigg, R. Shine and H. Ehmann. Sydney: Royal Zoological Society of New South Wales, pp. 117–23.

Lemos-Espinal, J. A., Smith, G. R. and Ballinger, R. E. (1997). Thermal ecology of the lizard, *Sceloporus gadoviae*, in an arid tropical scrub forest. *Journal of Arid Environments*, **35**, 311–19.

Lesage, V., Hammill, M.O. and Kovacs, K.M. (2001). Marine mammals and the community structure of the Estuary and Gulf of St Lawrence, Canada: evidence from stable isotope analysis. *Marine Ecology Progress Series*, **210**, 203–21.

Lettevall, E., Richter, C., Jaquet, N. *et al.* (2002). Social structure and residency in aggregations of male sperm whales. *Canadian Journal of Zoology*, **80**, 1189–96.

Leuzinger, Y. and Brossard, C. (1994). Répartition de *M. daubentonii* en fonction du sexes et de la période de l'année dans le Jura Bernois. *Résultats préliminaires. Mitteilunngen der Naturforschenden Gesellschaft Schaffhausen*, **39**, 135–43.

Lewis, S., Benvenuti, S., Dall'Antonia, L. *et al.* (2002). Sex-specific foraging behaviour in a monomorphic seabird. *Proceeding of the Royal Society of London*, **269**, 1687–93.

Lewis, S.E. (1992). Behaviour of Peter's tent-making bat, *Uroderma bilobatum*, at maternity roosts in Costa Rica. *Journal of Mammalogy*, **73**, 541–6.

(1996). Low roost-site fidelity in pallid bats: associated factors and effect on group stability. *Behavioural Ecology and Sociobiology*, **39**, 335–44.

Li, J., Luan, Y., Sun, I., Zhao, D. and Diao, Y. (1990). Studies on some problems of *Agkistrodon shedaoensis* population due to seasonal changes (in Chinese). In *From Water Onto Land*, ed. E. Zhao. China Forestry Press, pp. 273–6.

Lillegraven, J.A., Kielan-Jaworowska, Z. and Clemens, W.A. (1979). *Mesozoic Mammals: The First Two-Thirds of Mammalian History*. Berkeley: University of California Press.

Lillywhite, H.B. and Henderson, R.W. (1993). Behavioral and functional ecology of arboreal snakes. In *Snakes: Ecology and Behavior*, eds. R.A. Seigel and J.T. Collins. New York: McGraw-Hill, pp. 1–48.

Lindeman, P.V. (2003). Sexual difference in habitat use of Texas map turtles (Emydidae: *Graptemys versa*) and its relationship to size dimorphism and diet. *Canadian Journal of Zoology*, **81**, 1185–91.

Ling, J.K. (2002). Australian sea lion. In *Encyclopedia of Marine Mammals*, eds. W.F. Perrin, B. Wursig and J.G.M. Thewissen. San Diego: Academic Press.

Lingle, S. (2000). Seasonal variation in coyote feeding behaviour and mortality of white-tailed deer and mule deer. *Canadian Journal of Zoology*, **78**, 85–99.

(2001). Anti-predator strategies and grouping patterns in white-tailed deer and mule deer. *Ethology*, **107**, 295–314.

(2002). Coyote predation and habitat segregation of white-tailed deer and mule deer. *Ecology*, **83**, 2037–48.

Lingle, S. and Pellis, S.M. (2002). Fight or flight? Antipredator behavior and the escalation of coyote encounters with deer. *Oecologia*, **131**, 154–64.

Lingle, S. and Wilson, W.F. (2001). Detection and avoidance of predators in white-tailed deer (*Odocoileus virginianus*) and mule deer (*O. hemionus*). *Ethology*, **107**, 125–47.

Lister, B.C. (1976). The nature of niche expansion in West Indian *Anolis* lizards I. Ecological consequences of reduced competition. *Evolution*, **30**, 659–76.

Lister, B.C. and Aguayo, A.G. (1992). Seasonality, predation, and the behaviour of a tropical mainland anole. *Journal of Animal Ecology*, **61**, 717–33.

Loison, A., Gaillard, J.-M., Pelabon, C. and Yoccoz, N.G. (1999). What factors shape sexual size dimorphism in ungulates? *Evolutionary Ecology Research*, **1**, 611–33.

Long, J.A. and Pellegrini, A.D. (2003). Studying change in bullying and dominance with structural equation modeling. *School Psychology Review*, **32**, 401–17.

Lopez Ornat, A. and Greenberg, R. (1990). Sexual segregation by habitat in migratory warblers in Quintana Roo, Mexico. *Auk*, **107**, 539–43.

Lott, D.F. and Minta, S.C. (1983). Random individual association and social group instability in American bison (*Bison bison*). *Zeitschrift für Tierpsychology*, **61**, 153–72.

Lovern, M. B. (2000). Behavioral ontogeny in free-ranging juvenile male and female green anoles, *Anolis carolinensis*, in relation to sexual selection. *Journal of Herpetology*, **34**, 274–81.

Low, B. S. (2000). *Why Sex Matters: A Darwinian Look at Human Behavior*. Princeton: Princeton University Press.

Lowry, L. F., Burns, J. J. and Nelson, R. R. (1987). Polar bear, *Ursus maritimus*, predation on belugas, *Delphinapterus leucas*, in the Bering and Chukchi Seas. *Canadian Field-Naturalist*, **101**, 141–6.

Lowry, L. F., Frost, K. J. and Seaman, G. A. (1985). *Investigations of Beluga Whales in Coastal Waters of Western and Northern Alaska – III: Food Habits*. Fairbanks (Alaska): Alaska Department of Fish and Game, p. 24.

Lue, K. Y. and Chen, T. H. (1999). Activity, movement patterns, and home range of the yellow-margined box turtle (*Cuora flavomarginata*) in northern Taiwan. *Journal of Herpetology*, **33**, 590–600.

Luiselli, L. and Agrimi, U. (1991). Composition and variation of the diet of *Vipera aspis francisciredi* in relation to age and reproductive stage. *Amphibia-Reptilia*, **12**, 137–44.

Luiselli, L. and Angelici, F. M. (1998). Sexual size dimorphism and natural history traits are correlated with intersexual dietary divergence in royal pythons (*Python regius*) from the rainforests of southeastern Nigeria. *Italian Journal of Zoology*, **65**, 183–5.

Luiselli, L., Akani, G. C. and Capizzi, D. (1999). Is there any interspecific competition between dwarf crocodiles (*Osteolaemus tetraspis*) and Nile monitors (*Varanus niloticus ornatus*) in the swamps of central Africa? A study from southeastern Nigeria. *Journal of Zoology, London*, **247**, 127–31.

Luiselli, L., Capula, M. and Shine, R. (1996). Reproductive output, costs of reproduction, and ecology of the smooth snake, *Coronella austriaca*, in the eastern Italian Alps. *Oecologia*, **106**, 100–10.
(1997). Food habits, growth rates, and reproductive biology of grass snakes, *Natrix natrix* (Colubridae) in the Italian Alps. *Journal of Zoology, London*, **241**, 371–80.

Lunn, N. J. and Arnould, J. P. Y. (1997). Maternal investment in Antarctic fur seals: evidence for equality in the sexes. *Behavioral Ecology and Sociobiology*, **40**, 351–62.

Lunney, D. (1995). White-footed Dunnart. In *The Mammals of Australia*, ed. R. Strahan. Sydney: Reed New Holland, pp. 143–5.

Lunney, D. and Leary, T. (1989). Movement patterns of the white-footed dunnart, *Sminthopsis leucopus* (Marsupialia: Dasyuridae), in a logged, burnt forest on the south coast of New South Wales. *Australian Wildlife Research*, **16**, 207–16.

Lunney, D., Matthews, A. and Grigg, J. (2001). The diet of *Antechinus agilis* and *A. swainsonii* in unlogged and regenerating sites in Mumbulla State Forest, south-eastern New South Wales. *Wildlife Research*, **28**, 459–64.

Lunney, D., O'Connell, M. and Sanders, J. (1989). Habitat of the white-footed dunnart, *Sminthopsis leucopus* (Gray) (Dasyuridae: Marupialia), in a logged, burnt forest near Bega, New South Wales. *Australian Journal of Ecology*, **14**, 335–44.

Lyderson, C., Hammill, M. O. and Kovacs, K. M. (1994). Activity of lactating ice-breeding grey seals, *Halichoerus grypus*, from the Gulf of St. Lawrence, Canada. *Animal Behaviour*, **48**, 1417–25.

Lyle, J. M. (1983). Food and feeding habits of the lesser spotted dogfish, *Scyliorhinus canicula* (L.) in Isle of Man waters. *Journal of Fish Biology*, **23**, 725–38.

Lynch, J.F., Morton, E.S. and van der Voort, M.E. (1985). Habitat segregation between the sexes of wintering hooded warblers, *Wilsonia citrina. Auk,* **102,** 714–21.

MacArthur, R.A., Geist V. and Johnston, R.H. 1982. Cardiac and behavioral responses of mountain sheep to human disturbance. *Journal of Wildlife Management,* **46,** 351–8.

Maccoby, E.E. (1998). *The Two Sexes: Growing Up Apart, Coming Together.* Cambridge, MA, Harvard University Press.

MacFarlane, A.M. and Coulson G. (in press). Synchrony and timing of breeding influences sexual segregation in western grey and red Kangaroos (*Macropus Fuliginosus* and *M. rufus*). *Journal of Zoology, London.*

Macdonald, D. (2001). *The New Encyclopedia of Mammals.* Oxford: Oxford University Press.

Madsen, T. (1983). Growth rates, maturation and sexual size dimorphism in a population of grass snakes, *Natrix natrix,* in southern Sweden. *Oikos,* **40,** 277–82.

   (1984). Movements, home range size and habitat use of radio-tracked grass snakes (*Natrix natrix*) in southern Sweden. *Copeia,* **1984,** 707–13.

Madsen, T. and Shine, R. (1993a). Costs of reproduction in a population of European adders. *Oecologia,* **94,** 488–95.

   (1993b). Phenotypic plasticity in body sizes and sexual size dimorphism in European grass snakes. *Evolution,* **47,** 321–5.

   (1996). Seasonal migration of predators and prey: pythons and rats in tropical Australia. *Ecology,* **77,** 149–56.

   (2000). Energy versus risk: costs of reproduction in free-ranging pythons in tropical Australia. *Austral Ecology,* **25,** 670–5.

Madsen, T., Shine, R., Loman, J. and Håkansson, T. (1993). Determinants of mating success in male adders, *Vipera berus. Animal Behaviour,* **45,** 491–9.

Magnusson, W.E. (1993). Body temperatures of field-active Amazonian savanna lizards. *Journal of Herpetology,* **27,** 53–8.

Magurran, A.E. (1990). The adaptive significance of schooling as an antipredator defense in fish. *Annales Zoologici Fennici,* **27,** 51–66.

Magurran, A.E. and Garcia, M. (2000). Sex differences in behaviour as an indirect consequence of mating system. *Journal of Fish Biology,* **57,** 839–57.

Magurran, A.E., Seghers, B.H., Shaw, P.W. and Carvalho, G.R. (1994). Schooling preferences for familiar fish in the Guppy, *Poecilia reticulata. Journal of Fish Biology,* **45,** 401–6.

Mahmoud, I.Y. (1969). Comparative ecology of the kinosternid turtles of Oklahoma. *Southwestern Naturalist,* **14,** 31–66.

Main, M.B. (1994). Advantages of habitat selection and sexual segregation in mule and white-tailed deer. Ph.D. thesis, Corvallis, OR, USA: Oregon State University.

   (1998). Sexual segregation in ungulates: a reply. *Journal of Mammalogy,* **79,** 1410–15.

Main, M.B. and Coblentz, B.E. (1990). Sexual segregation among ungulates: a critique. *Wildlife Society Bulletin,* **18,** 204–10.

   (1996). Sexual segregation in Rocky Mountain mule deer. *Journal of Wildlife Management,* **60,** 97–507.

Main, M.B., Weckerly, F.W. and Bleich, V.C. (1996). Sexual segregation in ungulates: new directions for research. *Journal of Mammalogy,* **77,** 449–61.

Mann, J. and Barnett, H. (1999). Lethal tiger shark (*Galeocerdo cuvier*) attack on bottlenose dolphin (*Tursiops* sp.) calf: defense and reactions by the mother. *Marine Mammal Science,* **15,** 568–75.

Mann, J., Connor, R.C., Barre, L.M. and Heithaus, M.R. (2000a). Female reproductive success in wild bottlenose dolphins (*Tursiops* sp.): Life history, habitat, provisioning, and group size effects. *Behavioral Ecology*, **11**, 210–19.

Mann, J., Connor, R. C., Tyack, P. L. and Whitehead, H. (2000b). *Cetacean Societies: Field Studies of Dolphins and Whales*. Chicago: University of Chicago Press.

Mansergh, I. and Broome, L. (1994). *The Mountain Pygmy-Possum of the Australian Alps*. Kensington: New South Wales University Press.

Mansergh, I. and Scotts, D. (1989). Habitat continuity and social organizations of the Mountain Pygmy-possum restored by tunnel. *Journal of Wildlife Management*, **53**, 701–7.

(1990). Aspects of the life history and breeding biology of the Mountain Pygmy-possum, *Burramys parvus*, (Marsupialia: Burramyidae) in alpine Victoria. *Australian Mammalogy*, **13**, 179–91.

Mansergh, I., Baxter, B., Scotts, D., Brady, T. and Jolley, D. (1990). Diet of the mountain pygmy-possum, *Burramys parvus* (Marsupialia: Burramyidae) and other small mammals in the alpine environment at Mt. Higginbotham, Victoria. *Australian Mammalogy*, **13**, 167–77.

Marchal, C., Gerard, J.-F., Boisaubert, B. and Bideau, E. (1998). Instability and diurnal variation in size of winter groupings of field roe deer. *Revue d'Ecologie (Terre Vie)*, **53**, 59–68.

Marchant, S. and Higgins, P. J. (1993). *Handbook of Australian, New Zealand and Antarctic Birds*. vol. 2, Melbourne, Oxford University Press.

Marquiss, M. (1980). Habitat and diet of male and female hen harriers in Scotland in winter. *British Birds*, **73**, 555–60.

Marra, P. P. (2000). The role of behavioral dominance in structuring patterns of habitat occupancy in a migrant bird during the nonbreeding season. *Behavioural Ecology*, **11**, 299–308.

Marra, P. P. and Holberton, R. L. (1998). Corticosterone levels as indicators of habitat quality: effects of habitat segregation in a migratory bird during the non-breeding season. *Oecologia*, **116**, 284–92.

Marra, P. P. and Holmes, R. T. (1997). Avian removal experiments: do they test for habitat saturation or female availability? *Ecology*, **78**, 947–52.

(2001). Consequences of dominance-mediated habitat segregation in American redstarts during the nonbreeding season. *Auk*, **118**, 92–104.

Marra, P. P., Hobson, K. A. and Holmes, R. T. (1998). Linking winter and summer events in a migratory bird by using stable carbon isotopes. *Science*, **282**, 1884–6.

Marra, P. P., Sherry, T. W. and Holmes, R. T. (1993). Territorial exclusion by a long-distance migrant warbler in Jamaica: a removal experiment with American redstarts, *Setophaga ruticilla*. *Auk*, **110**, 565–72.

Marsden, S. J. and Sullivan, M. S. (2000). Intersexual differences in feeding ecology in a male-dominated wintering pochard *Aythya ferina* population. *Ardea*, **88**, 1–7.

Marsh, C. (1981). Time budget of Tana River red colobus. *Folia Primatologica*, **35**, 30–50.

Martin, A. R. and da Silva, V. M. F. (2004). River dolphins and flooded forest: seasonal habitat use and sexual segregation of botos (*Inia geoffrensis*) in an extreme cetacean environment. *Journal of Zoology, London*, **263**, 295–305.

Martin, J. and Lopez, P. (1999). Nuptial coloration and mate guarding affect escape decisions of male lizards (*Psammodromus algirus*). *Ethology*, **105**, 439–47.

Martin, R. and Handasyde, K. (1990). Population dynamics of the koala (*Phascolarctos cinereus*) in southeastern Australia. In *Biology of the Koala*, eds. A. K. Lee, K. A. Handasyde and G. D. Sanson. Chipping Norton, NSW: Surrey Beatty and Sons, pp. 75–84.

Martin, R.D. (1972). A preliminary field study of the lesser mouse lemur (*Microcebus murinus* J.F. Miller 1777). *Zietschrift für Tierpsycholgie*, **9**, 43–89.

Maruhashi, T. (1981). Activity patterns of a troop of Japanese macaques (*Macaca fuscata yakuii*) on Yakushima Island, Japan. *Primates*, **22**, 1–14.

Matthysen, E., Grubb, T.C. and Cimprich, D. (1991). Social control of sex-specific foraging behaviour in downy woodpeckers, *Picoides pubescens*. *Animal Behaviour*, **42**, 515–17.

Mautz, W.W. (1978). Sledding on a brushy hillside: the fat cycle in deer. *Wildlife Society Bulletin*, **6**, 88–90.

Maynard Smith, J. and Brown, R.L.W. (1986). Competition and body size. *Theoretical Population Biology*, **30**, 166–79.

McBride, A.F. and Kritzler, H. (1951). Observations on pregnancy, parturition, and post-natal behavior in the bottlenose dolphin. *Journal of Mammalogy*, **32**, 251–66.

McCann, T.S. (1980). Territoriality and breeding behaviour of adult male Antarctic Fur seal, *Arctocephalus gazella*. *Journal of Zoology London*, **192**, 295–310.

McCracken, G.F. (1984). Communal nursing in Mexican free-tailed bat maternity colonies. *Science*, **223**, 1090–1.

McCracken, G.F. and Bradbury, J.W. (1981). Social organisation and kinship in the polygynous bat *Phyllostomus hastatus*. *Behavioural Ecology and Sociobiology*, **8**, 11–34.

McCracken, G.F. and Wilkinson, G.S. (2000). Bat mating systems. In *Reproductive Biology of Bats*, eds. E.G. Crichton and P.H. Krutzsch, San Diego: Academic Press, pp. 321–62.

McCullough, D.R. (1979). *The George Reserve Deer Herd*. Ann Arbor, University of Michigan Press.

McCullough, D.R. and McCullough, Y. (2000). *Kangaroos in Outback Australia*. New York: Columbia University Press.

McCullough, D. R., Hirth, D.R. and Newhouse, S. J. (1989). Resource partitioning between sexes in white-tailed deer. *Journal of Wildlife Management*, **53**, 277–83.

McFarland-Symington, M. (1988). Demography, activity budgets, and ranging patterns of black spider monkeys (*Ateles belzebuth chamek*) in the Manu National Park. *American Journal of Primatology*, **15**, 45–67.

(1990). Fission-fusion social organization in *Ateles* and *Pan*. *International Journal of Primatology*, **11**, 47–61.

McIlhenny, E.A. (1937). Life history of the boat-tailed grackle in Louisiana. *Auk*, **54**, 274–95.

McKibben, J.N. and Nelson, D.R. (1986). Patterns of movement and grouping of gray reef sharks, *Carcharhinus amblyrhynchos*, at Enewetak, Marshall Islands. *Bulletin of Marine Science*, **38**, 89–110.

McKinnon, J. (1994). Feeding habits of the dusky dolphin, *Lagenorhynchus obscurus*, in the coastal waters of central Peru. *Fishery Bulletin*, **92**, 569–78.

McLaughlin, R.H. and O'Gower, A.K. (1971). Life history and underwater studies of a heterodont shark. *Ecological Monographs*, **41**, 271–89.

McLoyd, V. (1980). Verbally expressed modes of transformation in the fantasy and play of black preschool children. *Child Development*, **51**, 1133–9.

McNaughton, S.J. and Georgiadis, N.J. (1986) Ecology of African grazing and browsing mammals. *Annual Review of Ecology and Systematics*, **17**, 39–65.

McRobert, S.P. and Bradner, J. (1998). The influence of body coloration on shoaling preferences in fish. *Animal Behaviour*, **56**, 611–15.

Meaney, M.J. (1988). The sexual differentiation of social play. *Trends in Neurosciences*, **11**, 54–8.

Meaney, M.J., Stewart, J. and Beatty, W.W. (1985). Sex differences in social play: the socialization of sex roles. *Advances in the Study of Behaviour*, **15**, 1–58.

Menchen, F.C. and Winfield, I. (2000). Job search and sex segregation: Does sex of social contact matter? *Sex Roles*, **42**(9–10), 847–64.

Merchant, M.E., Shukla, S.S. and Akers, H.A. (1991). Lead concentrations in wing bones of the mottled duck. *Environmental Toxicology and Chemistry*, **10**, 1503–7.

Mesnick, S.L. (2001). Genetic relateness in sperm whales: evidence and cultural implications. *Behavior and Brain Science*, **26**, 346–7.

Metten, H. (1939a). Reproduction of the dogfish. *Nature*, **143**, 121–2.

(1939b). Studies on the reproduction of the dogfish. *Philosophical Transactions of the Royal Society of London B*, **230**, 217–38.

Michaud, R. (1993). Distribution estivale du béluga du Saint-Laurent; synthèse 1986 à 1992. *Rapport technique canadien des sciences halieutiques et aquatiques*, **1906**, vi–28.

(1999). Social organization of the St. Lawrence beluga, *Delphinapterus leucas*. *13th Conference on the Biology of Marine Mammals*, Maui, Hawaii, Society for marine mammalogy.

Michelena, P., Bouquet, P.M., Dissac, A. *et al.* (2004). An experimental test of hypotheses explaining social segregation in dimorphic ungulates. *Animal Behaviour*, **68**, 1371–80.

Miles, D.B., Snell, H.L. and Snell, H.M. (2001). Intrapopulation variation in endurance of Galapagos lava lizards (*Microlophus albemarlensis*): evidence for an interaction between natural and sexual selection. *Evolutionary Ecology Research*, **3**, 795–804.

Milinski, M. (1993). Predation risk and feeding behaviour. In *Behaviour of Teleost Fishes*, ed. T.J. Pitcher. London: Chapman and Hall, pp. 285–305.

Millar, J.S. and Hickling, G.J. (1992). The fasting endurance hypothesis revisited. *Functional Ecology*, **6**, 496–8.

Minchin, D. (1987). Fishes of the Lough Hyne marine reserve. *Journal of Fish Biology*, **31**, 343–52.

Miquelle, D.G., Peek, J.M. and Van Ballenberghe, V. (1992). Sexual segregation in Alaskan moose. *Wildlife Monographs*, **122**, 1–57.

Miranda, J.P. and Andrade, G.V. (2003). Seasonality in diet, perch use, and reproduction of the gecko *Gonatodes humeralis* from Eastern Brazilian Amazon. *Journal of Herpetology*, **37**, 433–8.

Mitani, J.C. and Amsler, S. (2003). Social and spatial aspects of male subgrouping in a community of wild chimpanzees. *Behaviour*, **140**, 869–84.

Mitani, J.C. and Watts, D.P. (2002). Why do male chimpanzees hunt and share meat? *Animal Behaviour*, **61**, 915–24.

Mitani, J.C., Gros-Louis, J. and Richards, A.F. (1996). Sexual dimorphism, the operational sex ratio, and the intensity of male competition in poygynous primates. *American Naturalist*, **147**, 966–80.

Mitani, J.C., Watts, D.P. and Lwanga, J. (2002). Ecological and social correlates of chimpanzee party size and composition. In *Behavioural Diversity in Chimpanzees and Bonobos*, eds., C. Boesch, G. Hohmann and L.F. Marchant. Cambridge: Cambridge University Press, pp. 102–11.

Mitchell, C. (1994). Migration alliances and coalitions among adult male South American squirrel monkeys. *Behaviour*, **130**, 169–89.

Mitchell, P. (1990). The home ranges and social activity of koalas – a quantitative analysis. In *Biology of the Koala*, eds. A.K. Lee, K.A. Handasyde and G.D. Sanson. Chipping Norton, NSW: Surrey Beatty and Sons, pp. 171–87.

Miyazaki, N. and Nishiwaki, M. (1978). School structure of the striped dolphins off the Pacific coast of Japan. *Scientific Report of the Whale Research Institute*, **30**, 65–115.

Moll, D. (1990). Population sizes and foraging ecology in a tropical freshwater stream turtle community. *Journal of Herpetology*, **24**, 48–53.

Moll, E. O. and Legler, J. M. (1971). The life history of a neotropical slider turtle, *Pseudemys scripta* (Schoepff) in Panama. Bulletin of the Los Angeles County Museum of Natural History, *Science*, **11**, 1–102.

Moore, C. L. (1985). Development of mammalian sexual behavior. In *The Comparative Development of Adaptative Skills: Evolutionary Implications*, ed. E. S. Gollin. Hillsdale: Lawrence Erlbaum Associate, pp. 19–56.

Morgantini, L. E. and Hudson, R. J. (1981). Sex differential in use of the physical environment by bighorn sheep (*Ovis canadensis*). *Canadian Field-Naturalist*, **95**, 69–74.

Mori, A. and Watanabe, K. (2003). Life history of male Japanese monkeys living on Koshima Islet. *Primates*, **44**, 119–26.

Moritz, C. (1993). The origin and evolution of parthenogenesis in the *Heteronotia binoei* complex: synthesis. *Genetica*, **90**, 269–80.

Morreale, S. J., Gibbons, J. W. and Congdon, J. D. (1984). Significance of activity and movement in the yellow-bellied slider turtle (*Pseudemys scripta*). *Canadian Journal of Zoology*, **62**, 1038–2.

Morrison, M. L. and With, K. A. (1987). Interseasonal and intersexual resource partitioning in hairy and white-headed woodpeckers. *Auk*, **104**, 225–33.

Morrissey, J. F. and Gruber, S. H. (1993). Habitat selection by juvenile lemon sharks, *Negaprion brevirostris*. *Environmental Biology of Fishes*, **38**, 311–19.

Morton, E. S. (1990). Habitat segregation by sex in the hooded warbler: experiments on proximate causation and discussion of its evolution. *American Naturalist*, **135**, 319–33.

Morton, E. S., van der Voort, M. E. and Greenberg, R. (1993). How a warbler chooses its habitat: field support for laboratory experiments. *Animal Behaviour*, **46**, 47–53.

Morton, S. (1978). An ecological study of *Sminthopsis crassicaudata* (Marsupialia: Dasyuridae) III. Reproduction and life history. *Australian Wildlife Research*, **5**, 183–211.

Morton, E. S., Lynch, J. F., Young, K. and Mehlhop, P. (1987). Do male hooded warblers exclude females from non-breeding territories in tropical forest? *Auk*, **104**, 133–5.

Moseby, K. E. and O'Donnell, E. (2003). Reintroduction of the greater bilby, *Macrotis lagotis* (Reid) (Marsupialia: Thylacomyidae), to northern South Australia: survival, ecology and notes on reintroduction protocols. *Wildlife Research*, **30**, 15–27.

Moskowitz, D. S., Suh, E. J. and Desaulniers, J. (1994). Situational influences on gender differences in agency and communion. *Journal of Personality and Social Psychology*, **66**(4), 753–61.

Moss, R. (1983). Gut size, body weight, and digestion of winter foods by grouse and ptarmigan. *Condor*, **85**, 185–93.

Mougin, J. L. (1968). Etude écologique de quatre espèces de pétrels antarctique. *Oiseau Revue Francophone Ornithology*, **38**, 1–52.

(1975). Ecologie comparée des Procellariidae Antarctiques et sub-Antarctiques. *CNFRA*, **36**, 1–195.

Mueller, H. C. (1990). The evolution of reversed sexual dimorphism in size in monogamous species of birds. *Biological Reviews of the Cambridge Philosophical Society*, **65**, 553–85.

Müller, A. E. and Thalmann, U. (2000). Origin and evolution of primate social organisation: a reconstruction. *Biological Reviews of the Cambridge Philosophical Society*, **75**, 405–35.

Murphy, M. T., Pierce, A., Shoen, J. *et al.* (2001). Population structure and habitat use by overwintering neotropical migrants on a remote oceanic island. *Biological Conservation*, **102**, 333–45.

Murray, M. G. and Illius, A. W. (2000). Vegetation modification and resource competition in grazing ungulates. *Oikos*, **89**, 501–8.

Murray, T. E., Bartle, J. A., Kalish, S. R. and Taylor, P. R. (1993). Incidental capture of seabirds by Japanese southern bluefin tuna longline vessels in New Zealand waters, 1988–1992. *Bird Conservation International*, **3**, 181–210.

Mushinsky, H. R., Hebrard, J. J. and Vodopich, D. S. (1982). Ontogeny of water snake foraging ecology. *Ecology*, **63**, 1624–9.

Myers, J. P. (1981). A test of three hypotheses for latitudinal segregation of the sexes in wintering birds. *Canadian Journal of Zoology*, **59**, 1527–34.

Myrberg, A. A. and Gruber, S. H. (1974). The behaviour of the bonnethead, *Sphyrna tiburo*. *Copeia*, **1974**, 358–74.

Myres, B. C. and Eells, M. M. (1968). Thermal aggregation in Boa constrictor. *Herpetologica*, **24**, 61–6.

Mysterud, A. (2000). The relationship between ecological segregation and sexual body size dimorphism in large herbivores. *Oecologia*, **124**, 40–54.

Nakagawa, N. (1989). Activity budget and diet of patas monkeys in Kala Maloue National Park, Cameroon. *Primates*, **30**, 27–34.

Naulleau, G. (1966). Etude complementaire de l'activitie de *Vipera aspis* dans la nature. *Vie et Milieu*, **17**, 461–509.

Nel, D. C., Ryan, P. G., Nel, J. L. *et al.* (2002). Foraging interactions between wandering albatrosses *Diomedea exulans* breeding on Marion Island and long-line fisheries in the southern Indian Ocean. *Ibis*, **144** (on-line), E141–E154.

Nelson, J. E. (1965). Behaviour of Australian Pteropodidae (Megachiroptera). *Animal Behaviour*, **13**, 544–57.

Nelson, M. E. and Mech, L. D. (1981). Deer social organization and wolf predation in northeastern Minnesota. *Wildlife Monographs*, **77**, 1–53.

Nelson, M. E. and Mech, L. D. (1984). Home-range formation and dispersal of deer in Northeastern Minnesota. *Journal of Mammalogy*, **65**, 567–75.

Neuhaus, P. and Ruckstuhl, K. E. (2002a). The link between sexual dimorphism, activity budgets, and group cohesion: the case of the plains zebra (*Equus burchelli*). *Canadian Journal of Zoology*, **80**, 1437–41.

(2002b). Foraging behaviour in Alpine ibex (*Capra ibex*): consequences of reproductive status, body size, age and sex. *Ecology, Ethology and Evolution*, **14**, 373–81.

Newsome, A. E. (1980). Differences in the diets of male and female red kangaroos in central Australia. *African Journal of Ecology*, **18**, 27–31.

Newton, I. (1986). *The Sparrowhawk*. Calton: T and AD Poyser.

(1998). *Population Limitation in Birds*. San Diego: Academic Press.

Newton-Fisher, N., Reynolds, V. and Plumtre, A. J. (2000). Food supply and chimpanzee (*Pan troglodytes schweinfurthii*) in the Budongo Forest Reserve, Uganda. *International Journal of Primatology*, **21**, 613–28.

Nicholls, D. G., Murray, M. D., Butcher, E. and Moors, P. (1997). Weather systems determine the non-breeding distribution of wandering albatrosses over southern oceans. *Emu*, **95**, 240–4.

Nichols, J. D. and Aramis, G. M. (1980). Sex-specific differences in winter distribution patterns of canvasbacks. *Condor*, **82**, 406–16.

Nievergelt, B. (1967). Die zusammensetzung der gruppen beim alpensteinbock. *Zeitschrift für Säugetierkunde*, **32**, 129–44.

(1981). *Ibexes in an African Environment*. Berlin: Springer Verlag.

Nikaido, M., Rooney, A. P. and Okada, N. (1999). Phylogenetic relationships among cetartiodactyls based on insertions of short and long interspersed elements: hippopotamuses are the closest extant relatives of whales. *Proceedings of the National Academy of Science*, **96**, 10261–6.

Nilsson, L. (1970). Food seeking activity of south Swedish diving ducks in the non-breeding season. *Oikos*, **21**, 145–54.

Nisbet, I. C. T. and Medway, L. (1972). Dispersion, population ecology and migration of eastern great reed warblers *Acrocephalus orientalis* wintering in Malaysia. *Ibis*, **114**, 451–94.

Nisbet, I. C. T., Montoya, J. P., Burger, J. and Hatch, J. J. (2002). Use of stable isotopes to investigate individual differences in diets and mercury exposures among common terns *Sterna hirundo* in breeding and wintering grounds. *Marine Ecology Progress Series*, **242**, 267–74.

Nishida, T. (1968). The social group of wild chimpanzees in the Mahale Mountains. *Primates*, **9**, 167–224.

Nishida, T. and Hosaka, K. (1996). Coalition strategies among adult male chimpanzees of the Mahale Mountains, Tanzania. In *Great Ape Societies*, eds. W. A. McGrew, L. A. Marchant, and T. Nishida. Cambridge: Cambridge University Press, pp. 114–34.

Nogueira, C., Sawaya, R. J. and Martins, M. (2003). Ecology of the pitviper, *Bothrops moojeni*, in the Brazilian Cerrado. *Journal of Herpetology*, **37**, 653–9.

Norbury, G. L., Coulson, G. M. and Walters, B. L. (1988). Aspects of the demography of the western grey kangaroo, *Macropus fuliginosus melanops*, in semiarid north-west Victoria. *Australian Wildlife Research*, **15**, 257–66.

Norris, K. S. (1994). Comparative view of cetacean social ecology, culture, and evolution. In *The Hawaiian Spinner Dolphin*, eds. K. S. Norris, B. Würsig, R. S. Well and M. Würsig. Berkeley: University of California Press, pp. 301–44.

Norris, K. S. and Dohl, T. P. (1980). The structure and functions of cetacean schools. In *Cetacean Behavior: Mechanisms and Functions*, ed. L. M. Herman. New York: John Wiley and Sons, pp. 211–59.

Norris, K. S. and Schilt, C. R. (1988). Cooperative societies in three-dimensional space: on the origins of aggregations, flocks, and schools, with special reference to dolphins and fish. *Ethology and Sociobiology*, **9**, 149–79.

Northcutt, R. G. (1977). Elasmobranch central nervous system organization and its possible evolutionary significance. *American Zoologist*, **17**, 411–29.

Novacek, M. J. (1992). Mammalian phylogeny: shaking the tree. *Nature, London*, **356**, 121–5.

Nowak, R. M. (1991). *Walker's Mammals of the World*, 5th edn. Baltimore: John Hopkins University Press.

Nunn, C. (1999). The number of males in primate social groups: a comparative test of the socioecological model. *Behavioral Ecology and Sociobiology*, **46**, 1–13.

O'Donnell, C. F. J. and Sedgeley, J. A. (1999). Use of roosts by the long-tailed bat, *Chalinolobus tuberculatus*, in temperate rain forest in New Zealand. *Journal of Mammalogy*, **80**, 913–23.

Oakwood, M. (2000). Reproduction and demography of the northern quoll, *Dasyurus hallucatus*, in the lowland savanna of northern Australia. *Australian Journal of Zoology*, **48**, 519–39.

(2002). Spatial and social organization of a carnivorous marsupial *Dasyurus hallucatus* (Marsupialia: Dasyuridae). *Journal of Zoology*, **257**, 237–48.

Ohguchi, O. (1978). Experiments on the selection against colour oddity of water fleas by three-spined stickelbacks. *Zeitschrift fur Tierpsychologie*, **47**, 254–67.

Olesiuk, P. F., Bigg, M. A. and Ellis, G. M. (1990). Life history and population dynamics of resident killer whales (Orcinus orca) in the coastal waters of British Columbia and Washington state. Report of the International Whaling Commission, Special Issue, 12, 209–405.

Oli, M. K. (1996). Seasonal patterns in habitat use of blue sheep Pseudois nayaur (Artiodactyla, Bovidae) in Nepal. Mammalia, 60, 187–93.

Oliver, W. R. B. (1955). New Zealand Birds. Wellington: A. H. and A. W. Reed.

Olsen, A. M. (1954). The biology, migration, and growth rate of the school shark, Galeorhinus australis (Macleay) (Carcharhinidae) in southeastern Australian waters. Australian Journal of Marine and Freshwater Research, 5, 353–410.

Ong, P. S. (1994). The Social Organization of the Common Ringtail Possum, Pseudocheirus peregrinus. Unpublished Ph.D. thesis, Monash University.

Ordway, L. L. and Krausman, P. R. (1986). Habitat use by mule deer. Journal of Wildlife Management, 50, 677–83.

Ortega, J. and Arita, H. T. (2000). Defence of females by dominant males of Artibeus jamaicensis (Chiroptera: Phyllostomidae). Ethology, 106, 395–407.

Ortega, J. and Arita, H. T. (2002). Subordinate males in harem groups of Jamaican fruit-eating bats (Artibeus jamaicensis): Satellites or sneaks? Ethology, 108, 1077–91.

Oswald, C., Fonken, P., Atkinson, D. and Palladino, M. (1993). Lactational water-balance and recycling in white-footed mice, red-backed voles, and gerbils. Journal of Mammalogy, 74, 963–70.

Owen, M. and Black, J. M. (1990). Waterfowl Ecology. Glasgow and London: Blackie.

(1991). The importance of migration mortality in non-passerine birds. In Bird Population Studies. Relevance to Conservation and Management, eds. C. M. Perrins, J.-D. Lebreton and G. J. M. Hirons. Oxford University Press.

Owen, M. and Dix, M. (1986). Sex ratios in some common British wintering ducks. Wildfowl, 37, 104–12.

Owens, I. P. F. and Bennet, P. M. (1994). Mortality costs of parental care and sexual dimorphism in birds. Proceedings of the Royal Society of London B, 257, 1–8.

Owen-Smith, N. (1993). Comparative mortality rates of male and female kudus: the costs of sexual size dimorphism. Journal of Animal Ecology, 62, 428–40.

Owen-Smith, R. N. (1988). Megaherbivores. The Influence of Very Large Body Size on Ecology. Cambridge: Cambridge University Press.

Ozoga, J. J. (1968). Variations in microclimate in a conifer swamp deeryard in Northern Michigan. Journal of Wildlife Management, 32, 574–81.

Ozoga, J. J. and Verme L. J. (1970). Winter feeding patterns of penned white-tailed deer. Journal of Wildlife Management, 34, 431–9.

Palombit, R. A. (1999). Infanticide and the evolution of pair bonds in nonhuman primates. Evolutionary Anthropology, 7, 117–28.

Palombit, R. A., Cheney, D. L., Fischer, J. et al. (2000). Male infanticide and defense of infants in chacma baboons. In Infanticide by Males and Its Implications, eds. C. P. van Schaik and C. H. Janson. Cambridge: Cambridge University Press, pp. 123–52.

Papastavrou, V., Smith, S. C. and Whitehead, H. (1989). Diving behaviour of the sperm whale, Physeter macrocephalus, off the Galapagos Islands. Canadian Journal of Zoology, 67, 839–46.

Parke, R. D. and Suomi, S. J. (1981). Adult male infant relationships: human and nonhuman primate evidence. In Behavioral Development, eds. K. Immelman, G. W. Barlow, L. Petronovitch, and M. Main. New York: Cambridge University Press, pp. 700–25.

Parker, K.L., Gillingham, M.P., Hanley, T.A. and Robbins, C.T. (1999). Energy and protein balance of free-ranging black-tailed deer in a natural forest environment. *Wildlife Monographs*, **143**, 1–48.

Parker, W.S. and Brown, W.S. (1980). Comparative ecology of two colubrid snakes, *Masticophis t. taeniatus* and *Pituophis melanoleucus deserticola*, in northern Utah. *Milwaukee Public Museum Publications in Biology and Geology*, **7**, 1–104.

Parmelee, D.F., Parmelee, J.M. and Fuller, M.R. (1985). Ornithological investigations at Palmer Station: the first long-distance tracking of seabirds by satellites. *Antarctic Journal of the United States*, 162–3.

Parmelee, J.R. and Guyer, C. (1995). Sexual differences in foraging behavior of an anoline lizard, *Norops humilis*. *Journal of Herpetology*, **29**, 619–21.

Parr, A.E. (1931). Sex dimorphism and schooling behavior among fishes. *American Naturalist*, **65**, 173–80.

Parrish, J.D. and Sherry, T.W. (1994). Sexual habitat segregation by American redstarts wintering in Jamaica: importance of resource seasonality. *Auk*, **111**, 38–49.

Parsons, K.N., Jones, G., Davidson-Watts, I. and Greenaway, F. (2003). Swarming of bats at underground sites in Britain – implications for conservation. *Biological Conservation*, **111**, 63–70.

Pasinelli, G. (2000). Sexual dimorphism and foraging niche partitioning in the middle spotted woodpecker *Dendrocopos medius*. *Ibis*, **2000**, 635–44.

Patterson, D.L. and Fraser, W.R. (2003). Satellite tracking southern giant petrels at Palmer station, Antarctica. *Feature Articles, Microwave Telemetry, Inc.*, **8**, 3–4.

Patterson, I.A.P., Reid, R.J., Wilson, B. *et al.* (1998). Evidence for infanticide in bottlenose dolphins: an explanation for violent interactions with harbour porpoises? *Proceedings of the Royal Society of London, Series B*, **265**, 1167–70.

Pearson, D., Shine, R. and How, R. (2002a). Sex-specific niche partitioning and sexual size dimorphism in Australian pythons (*Morelia spilota imbricata*). *Biological Journal of the Linnean Society*, **77**, 113–25.

Pearson, D., Shine, R. and Williams, A. (2002b). Geographic variation in sexual size dimorphism within a single snake species (*Morelia spilota*, Pythonidae). *Oecologia*, **131**, 418–26.

(2003). Thermal biology of large snakes in cool climates: a radio-telemetric study of carpet pythons (*Morelia spilota imbricata*) in south-western Australia. *Journal of Thermal Biology*, **28**, 117–31.

Pellegrini, A.D. (2003). Perceptions and possible functions of play and real fighting in early adolescence. *Child Development*, **74**, 1459–70.

Pellegrini, A.D. and Bartini, M. (2001). Dominance in early adolescent boys: affiliative and aggressive dimensions and possible functions. *Merrill-Palmer Quarterly*, **47**, 142–63.

Pellegrini, A.D. and Long, J.D. (2003). A sexual selection theory longitudinal analysis of sexual segregation and integration in early adolescence. *Journal of Experimental Child Psychology*, **85**, 257–78.

Pellegrini, A.D. and Perlmutter, J.C. (1989). Classroom contextual effects on children's play. *Developmental Psychology*, **25**, 289–96.

Pellegrini, A.D. and Smith, P.K. (1998). Physical activity play: the nature and function of a neglected aspect of play. *Child Development*, **69**, 577–98.

Pellegrini, A.D., Blatchford, P., Kato, K. and Baines, E. (2003). A short-time longitudinal study of children's playground games in primary school: implications for adjustment to school and social adjustment in the USA and the UK. *Social Development*, **13**, 107–23.

Pellegrini, A.D., Horvat, M. and Huberty, P.D. (1998). The relative cost of children's physical activity play. *Animal Behaviour*, **55**, 1053–106.

Pellew, R.A. (1983). The impacts of elephant, giraffe and fire upon the *Acacia tortilis* woodlands of the Serengeti. *African Journal of Ecology*, **21**, 41–74.

Pennycuick, C.J. (1987). Flight of seabirds. In *Seabirds: Feeding Ecology and Role in Marine Ecosystems*, ed. J.P. Croxall. Cambridge: Cambridge University Press, pp. 43–62.

Pepper, J.M., Mitani, J.C. and Watts, D.P. (1999). General gregariousness and specific partner preference among wild chimpanzees. *International Journal of Primatology*, **20**, 613–32.

Pereira, M.E. (1988). Effects of age and sex on intra-group spacing in juvenile savanna baboons, *Papio cynocephalus cynocephalus*. *Animal Behaviour*, **36**, 184–204.

(1989). Agonistic interactions of juvenile savanna baboons. II. Agonistic support and rank acquisition. *Ethology*, **80**, 152–71.

Pereira, M.E. and Fairbanks, L.A. (2002). Foreword 2002: family, friends, and the evolution of childood. In *Juvenile Primates*, eds. M.E. Pereira and L.A. Fairbanks, 2nd edn. Chicago: University of Chicago Press, pp. vii–xxiv.

Pereira, M.E. and Leigh, S.R. (2003). Modes of primate development. In *Primate Life Histories and Socioecology*, eds. P.M. Kappeler and M.E. Pereira. Chicago: University of Chicago Press, pp. 149–76.

Pérez-Barbería, F.J. and Gordon, I.J. (1998a). The influence of sexual dimorphism in body size and mouth morphology on diet selection and sexual segregation in cervids. *Acta Veterinaria Hungarica*, **46**, 357–67.

(1998b). Factors affecting food comminution during chewing in ruminants: a review. *Biological Journal of the Linnean Society*, **63**, 233–56.

(1999). Body size dimorphism and sexual segregation in polygynous ungulates: an experimental test with Soay sheep. *Oecologia*, **120**, 258–67.

Pérez-Barbería, F.J., Gordon, I.J. and Pagel, M. (2002). The origins of sexual dimorphism in body size in ungulates. *Evolution*, **56**, 1276–85.

Perez-Mellado, V. and de la Riva, I. (1993). Sexual size dimorphism and ecology: the case of a tropical lizard, *Tropidurus melanopleurus* (Sauria, Tropiduridae). *Copeia*, **1993**, 969–76.

Pernetta, J.C. (1977). Observations on the habits and morphology of the snake *Laticauda colubrina* in Fiji. *Canadian Journal of Zoology*, **55**, 1612–19.

Perret, M. (1998). Energetic advantage of nest sharing in a solitary primate, the lesser mouse lemur (*Microcebus murinus*). *Folia Primatologica*, **59**, 1–25.

Perrin, W.F. and Reilly, S.B. (1984). Reproductive parameters of dolphins and small whales of the family Delphinidae. *Report of the International Whaling Commission, Special Issue*, **6**, 97–134.

Perrin, W.F., Wilson, C.E. and Archer II, F.I. (1994). Striped dolphin – *Stenella coeruleoalba* (Meyen, 1833). In *Handbook of Marine Mammals: The First Book of Dolphins*, vol. 5, eds. S.H. Ridgway and R. Harrison. London: Academic Press, pp. 129–60.

Perry, G. (1996). The evolution of sexual dimorphism in the lizard *Anolis polylepis* (Iguania): evidence from intraspecific variation in foraging behavior and diet. *Canadian Journal of Zoology*, **74**, 1238–45.

Perry, G. and Brandeis, M. (1992). Variation in stomach contents of the gecko *Ptyodactylus hasselquistii guttatus* in relation to sex, season, and locality. *Amphibia-Reptilia*, **13**, 275–82.

Perry, G. and Garland, T. (2002). Lizard home ranges revisited: effects of sex, body size, diet, habitat, and phylogeny. *Ecology*, **83**, 1870–85.

Peter, H.-U., Kaiser, M. and Gebauer, A. (1988). Investigations on birds and seals at King George Island. *Geodätische und geophysikalische Veröffentlichungen*, **1**, 5–80.

Peters, W.D. and Grubb, T.C. (1983). An experimental analysis of sex-specific foraging in the downy woodpecker *Picoides pubescens*. *Ecology*, **64**, 1437–43.

Peterson, C. R., Gibson, A. R. and Dorcas, M. E. (1993). Snake thermal ecology: the causes and consequences of body-temperature variation. In *Snakes: Ecology and Behavior*, eds. R.A. Seigel and J.T. Collins. New York: McGraw-Hill, pp. 241–314.

Peterson, R.L. (1965). A review of the bats of the genus *Ametrida*, family Phyllostomidae. *Life Science Contribution, Royal Ontario Museum*, **65**, 1–13.

Pettit, K.E., Bishop, C.A. and Brooks, R.J. (1995). Home range and movements of the common snapping turtle, *Chelydra serpentina serpentina*, in a coastal wetland of Hamilton Harbor, Lake Ontario, Canada. *Canadian Field-Naturalist*, **109**, 192–200.

Peuhkuri, N. (1999). Size-assorted fish shoals and the majority's choice. *Behavioural Ecology and Sociobiology*, **46**, 307–12.

Pfeffer, P. (1967). Le mouflon de Corse (*Ovis ammon musimon* Schreber, 1782); position systématique, écologie, et éthologie comparées. *Mammalia*, **31**, 1–262.

Phelps, T.W. (1978). Seasonal movement of the snakes *Coronella austriaca, Vipera berus*, and *Natrix natrix* in southern England. *British Journal of Herpetology*, **5**, 775–81.

Phillips, J.A. (1995). Movement patterns and density of *Varanus albigularis*. *Journal of Herpetology*, **29**, 407–16.

Phillips, R.A., Silk, J.R.D., Croxall, J.P., Afanasyev, V. and Briggs, D.R. (2003). Accuracy of geolocation estimates for flying seabirds. *Marine Ecology Progress Series*, **266**, 265–72.

Phillips, R.A., Silk, J.R.D., Phalan, B., Catry, P. and Croxall, J.P. (2004). Seasonal sexual segregation in the two *Thalassarche* albatross species: competitive exclusion, reproductive role specialization or foraging niche divergence? *Proceeding of the Royal Society of London*, **271**, 1283–91.

Pianka, E.R. and Vitt, L.J. (2003). *Lizards: Windows to the Evolution of Diversity*. Berkeley, CA: University of California Press.

Pickering, S.P.C. (1989). Attendance patterns and behaviour in relation to experience and pair-bond formation in wandering albatrosses *Diomedea exulans* at South Georgia. *Ibis*, **131**, 183–95.

Pickering, S.P.C. and Berrow, S.D. (2002). Courtship behavior of the wandering albatross *Diomedea exulans* at Bird Island, South Georgia. *Marine Ornithology*, **29**, 29–37.

Pierce, B.M., Bleich, V.C. and Bowyer, R.T. (2000). Prey selection by mountain lions and coyotes: effects of hunting style, body size, and reproductive status. *Journal of Mammalogy*, **81**, 462–72.

Pierce, G.J. and Boyle, P.R. (1991). A review of methods for diet analysis in piscivorous marine mammals. *Oceanography and Marine Biology*, **29**, 409–86.

Pierotti, R. and Annett, C.A. (1991). Diet choice in the herring gull: constraints imposed by reproductive and ecological factors. *Ecology*, **72**, 319–28.

Pippard, L. (1985). Status of the St. Lawrence River population of beluga, *Delphinapterus leucas*. *Canadian Field-Naturalist*, **99**, 438–50.

Pitcher, E.G. and Schultz, L.H. (1983). *Boys and Girls at Play: The Development of Sex Roles*. South Hadley, MA: Bergin and Garvey.

Pitcher, T.J. and Parrish, J.K. (1993). Functions of shoaling behaviour in teleosts. In *Behaviour of Teleost Fishes*, ed. T.J. Pitcher. Chapman and Hall, London, pp. 363–439.

Pitcher, T.J., Magurran, A.E. and Edwards, J.I. (1985). Schooling mackerel and herring choose neighbours of similar size. *Marine Biology*, **86**, 319–22.

Pitman, R.L., Ballance, L.T., Mesnick, S.I. and Chivers, S.J. (2001). Killer whale predation on sperm whales: observations and implications. *Marine Mammal Science*, **17**, 494–507

Plummer, M.V. (1977). Activity, habitat, and population structure in the turtle *Trionyx muticus*. *Copeia*, **1977**, 431–440.

(1981a). Communal nesting of *Opheodrys aestivus* in the laboratory. *Copeia*, **1981**, 243–6.

(1981b). Habitat utilization, diet and movements of a temperate arboreal snake (*Opheodrys aestivus*). *Journal of Herpetology*, **15**, 425–32.

Plummer, M.V. and Farrar, D.B. (1981). Sexual dietary differences in a population of *Trionyx muticus*. *Journal of Herpetology*, **15**, 175–9.

Pluto, T.G. and Bellis, E.D. (1986). Habitat utilization by the turtle, *Graptemys geographica*, along a river. *Journal of Herpetology*, **20**, 22–31.

(1988). Seasonal and annual movements of riverine map turtles, *Graptemys geographica*. *Journal of Herpetology*, **22**, 152–8.

Poindron, P., Lévy, F. and Krehbiel, D. (1988). Genital, olfactory, and endocrine interactions in the development of maternal behaviour in the parturient ewe. *Psychoneuroendocrinology*, **13**, 99–125.

Polis, G.A. (1984). Age structure component of niche width and intra-specific resource partitioning: can age groups function as ecological species? *American Naturalist*, **123**, 541–64.

Pope, T. (1990). The reproductive consequences of male cooperation in the red howler monkey: paternity exclusion in multi-male and single-male troops using genetic markers. *Behavioral Ecology and Sociobiology*, **27**, 439–46.

(2000a). The evolution of male philopatry in neotropical monkeys. In *Primate Males*, ed. P. Kappeler. Cambridge: Cambridge University Press, pp. 219–35.

(2000b). Reproductive success increases with degree of kinship in cooperative coalitions of female red howler monkeys. *Behavioral Ecology and Sociobiology*, **48**, 253–67.

Porter, D.A. (1990). Feeding ecology of *Graptemys caglei* Haynes and McKown in the Guadalupe River, Dewitt County, Texas. M.Sc. thesis, Canyon, TX: West Texas State University.

Porter, R.H., Désiré, L., Bon, R., Orgeur, P. (2001). The role of familiarity in the development of social recognition by lambs. *Behaviour*, **134**, 207–19.

Porter, W.F. (1994). Family Meleagridae (Turkeys). In *Handbook of the Birds of the World*, vol. 2, eds. J. del Hoyo, A. Elliot and J. Sargatal. Barcelona: Lynx Edicions.

Post, D., Hausfater, G. and McKusky, S. (1981). Feeding behavior of yellow baboons (*Papio cynocephalus*): relationship to age, gender, and dominance rank. *Folia Primatologica*, **34**, 170–95.

Post, D.M., Armbrust, T.S., Horne, E.A. and Goheen, J.R. (2001). Sexual segregation results in differences in content and quality of bison (*Bos bison*) diets. *Journal of Mammalogy*, **82**, 407–13.

Post, E., Langvatn, R., Forchhammer, M.C. and Stenseth, N.C. (1999). Environmental variation shapes sexual dimorphism in red deer. *Proceedings of the National Academy of Sciences*, **96**, 4467–71.

Pough, F.H. (1973). Lizard energetics and diet. *Ecology*, **54**, 837–44.

(1980). The advantages of ectothermy for tetrapods. *American Naturalist*, **115**, 92–112.

Pough, F.H. and Groves, J.D. (1983). Specialization in the body form and food habits of snakes. *American Zoologist*, **23**, 443–54.

Pounds, J.A. and Jackson, J.F. (1983). Utilization of perch sites by sex and size classes of *Sceloporus undulatus undulatus*. *Journal of Herpetology*, **17**, 287–9.

Powell, G.L. and Russell, A.P. (1984). The diet of the eastern short-horned lizard (*Phrynosoma douglassi brevirostre*) in Alberta and its relationship to sexual size dimorphism. *Canadian Journal of Zoology*, **62**, 428–40.

Power, T.G. (2000). *Play and Exploration in Children and Animals*. Mahwah, NJ: Erlbaum.

Pratt, H.L. (1979). Reproduction in the blue shark, *Prionace glauca*. *U.S. Fishery Bulletin*, **77**, 445–70.

Pratt, H.L. and Carrier, J.C. (2001). A review of elasmobranch reproductive behaviour with a case study on the nurse shark, *Ginglymostoma cirratum*. *Environmental Biology of Fishes*, **60**, 157–88.

Preest, M.R. (1994). Sexual size dimorphism and feeding energetics in *Anolis carolinensis*: why do females take smaller prey than males? *Journal of Herpetology*, **28**, 292–8.

Preston, C.R. (1990). Distribution of raptor foraging in relation to prey biomass and habitat structure. *Condor*, **92**, 107–12.

Prestt, I. (1971). An ecological study of the viper *Vipera berus* in southern Britain. *Journal of Zoology, London*, **164**, 373–418.

Prince, P.A. and Francis, M.D. (1984). Activity budgets of foraging grey-headed albatrosses. *Condor*, **86**, 297–300.

Prince, P.A. and Morgan, R.A. (1987). Diet and feeding ecology of Procellariiformes. In *Seabirds: Feeding Ecology and Role in Marine Ecosystems*, ed. J.P. Croxall. Cambridge: Cambridge University Press, pp. 135–71.

Prince, P.A. and Walton, D.W.H. (1984). Automated measurement of feed size and feeding frequency in albatrosses. *Journal of Applied Ecology*, **21**, 789–94.

Prince, P.A., Croxall, J.P., Trathan, P.N. and Wood, A.G. (1998). The pelagic distribution of South Georgia albatrosses and their relationships with fisheries. In *Albatross Biology and Conservation*, eds. G. Robertson and R. Gales. Chipping Norton: Surrey Beatty and Sons, pp. 69–83.

Prince, P.A., Weimerskirch, H., Wood, A.G. and Croxall, J.P. (1999). Areas and scales of interactions between albatrosses and the marine environment: species, populations and sexes. In *Proceedings of the 22 International Ornithological Congress, Durban*, eds. N.J. Adams and R.H. Slotow. South Africa: Birdlife, pp. 2001–20.

Prins, H.H.T. (1987). The buffalo of Manyara. Ph.D. thesis, Rijksuniversiteit te Groningen.

(1989). Condition changes and choice of social environment in African buffalo bulls. *Behaviour*, **108**, 297–324.

(1996). *Ecology and Behaviour of the African Buffalo: Social Inequality and Decision Making*. London: Chapman & Hall.

Prins, H.H.T. and Iason, G.R. (1989). Dangerous lions and nonchalant buffalo. *Behaviour*, **108**, 262–95.

Promislow, D.E.L., Montgomerie, R. and Martin, T.E. (1992). Mortality costs of sexual dimorphism in birds. *Proceedings of the Royal Society of London B*, **250**, 143–50.

Provenza, F.D. and Balph, D.F. (1988). Development of dietary choice in livestock on rangelands and its implications for management. *Journal of Animal Science*, **66**, 2356–68.

Pusey, A.E. (1990). Behavioural changes at adolescence in chimpanzees. *Behaviour*, **115**, 203–46.

Pusey, A.E., Williams, J. and Goodall, J. (1997). The influence of dominance rank on the reproductive success of female chimpanzees. *Science*, **277**, 828–31.

Puttick, G. M. (1981). Sex-related differences in foraging behaviour of curlew sand-pipers. *Ornis Scandinavica*, **12**, 13–17.

Qualls, R., Shine, R., Donnellan, S. and Hutchinson, M. (1996). The evolution of viviparity within the Australian scincid lizard *Lerista bougainvillii*. *Journal of Zoology, London*, **237**, 13–26.

Quinn, J. S. (1990). Sexual size dimorphism and parental care patterns in a monomorphic and a dimorphic larid. *Auk*, **107**, 260–74.

Quintana, F. and Dell'Arciprete, O. P. (2002). Foraging grounds of southern giant petrels (*Macronectes giganteus*) on the Patagonian shelf. *Polar Biology*, **25**, 159–61.

Radespiel, U. (2000). Sociality in the grey mouse lemur (*Microcebus murinus*). *American Journal of Primatology*, **51**, 21–40.

Radespiel, U., Cepok, S., Zietemann, V. and Zimmermann, E. (1998). Sex specific usage patterns of sleeping sites in grey mouse lemurs (*Microcebus murinus*) in northwest Madagascar. *American Journal of Primatology*, **46**, 77–84.

Radespiel, U., Ehresmann, P. and Zimmermann, E. (2001a). Contest versus scramble competition for mates: the composition and spatial structure of a population of grey mouse lemurs (*Microcebus murinus*) in Madagascar. *Primates*, **42**, 207–20.

Radespiel, U., Sarikaya, Z., Zimmermann, E. and Bruford, M. (2001b). Sociogenetic structure in a free-living nocturnal primate population: Sex specific differences in the grey mouse lemur (*Microcebus murinus*). *Behavioral Ecology and Sociobiology*, **50**, 493–502.

Radford, A. N. and Plessis, M. A. (2003). Bill dimorphism and foraging niche partitioning in the green woodhoopoe. *Journal of Animal Ecology*, **72**, 258–69.

Rahmani, A. R. (1989). *The Great Indian Bustard*. Bombay: Bombay Natural History Society.

Rajporohit, L. S., Sommer, V. and Mohnot, S. M. (1995). Wanderers between harems and bachelor bands. Male hanuman langurs (*Presbytis entellus*) at Jodhpur in Rajasthan. *Behaviour*, **132**, 255–99.

Ralls, K. and Mesnick, S. L. (2002). Sexual dimorphism. In *Encyclopedia of Marine Mammals*, eds. W. F. Perrin, B. Wursig and J. G. M. Thewissen. San Diego: Academic Press.

Raman, T. R. S. (1997). Factors influencing seasonal and monthly changes in the group size of chital or axis deer in southern India. *Journal of Biosciences*, **22**, 203–18.

Ramírez Ávila, G. M., Guisset, J. L. and Deneubourg, J. L. (2003). Synchronization in light-controlled oscillators. *Physica*, **182**, 254–73.

Ramirez-Bautista, A. and Benabib, M. (2001). Perch height of the arboreal lizard *Anolis nebulosus* (Sauria: Polychrotidae) from a tropical dry forest of Mexico: effect of the reproductive season. *Copeia*, **2001**, 187–93.

Rands, S. A., Cowlishaw, G., Pettifor, R. A., Rowcliffe, J. M. and Johnstone, R. A. (2003). Spontaneous emergence of leaders and followers in foraging pairs. *Nature*, **423**, 432–4.

Ransome, R. D. (1990). *The Natural History of Hibernating Bats*. London: Christopher Helm.

Ranta, E. and Lindström, K. (1990). Assortative schooling in 3-spined sticklebacks. *Annales Zoologici Fennici*, **27**, 67–75.

Ranta, E., Peuhkuri, N. and Laurila, A. (1994). A theoretical exploration of antipredatory and foraging factors promoting phenotype-assorted fish schools. *Ecoscience*, **1**, 99–106.

Rau, G. H., Ainley, D. G., Bengtson, J. L., Torre, J. J. and Hopkins, T. L. (1992). 15N/14N and 13C/12C in Weddell Sea birds, seals, and fish: implications for diet and trophic structure. *Marine Ecology Progress Series*, **84**, 1–8.

Read, A. J. (1990). Reproductive seasonality in harbour porpoises, *Phocoena phocoena*, from the Bay of Fundy. *Canadian Journal of Zoology*, **68**, 284–8.

(1999). Harbour porpoise – *Phocoena phocoena* (Linnaeus, 1758). In *Handbook of Marine Mammals: The Second Book of Dolphins and the Porpoises*, vol. 6, eds. S. H. Ridgway and R. Harrison. London: Academic Press, pp. 323–56.

Read, A. J. and Hohn, A. A. (1995). Life in the fast lane: the life history of harbor porpoises from the Gulf of Maine. *Marine Mammal Science*, **11**, 423–40.

Read, A. J. and Tolley, K. A. (1997). Postnatal growth and allometry of harbour porpoises from the Bay of Fundy. *Canadian Journal of Zoology*, **75**, 122–30.

Read, A. J., Wells, R. S., Hohn, A. A. and Tolley, K. A. (1993). Patterns of growth in wild bottlenose dolphins, *Tursiops truncatus*. *Journal of Zoology, London*, **231**, 107–23.

Read, D. (1982). Observations of the movements of two arid zone planigales (Dasyuridae, Marsupialia). In *Carnivorous Marsupials*, ed. M. Archer. Sydney: Royal Zoological Society of New South Wales, pp. 227–31.

Read, D. G. (1984a). Movements and home ranges of three sympatric dasyurids, *Sminthopsis crassicaudata, Planigale gilesi* and *P. tenuirostris* (Marsupialia), in semiarid western New South Wales. *Australian Wildlife Research*, **11**, 223–34.

(1984b). Reproduction and breeding season of *Planigale gilesi* and *P. tenuirostris* (Marsupialia: Dasyuridae). *Australian Mammalogy*, **8**, 161–74.

(1989). Microhabitat separation and diel activity patterns of *Planigale gilesi* and *P. tenuirostris* (Marsupialia: Dasyuridae). *Australian Mammalogy*, **12**, 45–54.

(1995). Giles' Planigale. In *The Mammals of Australia*, ed. R. Strahan. Sydney: Reed New Holland, pp. 107–9.

Reaney, L. T. and Whiting, M. J. (2002). Life on a limb: ecology of the tree agama (*Acanthocercus a. atricollis*) in southern Africa. *Journal of Zoology, London*, **257**, 439–48.

Recchia, C. A. and Read, A. J. (1989). Stomach content of harbour porpoises, *Phoceona phoceona*, (L.), from the Bay of Fundy. *Canadian Journal of Zoology*, **67**, 2140–6.

Recher, H. F. and Holmes, R. T. (2000). The foraging ecology of birds of eucalypt forest and woodland. I. Differences between males and females. *Emu*, **100**, 205–15.

Reddy, E. and Fenton, M. B. (2003). Exploiting vulnerable prey: moths and red bats (*Lasiurus borealis*; Vespertilionidae). *Canadian Journal of Zoology*, **18**, 1553–60.

Reebs, S. G. and Saulnier, N. (1997). The effect of hunger on shoal choice in golden shiners (Pisces: Cyprinidae, *Notemigonus crysoleucas*). *Ethology*, **103**, 642–52.

Reed, R. N. and Shine, R. (2002). Lying in wait for extinction: ecological correlates of conservation status among Australian elapid snakes. *Conservation Biology*, **16**, 451–61.

Reid, K. (1995). The diet of Antarctic fur seals *Arctocephalus gazella* Peters 1875 during winter at South Georgia. *Antarctic Science*, **7**(3), 241–9.

Reid, K. and Arnould, J. P. Y. (1996). The diet of Antarctic fur seals *Arctocephalus gazella* during the breeding season at South Georgia. *Polar Biology*, **16**, 105–14.

Reijinders, P., Brasseur, S., van der Toorn, J. *et al.* (1993). *Seals, Fur Seals, Sea Lions, and Walrus. Status Survey and Conservation Action Plan.* Gland, Switzerland: IUCN.

Reilly, J. J. and Fedak, M. A. (1991). Rates of water turnover and energy-expenditure of free-living male common seals (*Phoca vitulina*). *Journal of Zoology*, **223**, 461–8.

Reinert, H. K. (1984). Habitat variation within sympatric snake populations. *Ecology*, **65**, 1673–82.

Reinert, H.K. and Kodrich, W.R. (1982). Movements and habitat utilization by the massasauga, *Sistrurus catenatus catenatus*. *Journal of Herpetology*, **16**, 162–71.

Reinert, H.K. and Zappalorti, R.T. (1988). Timber rattlesnakes (*Crotalus horridus*) of the Pine Barrens (New Jersey, USA): their movement patterns and habitat preference. *Copeia*, **1988**, 964–78.

Remis, M.J. (1999). Tree structure and sex differences in arboreality among western lowland gorillas (*Gorilla gorilla gorilla*) at Bai Hokou, Central African Republic. *Primates*, **40**, 383–96.

Rendell, L. and Whitehead, H. (2001). Culture in whales and dolphins. *Behavioral and Brain Sciences*, **24**, 309–82.

Renfree, M. B., Russell, E. M. and Wooller, R. D. (1984). Reproduction and life history of the Honey Possum, *Tarsipes rostratus*. In *Possums and Gliders*, eds. A.P. Smith and I.D. Hume. Chipping Norton, NSW: Surrey Beatty and Sons, pp. 427–37.

Rexstad, E.A. and Anderson, D.R. (1988). Effect of the point system on redistributing hunting pressure on mallards. *Journal of Wildlife Management*, **52**, 89–94.

Reynolds, J. D. and Jennings, S. (2000). The role of animal behavior in marine conservation. In *Behavior and Conservation*, eds. L.M. Gosling and W.J. Sutherland. Cambridge, United Kingdom: Cambridge University Press, pp. 238–57.

Reznick, D. (1996). Life history evolution in guppies: A model system for the empirical study of adaptation. *Netherlands Journal of Zoology*, **46**, 172–90.

Rhine, R., Bloland, P. and Lodwick, L. (1985). Progression of adult male chacma baboons (*Papio ursinus*) in the Moremi Wildlife Reserve. *International Journal of Primatology*, **6**, 115–22.

Rice, D.W. (1998). *Marine Mammals of the World, Systematics and Distribution.* Lawrence, KA, USA: Society of Marine Mammalogy.

(1989). Sperm whale – *Physeter macrocephalus* (Linnaeus, 1758). In *Handbook of Marine Mammals: River Dolphins and the Larger Toothed Whales*, vol. 4, eds. S.H. Ridgway and R. Harrison. London: Academic Press, pp. 177–233.

Richard, A., Rakotomanaga, P. and Schwartz, M. (1993). Dispersal by *Propithecus verrauxi* at Beza Mahafaly Reserve. *American Journal of Primatology*, **30**, 1–20.

Richard, C. and Pépin, D. (1990). Seasonal variation in intragroup-spacing behavior of foraging isards (*Rupicapra pyrenaica*). *Journal of Mammalogy*, **71**, 145–50.

Richard, K.R., Dillon, M.C., Whitehead, H. and Wright, J.M. (1996). Patterns of kinship in groups of free-living sperm whales (*Physeter macrocephalus*) revealed by multiple molecular analyses. *Proceedings of the Academy of Science of United States*, **93**, 8792–5.

Richard-Hansen, C. (1992). Association between individually marked isards (*Rupicapra pyrenaica*): seasonal and inter-annual variations. In *Ongulés/Ungulates 91*, eds. F. Spitz, G. Janeau, G. Gonzalez, and S. Aulagnier. Paris: S.F.E.P.M.-I.R.G.M., pp. 299–304.

Richards, A.F. (1996). Life history and behavior of female dolphins (*Tursiops* sp.) in Shark Bay, Western Australia. Ph.D thesis, Ann Arbor, Michigan: University of Michigan.

Richardson, A.J., Maharaj, G., Compagno, L.J.V. *et al.* (2000). Abundance, distribution, morphometrics, reproduction and diet of the Izak catshark. *Journal of Fish Biology*, **56**, 552–76.

Ridgway, S.H. and Harrison, R. (1989). *Handbook of Marine Mammals: River Dolphins and the Larger Toothed Whales*. London: Academic Press.

(1994). *Handbook of Marine Mammals: The First Book of Dolphins*. London: Academic Press.

(1999). *Handbook of Marine Mammals: The Second Book of Dolphins and the Porpoises.* London: Academic Press.

Riedman, M. (1990). *The Pinnipeds: Seals, Sea Lions, and Walruses.* Berkeley: University of California Press.

Rintamäki, P. T., Karvonen, E., Alatalo, R. V. and Lundberg, A. (1999). Why do black grouse males perform on lek sites outside the breeding season? *Journal of Avian Biology,* **30**, 359–66.

Ripley, W. E. (1946). The soupfin shark and the fishery. *California Fish Bulletin,* **64**, 7–37.

Rivas, J. and Burghardt, G. M. (2001). Understanding sexual size dimorphism in snakes: wearing the snake's shoes. *Animal Behaviour,* **62**, F1–F6.

Robert, K. A. and Thompson, M. B. (2001). Viviparous lizard selects sex of embryos. *Nature,* **412**, 698–9.

Roberts, C., Brotherton, P., Luschekina, A. A., Kholodova, M. V. and Milner-Gulland, E. J. (2001). Gazelles, dwarf antelopes and saigas. In *The New Encyclopedia of Mammals,* ed. D. Macdonald. Oxford: Oxford University Press, pp. 560–6.

Robichaud, D. and Rose, G. A. (2003). Sex differences in cod residency on a spawning ground. *Fisheries Research,* **60**, 33–43.

Rock, J. and Cree, A. (2003). Intraspecific variation in the effect of temperature on pregnancy in the viviparous gecko *Hoplodactylus maculatus. Herpetologica,* **59**, 8–22.

Rock, J., Andrews, R. M. and Cree, A. (2000). Effects of reproductive condition, season, and site on selected temperatures of a viviparous gecko. *Physiological and Biochemical Zoology,* **73**, 344–55.

Rodenhouse, N. L., Sherry, T. W. and Holmes, R. T. (1997). Site-dependent regulation of population size: a new synthesis. *Ecology,* **78**, 2025–42.

Rodhouse, P. G., Elvidge, C. D. and Trathan, P. N. (2001). Remote sensing of the global light-fishing fleet: an analysis of interactions with oceanography, other fisheries and predators. *Advances in Marine Biology,* **39**, 261–303.

Rodman, P. (1988). Diversity and consistency in ecology and behavior. In *Orang-Utan Biology,* ed. J. H. Schwartz. Oxford: Oxford University Press, pp. 31–51.

Rodman, P. S. (1979). Individual activity patterns and the solitary nature of orangutans. In *The Great Apes,* eds. D. A. Hamburg and E. R. McKown. Menlo Park: Benjamin Cummings, pp. 235–56.

Rodman, P. S. and Mitani, J. C. (1987). Orangutans: sexual dimorphism in a solitary species. In *Primate Societies,* eds. B. B. Smuts, D. L. Cheney, R. M. Seyfarth, R. W. Wrangham and T. T. Struhsaker. Chicago: University of Chicago Press, pp. 146–54.

Rodway, M. S., Regehr, H. M. and Cook, F. (2003). Sex and age differences in distribution, abundance, and habitat preferences of wintering harlequin ducks: implications for conservation and estimating recruitment rates. *Canadian Journal of Zoology,* **81**, 492–503.

Romeo, G., Lovari, S., Festa-Bianchet, M. (1997). Group leaving in mountain goats: are males ousted by adult females? *Behavioural Processes,* **40**, 243–6.

Rook, A. J. and Penning, P. D. (1991). Synchronization of eating, ruminating and idling activity by grazing sheep. *Applied Animal Behaviour Science,* **32**, 157–66.

Roosenburg, W. M., Halcy, K. L. and McGuire, S. (1999). Habitat selection and movements of diamondback terrapins, *Malaclemys terrapin,* in a Maryland estuary. *Chelonian Conservation Biology,* **3**, 425–9.

Rootes, W. L. and Chabreck, R. H. (1993). Cannibalism in the American alligator. *Herpetologica,* **49**, 99–107.

Rose, B.R. (1981). Factors affecting activity in *Sceloporus virgatus*. *Ecology*, **62**, 706–16.

Rose, F.L. and Judd, F.W. (1975). Activity and home range size of the Texas tortoise, *Gopherus berlandieri*, in South Texas. *Herpetologica*, **31**, 448–56.

Rose, L. (1994). Sex differences in diet and foraging in *Cebus capucinus*. *International Journal of Primatology*, **15**, 95–114.

Ross, C. (1992). Basal metabolic rate, body weight, and diet in primates: an evaluation of the evidence. *Folia Primatologica*, **58**, 7–23.

Ross, D.A. (1989). Population ecology of painted and Blanding's turtles (*Chrysemys picta* and *Emydoidea blandingii*) in central Wisconsin. *Transactions of the Wisconsin Academy of Science, Arts and Letters*, **77**, 77–84.

Ross, D.A., Brewster, K.N., Anderson, R.K., Ratner, N. and Brewster, C.M. (1991). Aspects of the ecology of wood turtles, *Clemmys insculpta*, in Wisconsin. *Canadian Field-Naturalist*, **105**, 363–7.

Ross, P.I., Jalkotzy, M.G. and Festa-Bianchet, M. (1997). Cougar predation on bighorn sheep in southwestern Alberta during winter. *Canadian Journal of Zoology*, **75**, 771–5.

Rothstein, A. and Griswold, J.G. (1991). Age and sex preferences for social partners by juvenile bison bulls, *Bison bison*. *Animal Behaviour*, **41**, 227–37.

Rowe, J.W. and Moll, E.O. (1991). A radiotelemetric study of activity and movements of the Blanding's turtle (*Emydoidea blandingii*) in northeastern Illinois. *Journal of Herpetology*, **25**, 178–85.

Rowell, T.E. (1973). Social organization of wild talapoin monkeys. *American Journal of Physical Anthropology*, **38**, 593–8.

Rubin, E.S., Boyce, W.M. and Bleich, V.C. (2000). Reproductive strategies of desert bighorn sheep. *Journal of Mammalogy*, **81**, 769–86.

Rubin, E.S., Boyce, W.M. and Caswell-Chen, E.P. (2002). Modeling demographic processes in an endangered population of bighorn sheep. *Journal of Wildlife Management*, **66**, 796–810.

Rubin, E.S., Boyce, W.M., Jorgensen, M.C. *et al.* (1998). Distribution and abundance of bighorn sheep in the Peninsular Ranges, California. *Wildlife Society Bulletin*, **26**, 539–51.

Ruby, D.E. and Baird, D.I. (1994). Intraspecific variation in behavior: comparisons between populations at different altitudes of the lizard *Sceloporus jarrovi*. *Journal of Herpetology*, **28**, 70–8.

Ruckstuhl, K.E. (1998). Foraging behaviour and sexual segregation in bighorn sheep. *Animal Behaviour*, **56**, 99–106.

(1999). To synchronise or not to synchronise: a dilemma for young bighorn males? *Behaviour*, **136**, 805–18.

Ruckstuhl, K.E. and Festa-Bianchet, M. (1998). Do reproductive status and lamb gender affect the foraging behavior of bighorn ewes? *Ethology*, **104**, 941–54.

(2001). Group choice by subadult male bighorn sheep: trade-offs between foraging efficiency and predator avoidance. *Ethology*, **107**, 161–72.

Ruckstuhl, K.E. and Kokko, H. (2002). Modelling sexual segregation in ungulates: effects of group size, activity budgets and synchrony. *Animal Behaviour*, **64**, 909–14.

Ruckstuhl, K.E. and Neuhaus, P. (2000). Sexual segregation in ungulates: a new approach. *Behaviour*, **137**, 361–77.

(2001). Behavioral synchrony in ibex groups: effects of age, sex, and habitat. *Behaviour*, **138**, 1033–46.

(2002). Sexual segregation in ungulates: a comparative test of three hypotheses. *Biological Review*, **77**, 77–96.

Ruckstuhl, K. E., Manica, A., MacColl, A. D. C., Pilkington, J. G., Clutton-Brock, T. H. (2005). The effects of castration, sex ratio and population density on social segregation and habitat use in Soay sheep. *Behavioral Ecology and Sociobiology* (in press).

Ruedas, L. A., Demboski, J. R. and Sison, R. V. (1994). Morphological and ecological variation in *Otopteropus cartilagonodus* Koch, 1969 (Mammalia: Chiroptera: Pteropodidae) from Luzon, Philippines. *Proceedings of the Biological Society of Washington*, **107**, 1–16.

Ruibal, R. and Philibosian, R. (1974). The population ecology of the lizard *Anolis acutus*. *Ecology*, **55**, 525–37.

Russell, E. M. (1979). The size and composition of groups in the red kangaroo, *Macropus rufus*. *Australian Wildlife Research*, **6**, 237–44.

(1986). Observations on the behaviour of the honey possum, *Tarsipes rostratus* (Marsupialia: Tarsipedidae) in captivity. *Australian Journal of Zoology*, **121**, 1–63.

Russo, D. (2002). Elevation affects the distribution of the two sexes in Daubenton's bats *Myotis daubentonii* (Chiroptera: Vespertilionidae) from Italy. *Mammalia*, **66**, 543–51.

Rutherford, P. L. and Gregory, P. T. (2003). How age, sex, and reproductive condition affect retreat-site selection and emergence, patterns in a temperate-zone lizard, *Elgaria coerulea*. *Ecoscience*, **10**, 24–32.

Ryan, P. G. and Boix-Hinzen, C. (1999). Consistent male-biased seabird mortality in the Patagonian toothfish longline fishery. *Auk*, **116**, 851–4.

Sabo, J. L. (2003). Hot rocks or no hot rocks: overnight retreat availability and selection by a diurnal lizard. *Oecologia*, **136**, 329–35.

Sachs, B. D. and Harris, V. S. (1978). Sex differences and developmental changes in selected juvenile activities (play) of domestic lambs. *Animal Behaviour*, **26**, 678–84.

Sailer, L. D., Gaulin, S. J. C., Boster, J. S. and Kurland, J. A. (1985). Measuring the relationship between dietary quality and body size in primates. *Primates*, **26**, 14–27.

Salamolard, M. and Weimerskirch, H. (1993). Relationship between foraging effort and energy requirement throughout the breeding season in the wandering albatross. *Functional Ecology*, **7**, 643–52.

Saltz, E., Dixon, D. and Johnson, J. (1977). Training disadvantaged preschoolers on various fantasy activities: Effects on cognitive functioning and impulse control. *Child Development*, **48**, 367–80.

Sanderson, R. A. (1974). Sexual dimorphism in the Barbour's map turtle, *Malaclemys barbouri* (Carr and Marchand). M.A. thesis, Tampa: University of South Florida.

Santos, X. and Llorente, G. A. (1998). Sexual and size-related differences in the diet of the snake *Natrix maura* from the Ebro Delta, Spain. *Herpetological Journal*, **8**, 161–5.

Santos, X., Gonzalez-Solis, J. and Llorente, G. A. (2000). Variation in the diet of the viperine snake *Natrix maura* in relation to prey availability. *Ecography*, **23**, 185–92.

Sato, K., Mitani, Y., Cameron, M. F. et al. (2002). Deep foraging dives in relation to the energy depletion of Weddell seal (*Leptonychotes weddellii*) mothers during lactation. *Polar Biology*, **25**, 696–702.

Savidge, J. A. (1988). Food habits of *Boiga irregularis*, an introduced preditor on Guam. *Journal of Herpetology*, **22**, 275–82.

Sayler, R. D. and Afton, A. D. (1981). Ecological aspects of common goldeneyes *Bucephala clangula* wintering on the Mississippi River, USA. *Ornis Scandinavica*, **12**, 99–108.

Schaefer, J.A. and Messier, F. (1995). Habitat selection as a hierarchy: the spatial scales of winter foraging by musk oxen. *Ecography*, **18**, 333–44.

Schaefer, R.J., Torres, S.G. and Bleich, V.C. (2000). Survivorship and cause-specific mortality in sympatric populations of mountain sheep and mule deer. *California Fish and Game*, **86**, 127–35.

Schmid, J. and Kappeler, P.M. (1998). Fluctuating sexual dimorphism and differential hibernation by sex in a primate, the grey mouse lemur. *Behavioral Ecology and Sociobiology*, **43**, 125–32.

Schmid-Nielson, K. (1989). *Scaling. Why is Animal Size so Important?* Cambridge: Cambridge University Press.

Schmidt-Nielsen, K. (1972). Locomotion: energy cost of swimming, flying, and running. *Science*, **177**, 222–8.

Schoener, T.W. (1967). The ecological significance of sexual dimorphism in size in the lizard *Anolis conspersus*. *Science*, **155**, 474–6.

(1968). The Anolis lizards of Bimini: resource partitioning in a complex fauna. *Ecology*, **49**, 704–26.

Schoener, T.W. and Gorman, G.C. (1968). Some niche differences in three lesser Antillean lizards of the genus *Anolis*. *Ecology*, **49**, 819–30.

Schoener, T.W., Slade, J.B. and Stinson, C.H. (1982). Diet and sexual dimorphism in the very catholic lizard genus, *Leiocephalus* of the Bahamas. *Oecologia*, **53**, 160–9.

Schreer, J.F. and Kovacs, K.M. (1997). Allometry of diving capacity in air-breathing vertebrates. *Canadian Journal of Zoology*, **75**, 339–58.

Schwaner, T.D. and Sarre, S.D. (1988). Body size of tiger snakes in southern Australia, with particular reference to *Notechis ater serventyi* (Elapidae) on Chappell Island. *Journal of Herpetology*, **22**, 24–33.

Schwartz, O.A., Bleich, V.C. and Holl, S.A. (1986). Genetics and the conservation of mountain sheep *Ovis canadensis nelsoni*. *Biological Conservation*, **37**, 179–90.

Schwarzkopf, L. (1994). Measuring trade-offs: a review of studies of costs of reproduction in lizards. In *Lizard Ecology: Historical and Experimental Perspectives*, eds. L.J. Vitt and E.R. Pianka. Princeton, NJ: Princeton University Press, pp. 7–30.

Schwarzkopf, L. and Shine, R. (1991). Thermal biology of reproduction in viviparous skinks, *Eulamprus tympanum*: why do gravid females bask more? *Oecologia*, **88**, 562–9.

(1992). Costs of reproduction in lizards: escape tactics and susceptibility to predation. *Behavioral Ecology and Sociobiology*, **31**, 17–25.

Scott, M.D., Wells, R.S. and Irvine, A.B. (1990). A long-term study of bottlenose dolphin on the west coast of Florida. In *The Bottlenose Dolphin*, eds. S. Leatherwood and R.R. Reeves. San Diego: Academic Press, pp. 235–44.

Scott, N.J., Jr, Wilson, D.E., Jones, C. and Andrews, R.M. (1976). The choice of perch dimensions by lizards of the genus *Anolis* (Reptilia, Lacertilia, Iguanidae). *Journal of Herpetology*, **10**, 75–84.

Scudder, R.M. and Burghardt, G.M. (1983). A comparative study of defensive behavior in three sympatric species of water snakes (*Nerodia*). *Zeitschrift für Tierpsychologie*, **63**, 17–26.

Searcy, W.A. and Yasukawa, K. (1981). Sexual size dimorphism and survival of male and female blackbirds (Icteridae). *Auk*, **98**, 457–65.

Secor, S.M. (1994). Ecological significance of movements and activity range for the sidewinder, *Crotalus cerastes*. *Copeia*, **1994**, 631–45.

Seebacher, F., Grigg, G.C. and Beard, L.A. (1999). Crocodiles as dinosaurs: behavioural thermoregulation in very large ectotherms leads to high and stable body temperatures. *Journal of Experimental Biology*, **202**, 77–86.

Seghers, B.H. (1973). *Zoology*. The University of British Columbia.

(1974). Schooling behaviour in the guppy (*Poecilia reticulata*): An evolutionary response to predation. *Evolution*, **28**, 486–9.

Seib, R.L. (1981). Size and shape in a neotropical burrowing colubrid snake, *Geophis nasalis*, and its prey. *American Zoologist*, **21**, 933.

Selander, R.K. (1966). Sexual dimorphism and differential niche utilization in birds. *Condor*, **68**, 113–51.

(1972). Sexual selection and dimorphism in birds. In *Sexual Selection and the Descent of Man 1971–1971*, ed. B. Campbell. Chicago: Heinemann, pp. 180–229.

Senft, R.L., Coughenour, M.B., Bailey, D.W. *et al.* (1987). Large herbivore foraging and ecological hierarchies. *Bioscience*, **37**, 789–99.

Sergeant, D.E. (1973). Biology of white whales (*Delphinapterus leucas*) in Western Hudson Bay. *Fisheries Research Board of Canada*, **30**, 1065–90.

Sergeant, D.E. and Brodie, P.F. (1969). Body size in white whales, *Delphinapterus leucas*. *Journal of the Fisheries Research Board of Canada*, **26**, 2561–80.

Sexton, O.J. (1964). Differential predation by the lizard, *Anolis carolinensis*, upon unicolored and polycolored insects after an interval of no contact. *Animal Behaviour*, **12**, 101–10.

Sexton, O.J., Bauman, J. and Ortleb, E. (1972). Seasonal food habits of *Anolis limifrons*. *Ecology*, **53**, 182–6.

Shackleton, D.M. (1991). Social maturation and productivity in bighorn sheep: are young males incompetent. *Applied Animal Behaviour Science*, **29**, 173–84.

Shaffer, S.A., Weimerskirch, H. and Costa, D.P. (2001). Functional significance of sexual dimorphism in wandering albatrosses, *Diomedea exulans*. *Functional Ecology*, **15**, 203–10.

Shah, B., Shine, R., Hudson, S. and Kearney, M. (2003). Sociality in lizards: why do thick-tailed geckos (*Nephrurus milii*) aggregate? *Behaviour*, **140**, 1039–52.

Shank, C.C. (1982). Age-sex differences in the diets of wintering Rocky Mountain bighorn sheep. *Ecology*, **63**, 627–33.

(1985). Inter- and intra-sexual segregation of chamois (*Rupicapra rupicapra*) by altitude and habitat during summer. *Zeitschrift für Säugetierkunde*, **50**, 117–25.

Shealer, D.A. (2002). Foraging behaviour and food of seabirds. In *Biology of Marine Birds*, eds. E.A. Schreiber and J. Burger. Boca Raton, Florida, USA: CRC Press, pp. 137–77.

Shealy, R.M. (1976). The natural history of the Alabama map turtle, *Graptemys pulchra* Baur, in Alabama. *Bulletin of the Florida State Museum, Biological Sciences Series*, **21**, 47–111.

Sherry, T.W. and Holmes, R.T. 1996. Winter habitat quality, population limitation and conservation of Neotropical-Nearctic migrant birds. *Ecology*, **77**, 36–48.

Shetty, S. and Shine, R. (2002a). Activity patterns of yellow-lipped sea kraits (*Laticauda colubrina*) on a Fijian island. *Copeia*, **2002**, 77–85.

(2002b). Sexual divergence in diets and morphology in Fijian sea snakes *Laticauda colubrina* (Laticaudinae). *Austral Ecology*, **27**, 77–84.

Shimoka, Y. (2003). Seasonal variation in association patterns of wild spider monkeys (*Ateles belzebuth belzebuth*) at La Macarena, Columbia. *Primates*, **44**, 83–90.

Shine, R. (1978). Sexual size dimorphism and male combat in snakes. *Oecologia*, **33**, 269–78.

(1979). Activity patterns in Australian elapid snakes (Squamata: Serpentes: Elapidae). *Herpetologica*, **35**, 1–11.

(1980a). 'Costs' of reproduction in reptiles. *Oecologia*, **46**, 92–100.

(1980b). Ecology of the Australian death adder, *Acanthophis antarcticus* (Elapidae): evidence for convergence with the Viperidae. *Herpetologica*, **36**, 281–9.

(1985). The evolution of viviparity in reptiles: an ecological analysis. In *Biology of the Reptilia*, vol. 15, eds. C. Gans and F. Billett. New York: John Wiley and Sons, pp. 605–94.

(1986). Sexual differences in morphology and niche utilization in an aquatic snake, *Acrochordus arafurae*. *Oecologia*, **69**, 260–7.

(1989a). Ecological causes for the evolution of sexual dimorphism: a review of the evidence. *Quarterly Review of Biology*, **64**, 419–60.

(1989b). Constraints, allometry and adaptation: food habits and reproductive biology of Australian brownsnakes (*Pseudonaja*, Elapidae). *Herpetologica*, **45**, 195–207.

(1991a). Intersexual dietary divergence and the evolution of sexual dimorphism in snakes. *American Naturalist*, **138**, 103–22.

(1991b). Why do larger snakes eat larger prey? *Functional Ecology*, **5**, 493–502.

(1994a). Allometric patterns in the ecology of Australian snakes. *Copeia*, **1994**, 851–67.

(1994b). Sexual size dimorphism in snakes revisited. *Copeia*, **1994**, 326–46.

(1999). Why is sex determined by nest temperature in many reptiles? *Trends in Ecology and Evolution*, **14**, 186–9.

Shine, R. and Crews, D. (1988). Why male garter snakes have small heads: the evolution and endocrine control of sexual dimorphism. *Evolution*, **42**, 1105–10.

Shine, R. and Fitzgerald, M. (1995). Variation in mating systems and sexual size dimorphism between populations of the Australian python *Morelia spilota* (Serpentes: Pythonidae). *Oecologia*, **103**, 490–8.

(1996). Large snakes in a mosaic rural landscape: the ecology of carpet pythons *Morelia spilota* (Serpentes: Pythonidae) in coastal eastern Australia. *Biological Conservation*, **76**, 113–22.

Shine, R. and Harlow, P. S. (1993). Maternal thermoregulation influences offspring viability in a viviparous lizard. *Oecologia*, **96**, 122–7.

Shine, R. and Wall, M. (2004). Why is intraspecific niche partitioning more common in snakes than in lizards? In *Foraging Modes in Lizards*, eds. D. B. Miles, S. M. Reilly and L. D. McBrayer. Cambridge University Press.

Shine, R., Barrott, E. G. and Elphick, M. J. (2002a). Some like it hot: effects of forest clearing on nest temperatures of montane reptiles. *Ecology*, **83**, 2808–15.

Shine, R., Branch, W. R., Harlow, P. S. and Webb, J. K. (1998a). Reproductive biology and food habits of horned adders, *Bitis caudalis* (Viperidae), from southern Africa. *Copeia*, **1998**, 391–401.

Shine, R., Cogger, H. G., Reed, R. N. S., Shetty, S. and Bonnet, X. (2002b). Aquatic and terrestrial locomotor speeds of amphibious sea-snakes (Serpentes, Laticaudidae). *Journal of Zoology, London*, **259**, 261–8.

Shine, R., Elphick, M. and Donnellan, S. (2002c). Co-occurrence of multiple, supposedly incompatible modes of sex determination in a lizard population. *Ecology Letters*, **5**, 486–9.

Shine, R., Haagner, G. V., Branch, W. R., Harlow, P. S. and Webb, J. K. (1996). Natural history of the African shieldnose snake *Aspidelaps scutatus* (Serpentes, Elapidae). *Journal of Herpetology*, **30**, 361–6.

Shine, R., Harlow, P. S., Keogh, J. S. and Boeadi (1998b). The influence of sex and body size on food habits of a giant tropical snake, *Python reticulatus*. *Functional Ecology*, **12**, 248–258.

Shine, R., Langkilde, T. and Mason, R.T. (2003b). Cryptic forcible insemination: male snakes exploit female physiology, anatomy and behavior to obtain coercive matings. *American Naturalist*, **162**, 653–67.

Shine, R., O'Connor, D., LeMaster, M.P. and Mason, R.T. (2001). Pick on someone your own size: ontogenetic shifts in mate choice by male garter snakes result in size-assortative mating. *Animal Behaviour*, **61**, 1–9.

Shine, R., Olsson, M.M., Lemaster, M.P., Moore, I.T. and Mason, R.T. (2000). Effects of sex, body size, temperature, and location on the antipredator tactics of free-ranging gartersnakes (*Thamnophis sirtalis*, Colubridae). *Behavioral Ecology*, **11**, 239–45.

Shine, R., Reed, R.N., Shetty, S. and Cogger, H.G. (2002d). Relationships between sexual dimorphism and niche partitioning within a clade of sea-snakes (Laticaudinae). *Oecologia*, **133**, 45–53.

Shine, R., Shine, T. and Shine, B.G. (2003a). Intraspecific habitat partitioning by the sea snake *Emydocephalus annulatus* (Serpentes, Hydrophiidae): the effects of sex, body size, and colour pattern. *Biological Journal of the Linnean Society*, **80**, 1–10.

Shine, R., Sun, L.X., Fitzgerald, M. and Kearney, M. (2003c). A radiotelemetric study of movements and thermal biology of insular Chinese pit-vipers (*Gloydius shedaoensis*, Viperidae). *Oikos*, **100**, 342–52.

Shine, R., Phillips, B., Langkilde, T., Lutterschmidt, D. and Mason, R.T. (2004). Mechanisms and consequences of sexual conflict in garter snakes (*Thamnophis sirtalis*, Colubridae). *Behavioral Ecology*, **15**, 654–60.

Shively, S.H. (1982). Factors limiting the upstream distribution of the Sabine map turtle. M.Sc. thesis, Lafayette, LA: University of Southwestern Louisiana.

Shively, S.H. and Jackson, J.F. (1985). Factors limiting the upstream distribution of the Sabine map turtle. *American Midland Naturalist*, **114**, 292–303.

Short, H.L. (1963). Rumen fermentation and energy relationships in white-tailed deer. *Journal of Wildlife Management*, **27**, 184–95.

Shumway, C.A. (1999). A neglected science: applying behavior to aquatic conservation. *Environmental Biology of Fishes*, **55**, 183–210.

Silk, J. (1987). Activities and feeding behavior of free-ranging pregnant female baboons. *International Journal of Primatology*, **8**, 596–613.

Sillet, T.S. and Holmes, R.T. (2002). Variation in survivorship of a migratory songbird throughout its annual cycle. *Journal of Animal Ecology*, **71**, 296–308.

Simon, C.A. (1975). Size selection of prey by the lizard *Sceloporus jarrovi*. *American Midland Naturalist*, **96**, 246–51.

Sims, D.W. (1996). The effect of body size on the standard metabolic rate of lesser spotted dogfish, *Scyliorhinus canicula*. *Journal of Fish Biology*, **48**, 542–4.

    (2003). Tractable models for testing theories about natural strategies: foraging behaviour and habitat selection of free-ranging sharks. *Journal of Fish Biology*, **63** (Supplement A), 53–73.

Sims, D.W. and Davies, S.J. (1994). Does specific dynamic action (SDA) regulate return of appetite in the lesser spotted dogfish, *Scyliorhinus canicula*? *Journal of Fish Biology*, **45**, 341–8.

Sims, D.W. and Quayle, V.A. (1998). Selective foraging behaviour of basking sharks on zooplankton in a small-scale front. *Nature*, **393**, 460–4.

Sims, D.W., Davies, S.J. and Bone, Q. (1993). On the diel rhythms in metabolism and activity of post-hatching lesser spotted dogfish, *Scyliorhinus canicula*. *Journal of Fish Biology*, **43**, 749–54.

Sims, D. W., Genner, M. J., Southward, A. J. and Hawkins, S. J. (2001b). Timing of squid migration reflects North Atlantic climate variability. *Proceedings of the Royal Society of London B*, **268**, 2607–11.

Sims, D. W., Nash, J. P. and Morritt, D. (2001a). Movements and activity of male and female dogfish in a tidal sea lough: alternative behavioural strategies and apparent sexual segregation. *Marine Biology*, **139**, 1165–75.

Sinclair, A. R. E. (1977). *The African Buffalo. A Study of Resource Limitation of Populations.* Chicago: University of Chicago Press.

Singleton, I. and van Schaik, C. P. (2001). Orangutan home range size and its determinants. *International Journal of Primatology*, **22**, 877–912.

Siniff, D. B., DeMaster, D. P., Hofman, R. J. and Eberhardt, L. L. (1977). An analysis of the dynamics of a Weddell seal population. *Ecological Monographs*, **47**, 319–35.

Slip, D. J. and Shine, R. (1988a). Feeding habits of the diamond python, *Morelia s. spilota*: ambush predation by a boid snake. *Journal of Herpetology*, **22**, 323–30.

(1988b). Habitat use, movements, and activity patterns of free-ranging diamond pythons, *Morelia spilota spilota* (Serpentes, Boidae): a radiotelemetric study. *Australian Wildlife Research*, **15**, 515–31.

Slip, D. J., Hindell, M. A. and Burton, H. R. (1994). Diving behaviour of southern elephant seals from Macquarie Island: An overview. In *Elephant Seals: Population Ecology, Behaviour and Physiology*, eds. B. J. Le Boeuf and R. M. Laws. Berkley: University of California Press, pp. 253–70.

Slotow, R., van Dyk, G., Poole, J., Page, B. and Klocke, A. (2000). Older bull elephants control young males. *Nature*, **408**, 425–6.

Smith, A. P. and Broome, L. (1992). The effects of season, sex and habitat on the diet of the mountain pygmy-possum (*Burramys parvus*). *Wildlife Research*, **19**, 755–68.

Smith, C. (1977). Feeding and ranging behavior of mantled howler monkeys (*Alouatta palliata*). In *Primate Ecology*, ed. T. H. Clutton-Brock. Cambridge: Cambridge University Press, pp. 183–222.

Smith, G. J. D. and Gaskin, D. E. (1983). An environmental index for habitat utilization by female harbour porpoises with calves near Deer Island, Bay of Fundy. *Ophelia*, **22**, 1–13.

Smith, G. R. (1996). Habitat use and fidelity in the striped plateau lizard *Sceloporus virgatus*. *American Midland Naturalist*, **135**, 68–80.

Smith, G. R. and Ballinger, R. E. (1994). Thermal ecology of *Sceloporus virgatus* from southeastern Arizona, with comparison to *Urosaurus ornatus*. *Journal of Herpetology*, **28**, 65–9.

Smith, G. R., Ballinger, R. E. and Congdon, J. D. (1993). Thermal ecology of the high-altitude bunch grass lizard, *Sceloporus scalaris*. *Canadian Journal of Zoology*, **71**, 2152–5.

Smith, M. S. R. (1966). Injuries as an indication of social behaviour in the Weddell seal (*Leptonychotes wedellii*). *Mammalia*, **30**, 241–6.

Smith, P. C. and Evans, P. R. (1973). Studies of shorebirds at Lindisfarne, Northumberland. 1. Feeding ecology and behaviour of the bar-tailed godwit. *Wildfowl*, **24**, 135–9.

Smith, R. J. and Jungers, W. L. (1997). Body mass in comparative primatology. *Journal of Human Evolution*, **32**, 523–59.

Smith, T. G., Hammill, M. O. and Martin, A. R. (1994). Herd composition and behaviour of white whales (*Delphinapterus leucas*) in two Canadian arctic estuaries. *Bioscience*, **39**, 175–84.

Snell, H.L., Jennings, R.D., Snell, H.M. and Harcourt, S. (1988). Intrapopulation variation in predator avoidance performance of Galapagos lava lizards: the interaction of sexual and natural selection. *Evolutionary Ecology*, **2**, 353–69.

Snelson, F.F., Mulligan, T.J. and Williams, S.E. (1984). Food habits, occurrence, and population structure of the bull shark, *Carcharhinus leucas*, in Florida coastal lagoons. *Bulletin of Marine Science*, **34**, 71–80.

Southwell, C.J. (1984). Variability in grouping in the eastern grey kangaroo, *Macropus giganteus* I. Group density and group size. *Australian Wildlife Research*, **11**, 423–35.

Spaeth, D.F., Bowyer, R.T., Stephenson, T.R. and Barboza, P.S. (2004). Sexual segregation in moose *Alces alces*: an experimental manipulation of foraging behaviour. *Wildlife Biology*, **10**, 59–72.

Speakman, J.R. and Thomas, D.W. (2003). Physiological ecology and energetics of bats. In *Bat Ecology*, eds. T.H. Kunz and M.B. Fenton. Chicago: University of Chicago Press, pp. 430–90.

Speakman, J.R., Irwin, N., Tallach, N. and Stone, R. (1999). Effect of roost size on the emergence behaviour of pipistrelle bats. *Animal Behaviour*, **58**, 787–95.

Sprague, D.S., Suzuki, S., Takahashi, H. and Sato, S. (1998). Male life histories in natural populations of Japanese macaques: migration, dominance rank, and troop participation of males in two habitats. *Primates*, **39**, 351–63.

Springer, M.S., Kirsch, J.A.W. and Case, J.A. (1997). The chronicle of marsupial evolution. In *Molecular Evolution and Adaptive Radiation*, eds. T.J. Givnish and K.J. Sytsma. Cambridge: Cambridge University Press, pp. 129–61.

Springer, S. (1967). Social organization of shark populations. In *Sharks, Skates and Rays*, eds. P.W. Gilbert, R.F. Mathewson and D.P. Rall. Baltimore MD: Johns Hopkins University Press, pp. 149–74.

Sroufe, L.A., Bennett, C., Englund, M., Urban, J. and Shulman, S. (1993). The significance of cross-gender boundaries in preadolescence: contemporary correlates and antecedents of boundary violation and maintenance. *Child Development*, **64**, 455–66.

St Aubin, D.J., Smith, T.G. and Geraci, J.R. (1990). Seasonal epidermal molt in beluga whales, *Delphinapterus leucas*. *Canadian Journal of Zoology*, **68**, 359–67.

Stahl, J.C. and Sagar, P.M. (2000). Foraging strategies of southern Buller's albatrosses *Diomedea b. bulleri* breeding on The Snares, New Zealand. *Journal of the Royal Society of New Zealand*, **30**, 299–318.

Staines, B.W. (1976). The use of natural shelter by red deer (*Cervus elaphus*) in relation to weather in North-East Scotland. *Journal of Zoology*, **180**, 1–8.

(1977). Factors affecting the seasonal distribution of red deer (*Cervus elaphus*, L.) in Glen Dye, North-East Scotland. *Annals of Applied Biology*, **87**, 495–512.

Staines, B.W., Crisp, J.M. and Parish, T. (1982). Differences in the quality of food eaten by red deer (*Cervus elaphus*) stags and hinds in winter. *Journal of Applied Ecology*, **19**, 65–77.

Stammbach, E. (1987). Desert, forest, and montane baboons: multilevel societies. In *Primate Societies*, eds. B.B. Smuts, D.L. Cheney, R.M. Seyfarth, R.W. Wrangham and T.T. Struhsaker. Chicago: University of Chicago Press, pp. 112–20.

Stamps, J.A. (1977). The function of the survey posture in Anolis lizards. *Copeia*, **1977**, 756–8.

(1983). Sexual selection, sexual dimorphism, and territoriality. In *Lizard Ecology: Studies of a Model Organism*, eds. R.B. Huey, E.R. Pianka and T.W. Schoener. Cambridge, MA: Harvard University Press, pp. 169–204.

Staniland, I.J. and Boyd, I.L. (2003). Variation in the foraging location of Antarctic fur seals (*Arctocephalus gazella*), the effects on diving behaviour. *Marine Mammal Science*, **19**, 331–43.

Steenbeek, R., Sterck, E. H. M., de Vries, H. and van Hooff, J. A. R. A. M. (2000). Costs and benefits of the one-male, age-graded, and all-male phase in wild Thomas' langurs groups. In *Primate Males*, ed. P. Kappeler. Cambridge: Cambridge University Press, pp. 130–45.

Sterck, E. H., Watts, D. P. and van Schaik, C. P. (1997). The evolution of social relationships in female primates. *Behavioral Ecology and Sociobiology*, **41**, 291–309.

Stevens, J. D. (1974). The occurrence and significance of tooth cuts on the blue shark (*Prionace glauca* L.) from British waters. *Journal of the Marine Biological Association of the United Kingdom*, **54**, 373–8.

(1976). First results of shark tagging in the north-east Atlantic, 1972–1975. *Journal of the Marine Biological Association of the United Kingdom*, **56**, 929–37.

Stevick, P. T., Allen, J., Berube, M. *et al.* (2003). Segregation of migration by feeding ground origin in North Atlantic humpback whales (*Megaptera novaeangliae*). *Journal of Zoology, London*, **259**, 231–7.

Stewart, B. S. and Delong, R. L. (1994). Postbreeding foraging migrations of northern elephant seals. In *Elephant Seals: Population Ecology, Behaviour, and Physiology*, eds. B. J. Le Boeuf and R. M. Laws. Berkley: University of California Press, pp. 290–309.

Stewart, B. S., Craig, M. P. and Antonelis, G. A. (1998). Characterization of Hawaiian Monk Seal (*Monachus schauinslandi*) pelagic habitat, home range and diving behaviour. Honolulu: National Marine Fisheries Service, Southwest Fisheries Science Center Honolulu Laboratory, 2570 Dole Street, Honolulu, Hawaii 96822-2902.

Stirling, I. (1969). Ecology of the Weddell seal in McMurdo Sound, Antarctica. *Ecology*, **50**, 573–86.

(1983). The evolution of mating systems in pinnipeds. In *Advances in the Study of Mammalian Behaviour*, eds. J. F. Eisenberg and D. G. Kleinmann. Lawrence, Kansas: Allen Press, pp. 489–527.

Stoddart, D. and Braithwaite, R. (1979). A strategy for utilization of regenerating heathland habitat by the brown bandicoot (*Isoodon obesulus*; Marsupialia, Peramelidae). *Journal of Animal Ecology*, **48**, 165–79.

Stokke, S. (1999). Sex differences in feeding-patch choice in a megaherbivore: elephants in Chobe National Park, Botswana. *Canadian Journal of Zoology*, **77**, 1723–32.

Stokke, S. and du Toit, J. T. (2000). Sex and size related differences in the dry season feeding patterns of elephants in Chobe National Park, Botswana. *Ecography*, **23**, 70–80.

(2002). Sexual segregation in habitat use by elephants in Chobe National Park, Botswana. *African Journal of Ecology*, **40**, 360–71.

Storz, J. F. and Williams, C. F. (1996). Summer population structure of sub-alpine bats in Colorado. *Southwestern Naturalist*, **41**, 322–4.

Storz, J. F., Bhat, H. R. and Kunz, T. H. (2000). Social structure of a polygynous tent-making bat, *Cynopterus sphinx* (Megachiroptera). *Journal of Zoology, London*, **251**, 151–65.

Strahan, R. (1995). *The Mammals of Australia*, 2nd edn. Sydney: Reed New Holland.

Straits, B. C. (1998). Occupational sex segregation: the role of personal ties. *Journal of Vocational Behavior*, **52**(2), 191–207.

Strikwerda, T. E., Fuller, M. R., Seegar, W. S., Howey, P. W. and Black, H. D. (1986). Bird-borne satellite transmitter and location program. *Johns Hopkins APL Technical Digest*, **7**, 203–8.

Suhonen, J. and Kuitunen, M. (1991). Intersexual foraging niche differentiation within the breeding pair in the common treecreeper *Certhia familiaris*. *Ornis Scandinavica*, **22**, 313–18.

Sukumar, R. and Gadgil, M. (1988). Male-female differences in foraging on crops by Asian elephants. *Animal Behaviour*, **36**, 1233–5.

Summers, R.W., Westlake, G.E. and Feare, C.J. (1987). Differences in the ages, sexes and physical condition of starlings *Sturnus vulgaris* at the centre and periphery of roosts. *Ibis*, **129**, 96–102.

Sun, L.X., Shine, R., Zhao, D.B. and Tang, Z.R. (2002). Low costs, high output: reproduction in an insular pit-viper (*Gloydius shedaoensis*, Viperidae) from north-eastern China. *Journal of Zoology, London*, **256**, 511–21.

Sund, O. (1943). Et brugdebarsel. *Naturen*, **67**, 285–6.

Swennen, C. (1984). Differences in quality of roosting flocks of oystercatchers. In *Coastal Waders and Wildfowl in Winter*, eds. P.R. Evans, J.D. Goss-Custard and W.G. Hale. Cambridge: Cambridge University Press.

Swennen, C., de Bruijn, L.L.M., Duiven, P., Leopold, M.F. and Marteijn, E.C.L. (1983). Differences in bill form of the oystercatcher *Haematopus ostralegus*; a dynamic adaptation to specific foraging techniques. *Netherlands Journal of Sea Research*, **17**, 57–83.

Swingland, I.R. and Lessells, C.M. (1979). The natural regulation of giant tortoise populations on Aldabra Atoll: movement polymorphism, reproductive success and mortality. *Journal of Animal Ecology*, **48**, 639–54.

Sydeman, W.J. and Nur, N. (1994). Life history strategies of female northern elephant seals. In *Elephant Seals: Population Ecology, Behaviour and Physiology*, eds. B.J. Le Boeuf and R.M. Laws. Berkley: University of California Press, pp. 137–53.

Talbot, J.J. (1979). Time budget, niche overlap, inter- and intraspecific aggression in *Anolis humilis* and *A. limifrons* from Costa Rica. *Copeia*, **1979**, 472–81.

Tamisier, A. (1985). Hunting as a key environmental parameter for the Western Palearctic duck populations. *Wildfowl*, **36**, 95–103.

Taylor, R.J. (1981). The comparative ecology of the Eastern grey kangaroo and wallaroo in the New England tablelands of New South Wales. Ph.D. thesis, University of New England.

(1983). Association of social classes of the wallaroo, *Macropus robustus* (Marsupialia: Macropodidae). *Australian Wildlife Research*, **10**, 39–45.

Tedman, R.A. and Bryden, M.M. (1979). Cow-pup behaviour of the Weddell seal, *Leptonychotes weddelli* (Pinnipedia), in McMurdo Sound, Antarctica. *Australian Journal of Wildlife Research*, **6**, 19–37.

Terborgh, J. (1990). *Where Have All the Birds Gone?* Princeton: Princeton University Press.

Terranova, M.L. and Laviola, G. (1995). Individual differences in mouse behavioural development: effects of precious weaning and ongoing sexual segregation. *Animal Behaviour*, **50**, 1261–71.

Testa, J.W. (1994). Over winter movements and diving behaviour of female Weddell seal (*Leptonychotes weddellii*) in the Southwestern Ross Sea, Antarctica. *Canadian Journal of Zoology*, **72**, 1700–10.

Theodorakis, C.W. (1989). Size segregation and the effects of oddity on predation risk in minnow schools. *Animal Behaviour*, **38**, 496–502.

Thill, R.E., Martin, A., Jr, Morris, H.F., Jr and McCune E.D. (1987). Grazing and burning impacts on deer diets on Louisiana pine-bluestem range. *Journal of Wildlife Management*, **51**, 873–80.

Thomas, C.D., Baguette, M. and Lewis, O.T. (2000). Butterfly movement and conservation in patchy landscapes. In *Behavior and Conservation*, eds. L.M. Gosling and W.J. Sutherland. Cambridge, United Kingdom: Cambridge University Press, pp. 85–104.

Thomas, D.W. and Cloutier, D. (1992). Evaporative water loss by hibernating little brown bats, *Myotis lucifugus*. *Physiological Ecology*, **65**, 433–56.

Thomas, L.N. (1987). The effects of stress on some aspects of the demography and physiology of *Isoodon obesulus* (Shaw and Nodder). Unpublished Masters thesis, University of Western Australia.

Thompson, P.M., Fedak, M.A., McConnell, B.J. and Nicholas, K.S. (1989). Seasonal and sex-related variation in the activity patterns of common seals (*Phoca vitulina*). *Journal of Applied Ecology*, **27**, 521–35.

Thompson, P.M., Mackay, A., Tollit, D.J., Enderby, S. and Hammond, P.S. (1998). The influence of body size and sex on the characteristics of harbour seal foraging trips. *Canadian Journal of Zoology – Revue Canadienne De Zoologie*, **76**, 1044–53.

Thompson, P.M., Tollit, D.J., Wood, D. *et al.* (1997). Estimating harbour seal abundance and distribution in an esturine habitat in N.E. Scotland. *Journal of Applied Ecology*, **34**, 43–52.

Thorhallsdottir, A.G., Provenza, F.D. and Balph, D.F. (1990). Ability of lambs to learn about novel foods while observing or participating with social models. *Applied Animal Behaviour Science*, **25**, 25–33.

Tickell, W.L.N. (1968). The biology of the great albatrosses *Diomedea exulans* and *Diomedea epomophora*. *Antarctic Research Series*, **12**, 1–55.

(2000). *Albatrosses*. East Sussex, England: Pica Press.

Tieszen, L.L. and Imbamba, S.K. (1980). Photosynthetic systems, carbon isotope discrimination and herbivore selectivity in Kenya. *African Journal of Ecology*, **18**, 237–42.

Tolley, K.A., Read, A.J., Well, R.S. *et al.* (1995). Sexual dimorphism in wild bottlenose dolphins (*Tursiops truncatus*) from Sarasota, Florida. *Journal of Mammalogy*, **76**, 1190–8.

Townshend, D.J. (1981). The importance of field feeding to the survival of wintering male and female curlews *Numenius arquata* on the Tees estuary. In *Feeding and Survival Strategies of Estuarine Organisms*, eds. N.V. Jones and W.J. Wolff. New York: Plenum Press.

Tricas, T.C. and Le Feuvre, E.M. (1985). Mating in the reef white-tip shark *Triaenodon obesus*. *Marine Biology*, **84**, 233–7.

Trivers, R. (1971). The evolution of reciprocal altruism. *Quarterly Review of Biology*, **46**, 35–57.

Trivers, R.L. (1972). Parental investment and sexual selection. In *Sexual Selection and the Descent of Man*, ed. B. Campbell. Chicago: Aldine, pp. 1871–971.

Tuck, G.N., Polacheck, T. and Bulman, C. (2003). Spatio-temporal trends of long-line fishing effort in the Southern Ocean and implications for seabird by-catch. *Biological Conservation*, **114**, 1–27.

Tucker, A.D., FitzSimmons, N.N. and Gibbons, J.W. (1995). Resource partitioning by the estuarine turtle *Malaclemys terrapin*: trophic, spatial, and temporal foraging constraints. *Herpetologica*, **51**, 167–81.

Tucker, A.D., Limpus, C.J., McCallum, H.I. and McDonald, K.R. (1997). Movements and home ranges of *Crocodylus johnstoni* in the Lynd River, Queensland. *Wildlife Research*, **24**, 379–96.

Tucker, A.D., McCallum, H.I., Limpus, C.J. and McDonald, K.R. (1998). Sex-biased dispersal in a long-lived polygynous reptile (*Crocodylus johnstoni*). *Behavioral Ecology and Sociobiology*, **44**, 85–90.

Tuttle, M.D. (1979). Status, causes of decline, and management of endangered grey bats. *Journal of Wildlife Management*, **43**, 1–17.

Twente, J.W. (1955). Some aspects of habitat selection and other behaviour of cavern-dwelling bats. *Ecology*, **36**, 706–32.

Tyndale-Biscoe, H. and Renfree, M. (1987). *Reproductive Physiology of Marsupials*. Cambridge: Cambridge University Press.

Uetz, P. (2000). How many reptile species? *Herpetological Review*, **31**, 13–15.

Van Marken Lichtenbelt, W.D., Wesselingh, R.A., Vogel, J.T. and Albers, K.B.M. (1993). Energy budgets in free-living green iguanas in a seasonal environment. *Ecology*, **74**, 1157–72.

van Noordwijk, A.J. (1994). The interaction of inbreeding depression and environmental stochasticity in the risk of extinction of small populations. In *Conservation Genetics*, eds. V. Loeschcke, J. Tomiuk and S.K. Jain. Basel, Switzerland: Birkhauser Verlag, pp. 131–46.

van Noordwijk, M. and van Schaik, C.P. (2002). Career moves: transfers and rank challenge decisions by male long-tailed macaques. *Behaviour*, **138**, 359–95.

van Noordwijk, M., Hemelrijk, C. K., Herremans, L. A. M. and Sterck, E. H. M. (1993). Spatial position and behavioral sex differences in juvenile long-tailed macaques. In *Juvenile Primates*, eds. M.E. Pereira and L.A. Fairbanks. Oxford: Oxford University Press, pp. 77–85.

van Schaik, C.P. (1983). Why are diurnal primates living in groups? *Behaviour*, **37**, 120–44.

(1989). The ecology of social relationships amongst female primates. In *Comparative Socioecology*, eds. V. Standen and R. Foley. London: Blackwell, pp. 195–218.

(1999). The fission-fusion social system of orangutans. *Primates*, **40**, 69–86.

(2000a). Vulnerability to infanticide by males: patterns among mammals. In *Infanticide by Males and Its Implications*, eds. C.P. van Schaik and C.H. Janson. Cambridge: Cambridge University Press, pp. 61–71.

(2000b). Infanticide by male primates: the sexual selection hypothesis revisited. In *Infanticide by Males and Its Implications*, eds. C.P. van Schaik and C.H. Janson. Cambridge: Cambridge University Press, pp. 27–60.

van Schaik, C.P. and Kappeler, P.M. (1997). Infanticide risk and the evolution of male-female association in primates. *Proceedings of the Royal Society, London, B*, **264**, 1687–94.

van Schaik, C.P. and van Noordwijk, M.A. (1989). The special role of male Cebus monkeys in predation avoidance and its effects on group composition. *Behavioral Ecology and Sociobiology*, **24**, 265–76.

Van Soest, P.J. (1982). *Nutritional Ecology of the Ruminant*. Corvallis: O and B Books.

(1994). *Nutritional Ecology of the Ruminant*, 2nd edn. Ithaca: Cornell University Press.

(1996). Allometry and ecology of feeding behavior and digestive capacity in herbivores. *Zoo Biology*, **15**, 455–79.

Vaughan, T.A. (1976). Nocturnal behaviour of the African false vampire bat (*Cardioderma cor*). *Journal of Mammalogy*, **57**, 227–48.

Vaughan, T.A. and O'Shea, T.J. (1976). Roosting ecology of the pallid bat, *Antrozous pallidus*. *Journal of Mammalogy*, **57**, 227–48.

Vaughan, T.A. and Vaughan, R.P. (1986). Seasonality and the behaviour of the African yellow-winged bat (*Lavia frons*). *Journal of Mammalogy*, **67**, 91–102.

(1987). Parental behaviour in the African yellow-winged bat (*Lavia frons*). *Journal of Mammalogy*, **68**, 217–23.

Vehrencamp, S.L., Stiles, F.G. and Bradbury, J.W. (1977). Observations on the foraging behaviour and avian prey of the neotropical carnivorous bat, *Vampyrum spectrum*. *Journal of Mammalogy*, **58**, 469–78.

Verme, L.J. (1988). Niche selection by male white-tailed deer: an alternative hypothesis. *Wildlife Society Bulletin*, **16**, 448–51.

Vernes, K. and Pope, L.C. (2001). Stability of nest range, home range and movements of the northern bettong (Bettongia tropica) following moderate-intensity fire in a tropical woodland, north-eastern Queensland. Wildlife Research, 28, 141–50.

(2002). Fecundity, pouch young survivorship and breeding season of the northern bettong (Bettongia tropica) in the wild. Australian Mammalogy, 23, 95–100.

Vidal, R.M., Macias-Caballero, C. and Duncan, C.D. (1994). The occurrence and ecology of the golden-cheeked warbler in the highlands of Northern Chiapas, Mexico. Condor, 96, 684–91.

Viitanen, P. (1967). Hibernation and seasonal movements of the viper, Vipera berus berus (L.) in southern Finland. Annales Zoologici Fennici, 4, 472–546.

Villard, M.-A., Martin, P.R. and Drummond, C.G. (1993). Habitat fragmentation and pairing success in the ovenbird, Seiurus aurocapillus. Auk, 110, 759–68.

Villaret, J.C. and Bon, R. (1995). Social and spatial segregation in alpine ibex (Capra ibex) in Bargy, French Alps. Ethology, 101, 291–300.

(1998). Sociality and relationships in Alpine ibex (Capra ibex). Revue d'Ecologie (Terre Vie), 53, 153–70.

Villaret, J.C., Rivet, A. and Bon, R. (1997). Sexual segregation of habitat by the alpine ibex in the French Alps. Journal of Mammalogy, 78, 1273–81.

Vincent, A. and Sadovy, Y. (1998). Reproductive ecology in the conservation and management of fishes. In Behavioral Ecology and Conservation Biology, ed. T. Caro. New York: Oxford University Press, pp. 209–45.

Vincent, S.E., Herrel, A. and Irschick, D.J. (2004). Ontogeny of intersexual head shape and prey selection in the pitviper Agkistrodon piscivorus. Biological Journal of the Linnean Society, 81, 151–9.

Visagie, L. (2001). Grouping behaviour in the armadillo lizard, Cordylus cataphractus. M.Sc. thesis, Matieland, South Africa: University of Stellenbosch.

Vitt, L.J. and Cooper, W.E. (1986). Foraging and diet of a diurnal predator (Eumeces laticeps) feeding on hidden prey. Journal of Herpetology, 20, 408–15.

Vitt, L.J. and Zani, P.A. (1996). Ecology of the elusive tropical lizard Tropidurus [equals Uracentron] flaviceps (Tropiduridae) in lowland rain forest of Ecuador. Herpetologica, 52, 121–32.

Vladykov, V.D. (1946). Nourriture du marsouin blanc ou béluga (Delphinapterus leucas) du fleuve Saint-Laurent. Etudes sur les mammifères aquatiques (IV). Québec: Département des pêcheries, Province de Québec, pp. 129.

Vogt, R.C. (1981). Food partitioning in three sympatric species of map turtle, genus Graptemys (Testudinata, Emydidae). American Midland Naturalist, 105, 102–11.

Voigt, C.C. and Streich, W.J. (2003). Queuing for harem access in colonies of the greater sac-winged bat. Animal Behaviour, 65, 149–56.

Voisin, J. (1968). Les pétrels géants (Macronectes halli et Macronectes giganteus) de l'île de la Possession. L'Oiseau, 38, 7–122.

(1991). Sur le régime et l'écologie alimentaires des pétrels géants Macronectes halli et M. giganteus de l'archipel Crozet. L'Oiseau, 61, 39–49.

Voisin, J. and Bester, M.N. (1981). The specific status of Giant petrels Macronectes at Gough Island. In Proceedings of the Symposium on Birds on the Sea and Shore, 1979, ed. J. Cooper. Cape Town: African Seabird Group, pp. 215–22.

Volkman, N.J., Presler, P. and Trivelpiece, W. (1980). Diets of pygoscelid penguins at King George Island, Antarctica. Condor, 82, 373–8.

Wagenknecht, E. (1986). Rotwild. Malsungen: Neumann-Neudamm Verlag.

Wagner, R.H. (1997). Differences in prey species delivered to nestlings by male and female razorbills Alca torda. Seabird, 19, 58–9.

Walker, B. G. and Bowen, W. D. (1993). Changes in body-mass and feeding-behavior in male harbor seals, *Phoca vitulina*, in relation to female reproductive status. *Journal of Zoology*, **231**, 423–36.

Wanless, S., Corfield, T., Harris, M. P., Buckland, S. T. and Morris, J. A. (1993). Diving behaviour of the shag (Aves: Pelecaniformes) in relation to water depth and prey size. *Journal of Zoology, London*, **231**, 11–25.

Wanless, S., Finney, S. K., Harris, M. P. and McCafferty, D. J. (1999). Effect of the diel light cycle on the diving behaviour of two bottom feeding marine birds: the blue-eyed shag *Phalacrocorax atriceps* and the European shag *P. aristotelis*. *Marine Ecology Progress Series*, **188**, 219–24.

Wanless, S., Harris, M. P., Morris, J. A. (1995). Factors affecting daily activity budgets of South-Georgian shags during chick rearing at Bird Island, South Georgia. *Condor*, **97**, 550–8.

Wapstra, E., Olsson, M., Shine, R. *et al.* (2004). Maternal basking behaviour determines offspring sex in a viviparous reptile. *Biology Letters*, **271**, S230–S232.

Ward, A. J. W. and Hart, P. J. B. (2003). The effects of kin and familiarity on interactions between fish. *Fish and Fisheries*, **4**, 348–58.

Ward, S. J. (1990). Life history of the feathertail glider, *Acrobates pygmaeus* (Acrobatidae: Marsupialia) in south-eastern Australia. *Australian Journal of Zoology*, **38**, 503–17.

Ward, S. J. and Renfree, M. B. (1988). Reproduction in females of the feathertail glider, *Acrobates pygmaeus* (Marsupialia). *Journal of Zoology, London*, **216**, 225–39.

Wardle, C. S. (1993). Fish behaviour and fishing gear. In *Behaviour of Teleost Fishes*, ed. T. J. Pitcher. London: Chapman and Hall, pp. 607–43.

Warham, J. (1990). *The Petrels: Their Ecology and Breeding Systems*. London: Academic Press.

Waser, P. M. (1977). Feeding, ranging, and group size in the mangabey, *Cercocebus albigena*. In *Primate Ecology*, ed. T. H. Clutton-Brock. Cambridge: Cambridge University Press, pp. 183–222.

(1985). Spatial structure in mangabey groups. *International Journal of Primatology*, **6**, 569–80.

Waterman, J. M. (1995). The social organization of the Cape ground squirrel (*Xerus inauris*; Rodentia: Sciuridae). *Ethology*, **101**, 130–47.

(1997). Why do male Cape ground squirrels live in groups? *Animal Behaviour*, **53**, 809–17.

Waters, J. C. (1974). The biological significance of the basking habit in the black-knobbed sawback, *Graptemys nigronoda* Cagle. M.Sc. thesis, Auburn, AL: Auburn University.

Watkins, J. L., Buchholz, F., Priddle, J., Morris, D. J. and Ricketts, C. (1992). Variation in reproductive status of Antarctic krill swarms evidence for a size-related sorting mechanism? *Marine Ecology Progress Series*, **82**, 163–74.

Watson, A. and Staines, B. W. (1978). Differences in the quality of wintering areas used by male and female red deer (*Cervus elaphus*) in Aberdeenshire. *Journal of Zoology, London*, **186**, 544–50.

Watt, E. M. and Fenton, M. B. (1995). DNA fingerprinting provides evidence of discriminant suckling and non-random mating in little brown bats, *Myotis lucifugus*. *Molecular Ecology*, **4**, 261–4.

Watts, D. P. (1984). Composition and variability of mountain gorilla diets in the central Virungas. *American Journal of Primatology*, **7**, 325–56.

(1988). Environmental influences on mountain gorilla time budgets. *American Journal of Primatology*, **15**, 295–312.

(1991). Ecology of gorillas and its relationship to female transfer in mountain gorillas. *International Journal of Primatology*, **11**, 21–45.

Watts, D. P. and Mitani, J. C. (2001). Boundary patrols and intergroup encounters in wild chimpanzees. *Behaviour*, **138**, 299–327.

Watts, P. S., Hansen, S. and Lavigne, D. M. (1993). Models of heat loss by marine mammals: Thermoregulation below the zone of irrelevance. *Journal of Theoretical Biology*, **163**, 505–25.

Webb, G. J. W. and Smith, A. M. A. (1984). Sex ratio and survivorship in the Australian freshwater crocodile, *Crocodylus johnstoni*. *Symposia of the Zoological Society of London*, **52**, 319–55.

Webb, G. J. W., Buckworth, R. and Manolis, S. C. (1983). *Crocodylus johnstoni* in the McKinlay River area, N. T. III. Growth, movement and the population age structure. *Australian Wildlife Research*, **10**, 383–401.

Webb, J. K. and Shine, R. (1997). A field study of spatial ecology and movements of a threatened snake species, *Hoplocephalus bungaroides*. *Biological Conservation*, **82**, 203–17.

Webb, J. K., Brown, G. P. and Shine, R. (2001). Body size, locomotor speed and antipredator behaviour in a tropical snake (*Tropidonophis mairii*, Colubridae): the influence of incubation environments and genetic factors. *Functional Ecology*, **15**, 561–8.

Webb, R. G. (1961). Observations on the life histories of turtles (genus *Pseudemys* and *Graptemys*) in Lake Texoma, Oklahoma. *American Midland Naturalist*, **65**, 193–214.

Weckerly, F. W. (1993). Intersexual resource partitioning in black-tailed deer: a test of the body size hypothesis. *Journal of Wildlife Management*, **57**, 475–94.

(1998). Sexual-size dimorphism: influence of mass and mating systems in the most dimorphic mammals. *Journal of Mammalogy*, **79**, 33–52.

Weckerly, F. W., Ricca, M. A. and Meyer, K. P. (2001). Sexual segregation in Roosevelt elk: cropping rates and aggression in mixed-sex groups. *Journal of Mammalogy*, **82**, 825–35.

Weeden, R. B. (1964). Spatial separation of sexes in rock and willow ptarmigan in winter. *Auk*, **81**, 534–41.

Weidt, A., Hagenah, N., Randrianambinina, B., Radespiele, U. and Zimmerman, E. (2004). Social organization of the golden brown mouse lemur (*Microcebus ravelobensis*). *American Journal of Physical Anthropology*, **123**, 40–51.

Weihs, D. and Webb, P. W. (1983). Optimisation of locomotion. In *Fish Biomechanics*, eds. P. W. Webb and D. Weihs. New York: Praeger, pp. 339–71.

Weilgart, L. S. and Whitehead, H. (1986). Observations of a sperm whale (*Physeter catodon*) birth. *Journal of Mammalogy*, **67**, 399–401.

Weilgart, L., Whitehead, H. and Payne, K. (1996). A colossal convergence. *American Scientist*, **84**, 278–87.

Weimerskirch, H. (1991). Sex-specific differences in molt strategy in relation to breeding in the wandering albatross. *Condor*, **93**, 731–7.

(1992). Reproductive effort in long-lived birds: age-specific patterns of condition, reproduction and survival in the wandering albatross. *Oikos*, **64**, 464–73.

(1995). Regulation of foraging trips and incubation routine in male and female wandering albatrosses. *Oecologia*, **102**, 37–43.

(1998). Foraging strategies of Indian Ocean albatrosses and their relationships with fisheries. In *Albatross Biology and Conservation*, eds. G. Robertson and R. Gales. Chipping Norton, Australia: Surrey Beatty and Sons, pp. 168–79.

Weimerskirch, H. and Jouventin, P. (1987). Population dynamics of the wandering albatross, *Diomedea exulans*, of the Crozet Islands: causes and consequences of the population decline. *Oikos*, **49**, 315–22.

(1998). Changes in population sizes and demographic parameters of six albatross species breeding on the French sub-Antarctic islands. In *Albatross Biology and Conservation*, eds. G. Robertson and R. Gales. Chipping Norton, Australia: Surrey Beatty and Sons, pp. 84–91.

Weimerskirch, H. and Wilson, R.P. (2000). Oceanic respite for wandering albatrosses. *Nature*, **406**, 955–6.

Weimerskirch, H., Barbraud, C. and Lys, P. (2000). Sex differences in parental investment and chick growth in Wandering Albatrosses: Fitness consequences. *Ecology*, **81**, 309–18.

Weimerskirch, H., Bonadonna, F., Bailleul, F. *et al.* (2002). GPS tracking of foraging albatrosses. *Science*, **295**, 1259–69.

Weimerskirch, H., Brothers, N. and Jouventin, P. (1997a). Population dynamics of wandering albatross *Diomedea exulans* and the Amsterdam albatross *D. amsterdamensis* in the Indian Ocean and their relationships with long-line fisheries: conservation implications. *Biological Conservation*, **79**, 257–70.

Weimerskirch, H., Cherel, Y., Cuenot-chaillet, F. and Ridoux, V. (1997b). Alternative foraging strategies and resource allocation by male and female wandering albatrosses. *Ecology*, **78**, 2051–63.

Weimerskirch, H., Lequette, B. and Jouventin, P. (1989). Development and maturation of plumage in the wandering albatross *Diomedea exulans*. *Journal of Zoology, London*, **219**, 411–21.

Weimerskirch, H., Salamolard, M., Sarrazin, F. and Jouventin, P. (1993). Foraging strategy of wandering albatrosses through the breeding season: a study using satellite telemetry. *Auk*, **110**, 325–42.

Wells, R. S. (2003). Dolphin social complexity: lessons from long-term study and life history. In *Animal Social Complexity: Intelligence, Culture, and Individualized Societies*, eds. F.B.M. De Waal and P.L. Tyack. Cambridge, Massachusetts: Harvard University Press, pp. 32–56.

Wells, R. S. and Scott, M. S. (1999). Bottlenose dolphin – *Tursiops truncatus* (Montagu, 1821). In *Handbook of Marine Mammals: The Second Book of Dolphins and Porpoises*, vol. 6, eds. S.H. Ridgway and R. Harrison. London: Academic Press, pp. 137–82.

Wells, R. S., Irvine, A. B. and Scott, M. D. (1980). The social ecology of inshore odontocetes. In *Cetacean Behavior: Mechanisms and Functions*, ed. L.M. Herman. New York: John Wiley and Sons, pp. 263–317.

Wells, R. S., Scott, M. D. and Irvine, A. B. (1987). The social structure of free-ranging Bottlenose dolphins. In *Current Mammalogy*, vol. 1, ed. H.H. Genoways. New York: Plenum Press, pp. 247–305.

Wells, R.T. (1978). Field observations of the hairy-nosed wombat, *Lasiorhinus latifrons* (Owen). *Australian Wildlife Research*, **5**, 299–303.

Wetherbee, B. M., Gruber, S. H. and Cortes, E. (1990). Diet, feeding habits, digestion and consumption in sharks, with special reference to the lemon shark, *Negaprion brevirostris*. In *Elasmobranchs as Living Resources: Advances in the Biology, Ecology, Systematics and Status of the Fisheries, NOAA Technical Report 90*, eds. H.L. Pratt, S.H. Gruber and T. Taniuchi. Seattle, WA: National Oceanic and Atmospheric Administration, pp. 29–47.

Whitaker, P. B. and Shine, R. (2001). Thermal biology and activity patterns of the eastern brownsnake (*Pseudonaja textilis*): a radiotelemetric study. *Herpetologica*, **58**, 436–62.

(2003). A radiotelemetric study of movements and shelter-site selection by free-ranging brownsnakes (*Pseudonaja textilis*, Elapidae). *Herpetological Monographs*, **17**, 130–44.

White, M. and Kolb, J.A. (1974). A preliminary study of *Thamnophis* near Sagehagen Creek, California. *Copeia*, **1974**, 126–36.

Whitehead, H. (1993). The behaviour of mature male sperm whales on the Galapagos Islands breeding grounds. *Canadian Journal of Zoology*, **71**, 689–99.

(1996). Babysitting, dive synchrony, and indications of alloparental care in sperm whales. *Behavioral Ecology and Sociobiology*, **38**, 237–44.

(2003). *Sperm Whales: Social Evolution in the Ocean*. Chicago: The University of Chicago Press.

Whitehead, H. and Mann, J. (2000). Female reproductive strategies of cetaceans: Life histories and calf care. In *Cetacean Societies: Field Studies of Dolphins and Whales*, eds. J. Mann, R.C. Connor, P.L. Tyack and H. Whitehead. Chicago: University of Chicago Press, pp. 219–47.

Whitehead, H. and Weilgart, L. (2000). The sperm whale: Social females and roving males. In *Cetacean Societies: Field Studies of Dolphins and Whales*, eds. J. Mann, R.C. Connor, P.L. Tyack and H. Whitehead. Chicago: University of Chicago Press, pp. 154–73.

Whitehead, H., Waters, S. and Lyrholm, T. (1991). Social organization of female sperm whales and their offspring: constant companions and casual acquaintances. *Behavioral Ecology and Sociobiology*, **29**, 385–9.

Whiting, B. and Edwards, C. (1973). A cross-cultural analysis of sex-differences in the behavior of children age 3 through 11. *Journal of Social Psychology*, **91**, 171–88.

Whiting, M.J. and Greeff, J.M. (1997). Facultative frugivory in the Cape flat lizard, *Platysaurus capensis* (Sauria: Cordylidae). *Copeia*, **1997**, 811–18.

Wickings, E.J. and Dixon, A.F. (1992). Testicular function, secondary sexual development, and social status in male mandrills (*Mandrillus sphinx*). *Physiology and Behavior*, **52**, 909–16.

Wikelski, M. and Trillmich, F. (1994). Foraging strategies of the Galapagos marine iguana (*Amblyrhynchus cristatus*): adapting behavioral rules to ontogenic size change. *Behaviour*, **128**, 255–79.

Wilkinson, G.S. (1992). Information transfer at evening bat colonies. *Animal Behaviour*, **44**, 501–18.

(1995). Information transfer in bats. In *Ecology, Evolution and Behaviour of Bats*, eds. P.A. Racey and S.M. Swift. Symposium of the Zoological Society of London 67. Oxford: Clarendon Press, pp. 345–360.

Wilkinson, L.C. and Barclay, R.M.R. (1997). Differences in the foraging behaviour of male and female big brown bats (*Eptesicus fuscus*) during the reproductive period. *Ecoscience*, **4**, 279–85.

Wilkinson, G.S. and Boughman, J.W. (1998). Social calls co-ordinate foraging in greater spear-nosed bats. *Animal Behaviour*, **55**, 337–50.

Williams, A.J. (1982). Sexual size-dimorphism in the growth of *Macronectes*. *Acta XVIII Congressus Internationalis Ornithologici*, **2**, 1190–1.

Williams, J.M., Oehlert, G.W., Carlis, J.V. and Pusey, A.E. (2004). Why do male chimpanzees defend a group range? *Animal Behaviour*, **68**, 523–32.

Williams, J.M., Pusey, A.E., Carlis, J.V., Goodall. J. and Farm, B.E. (2002). Space use and group membership in female chimpanzees: alternative strategies. *Animal Behaviour*, **63**, 347–60.

Williams, T.A. and Christiansen, J.L. (1981). The niches of two sympatric softshell turtles, *Trionyx muticus* and *Trionyx spiniferus*, in Iowa. *Journal of Herpetology*, **15**, 30–8.

Williams, T. C., Ireland, L. C. and Williams, J. M. (1973). High altitude flights of the free-tailed bat, *Tadarida brasiliensis* observed with radar. *Journal of Mammalogy*, **54**, 807–21.

Williams, T. D. (1991). Foraging ecology and diet of gentoo penguins *Pygoscelis papua* at South Georgia during winter and an assessment of their winter prey consumption. *Ibis*, **133**, 3–13.

Williams, T. M. (1999). The evolution of low cost efficient swimming in marine mammals: limits to energetic optimization. *Philosophical Transactions of the Royal Society of London*, **354**, 193–201.

Wilson, B. (1995). The ecology of bottlenose dolphin in Moray Firth, Scotland: a population at the northern extreme of the species' range. Ph.D. thesis, Scotland: University of Aberdeen.

Wilson, E. O. (1980). *Sociobiology: The Abridged Edition*. Cambridge: Belknap Press.

Wilson, M. and Wrangham, R. W. (2003). Intergroup relations in chimpanzees. *Annual Review of Anthropology*, **32**, 363–92.

Wilson, R. P., Copper, J. and Plotz, J. (1992). Can we determine when marine endotherms feed? A case study of seabirds. *Journal of Experimental Biology*, **167**, 267–75.

Wilson, M. L., Wallauer, W. R. and Pusey, A. E. (2004). Intergroup violence in chimpanzees: new cases from Gombe National Park, Tanzania. *International Journal of Primatology*, **25**, 523–50.

Wirtz, P. and Morato, T. (2001). Unequal sex ratios in longline catches. *Journal of Marine Biology*, **81**, 1–2.

Witter, M. S. and Cuthill, I. C. (1993). The ecological costs of avian fat storage. *Philosophical Transactions of the Royal Society of London B*, **340**, 73–90.

Wolf, N. G. (1985). Odd fish abandon mixed-species groups when threatened. *Behavioral Ecology and Sociobiology*, **17**, 47–52.

Wood, A. G., Naef-Daenzer, B., Prince, P. A. and Croxall, J. P. (2000). Quantifying habitat use in satellite-tracked pelagic seabirds: application of kernel estimation to albatross locations. *Journal of Avian Biology*, **31**, 278–86.

Wood, C. M. and McDonald, D. G. (1997). *Global Warming: Implications for Freshwater and Marine Fish*. Cambridge: Cambridge University Press.

Woodroffe, R. and Ginsberg, J. R. (2000). Ranging behaviour and vulnerability to extinction in carnivores. In *Behavior and Conservation*, eds. L. M. Gosling and W. J. Sutherland. Cambridge, United Kingdom: Cambridge University Press, pp. 125–40.

Woolington, D. W. (1993). Sex ratios of canvasbacks wintering in Louisiana. *Journal of Wildlife Management*, **57**, 751–8.

Wooller, R. D., Richardson, K. C., Garavanta, C. A. M., Saffer, V. M. and Bryant, K. A. (2000). Opportunistic breeding in the polyandrous honey possum, *Tarsipes rostratus*. *Australian Journal of Zoology*, **48**, 669–80.

Woolnough, A. P. and du Toit, J. T. (2001). Vertical zonation of browse quality in tree canopies exposed to a size-structured guild of African browsing ungulates. *Oecologia*, **129**, 585–90.

Wootton, R. J. (1976) *The Biology of the Sticklebacks*. London: Academic Press.

(1998) *Ecology of Teleost Fishes*. Dordrecht, The Netherlands: Kluwer Academic Pulishers.

Worrell, E. (1958). *Song of the Snake*. Sydney: Angus and Robertson.

Wourms, J. P. and Demski, L. S. (1993). The reproduction and development of sharks, skates, rays and ratfishes: introduction, history, overview, and future prospects. *Environmental Biology of Fishes*, **38**, 7–21.

Wrangham, R. W. (1979). On the evolution of ape social systems. *Social Science Information*, **18**, 335–68.

(1999). Evolution of coalitionary killing. *Yearbook of Physical Anthropology*, **42**, 1–30.

(2000). Why are male chimpanzees more gregarious than mothers? A scramble competition hypothesis. In *Primate Males*, ed. P. M. Kappeler. Cambridge: Cambridge University Press, pp. 248–58.

Wrangham, R. W. and Rubenstein, D. I. (1986). Social evolution in birds and mammals. In *Ecological Aspects of Social Evolution*, ed. R. Wrangham. Princeton, New Jersey: Princeton University Press, pp. 452–71.

Wrangham, R. W. and Smuts, B. B. (1980). Sex differences in the behavioural ecology of chimpanzees in the Gombe National Park, Tanzania. *Journal of Reproduction and Fertility*, Suppl. **28**, S13–S31.

Wrangham, R. W., Chapman, C. A., Clark-Arcadi, A. P. and Isabirye-Basuta, G. (1996). Social ecology of Kanyawara chimpanzees: implications for understanding the costs of great ape groups. In *Great Ape Societies*, eds. W. C. McGrew, L. F. Marchant and N. Toshisada. Cambridge: Cambridge University Press, pp. 45–57.

Wright, P. H. (1988). Interpreting research on gender differences in friendship: A case for moderation and plea for caution. *Journal of Social and Personal Relationships*, **5**(3), 367–73.

Wright, P. H. and Scanlon, M. B. (1991). Gender role orientations and friendship: some attenuation, but gender differences abound. *Sex Roles*, **24**(9–10), 551–66.

Wrona, F. J. and Dixon W. J. (1991). Group size and predation risk: a field analysis of encounter and dilution effects. *American Naturalist*, **137**, 186–201.

Wunderle, J. M. (1995). Population characteristics of black-throated blue warblers wintering in three sites on Puerto Rico. *Auk*, **112**, 931–46.

Würsig, B. and Bastida, R. (1986). Long-range movement and individual associations of two dusky dolphins (*Lagenorhynchus obscurus*) off Argentina. *Journal of Mammalogy*, **67**, 773–4.

Würsig, B. and Würsig, M. (1980). Behavior and ecology of the Dusky dolphin, *Lagenorhynchus obscurus*, in the South Atlantic. *Fishery Bulletin*, **77**, 871–90.

Xavier, J. C., Croxall, J. P., Trathan, P. N. and Rodhouse, P. G. (2003a). Inter-annual variation in the cephalopod component of the diet of wandering albatrosses *Diomedea exulans* breeding at Bird Island, South Georgia. *Marine Biology*, **142**, 611–22.

Xavier, J. C., Croxall, J. P. and Reid, K. (2003b). Interannual variation in the diets of two albatross species breeding at South Georgia: implications for breeding performance. *Ibis*, **145**, 593–610.

Xavier, J. C., Croxall, J. P., Trathan, P. N. and Wood, A. G. (2003c). Feeding strategies and diets of breeding grey-headed and wandering albatrosses at South Georgia. *Marine Biology*, **143**, 221–32.

Xavier, J. C., Trathan, P. N., Croxall, J. P. *et al.* (2004). Foraging ecology and interactions with fisheries of wandering albatrosses (*Diomedea exulans*) breeding at South Georgia. *Fisheries Oceanography*, **13**(5), 324–44.

Yamagiwa, J. (1987). Intra- and inter-group interactions of an all-male group of Virunga mountain gorillas (*Gorilla gorilla beringei*). *Primates*, **28**, 1–30.

Yearsley, J. M. and Pérez-Barbería, F. J. (2005). Does the activity budget hypothesis explain sexual segregation in ungulates? *Animal Behaviour*, **69**, 257–67.

Young, D. D. and Cockcroft, V. G. (1994). Diet of common dolphins (*Delphinus delphis*) off the south-east of southern Africa: opportunism or specialization? *Journal of Zoology, London*, **234**, 41–53.

Young, T. P. and Isbell, L. A. (1991). Sex-differences in giraffe feeding ecology – energetic and social constraints. *Ethology*, **87**, 79–89.

Yurk, H., Barrett-Lennard, L., Ford, J.K.B. and Matkins, C.O. (2002). Cultural transmission within maternal lineages: vocal clans in resident killer whales in southern Alaska. *Animal Behaviour*, **63**, 1103–19.

Zari, T.A. (1987). The energetics and thermal physiology of the Wiegmann's skink, *Mabuya brevicollis*. Ph.D. thesis, Nottingham, UK: University of Nottingham.

(1998). Effects of sexual condition on food consumption and temperature selection in the herbivorous desert lizard, *Uromastyx philbyi*. *Journal of Arid Environments*, **38**, 371–7.

Zharikov, Y. and Skilleter, G.A. (2002). Sex-specific intertidal habitat use in subtropically wintering bar-tailed godwits. *Canadian Journal of Zoology*, **80**, 1918–29.

Zinner, H. (1985). On behavioral and sexual dimorphism of *Telescopus dhara* Forscal 1776 (Reptilia: Serpentes, Colubridae). *Journal of the Herpetological Association of Africa*, **31**, 5–6.

Zuberbühler, J.K.D. and Bshary, R. (1999). The predator deterrence function of primate alarm calls. *Ethology*, **105**, 477–90.

Zucker, N. (1986). Perch height preferences of male and female tree lizards, *Urosaurus ornatus*: a matter of food competition or social role? *Journal of Herpetology*, **20**, 547–53.

Zuk, M. and McKean, K.A. (1996). Sex differences in parasite infections: patterns and processes. *International Journal of Parasitology*, **26**, 1009–24.

Zwarts, L. (1988). Numbers and distribution of coastal waders in Guinea-Bissau. *Ardea*, **76**, 42–55.

Index